D0988454

Building Regulations in Brief

25.

14. MAY

25. M

20.

About the author

Ray Tricker (MSc, IEng, FIIE (elec), FCIM, FIQA, MIRSE) is the Principal Consultant of Herne European Consultancy Ltd – a company specializing in Integrated Management Systems – is also an established Butterworth-Heinemann author. He served with the Royal Corps of Signals (for a total of 37 years) during which time he held various managerial posts culminating in being appointed as the Chief Engineer of NATO's Communication Security Agency (ACE COMSEC).

Most of Ray's work since joining Herne has centred on the European railways. He has held a number of posts with the Union International des Chemins de fer (UIC), e.g. Quality Manager of the European Train Control System (ETCS), European Union (EU) T500 Review Team Leader, European Rail Traffic Management System (ERTMS) Users Group Project Co-ordinator, HEROE (Harmonization of European Rail Rules) Project Co-ordinator and currently (as well as writing books for Butterworth-Heinemann!) he is busy assisting small businesses from around the world (usually on a no cost basis) produce their own auditable Quality Management Systems to meet the requirements of ISO 9001:2000. He is also consultant to the Association of American Railroads (AAR) advising them on ISO 9001:2000 compliance and has recently been appointed as UICAS Technical Specialist for the assessment of Notified Bodies for the Harmonisation of the trans-European high speed rail system.

To my friends Ken and Val Judge whose DIY skills in converting a tumbledown barn into a luxury home were the inspiration for writing this book.

Building Regulations in Brief

Ray Tricker

RECEIVED 3 1 OCT 2003

AMSTERDAM BOSTON HEIDELBERG LONDON NEW YORK OXFORD
PARIS SAN DIEGO SAN FRANCISCO SINGAPORE SYDNEY TOKYO

Butterworth-Heinemann
An imprint of Elsevier
Linacre House, Jordan Hill, Oxford OX2 8DP
200 Wheeler Road, Burlington, MA 01803

First published 2003

Copyright © 2003, Ray Tricker. All rights reserved

The right of Ray Tricker to be identified as the author of this work
has been asserted in accordance with the Copyright, Designs and
Patents Act 1988

No part of this publication may be
reproduced in any material form (including
photocopying or storing in any medium by electronic
means and whether or not transiently or incidentally
to some other use of this publication) without the
written permission of the copyright holder except
in accordance with the provisions of the Copyright,
Designs and Patents Act 1988 or under the terms of a
licence issued by the Copyright Licensing Agency Ltd,
90 Tottenham Court Road, London, England W1T 4LP.
Applications for the copyright holder's written permission
to reproduce any part of this publication should be addressed
to the publisher

Permissions may be sought directly from Elsevier's Science
and Technology Rights Department in Oxford, UK:
phone: (+44) (0) 1865 843830; fax: (+44) (0) 1865 853333;
e-mail: permissions@elsevier.co.uk. You may also complete
your request on-line via the Elsevier Science homepage
(http://www.elsevier.com), by selecting 'Customer Support'
and then 'Obtaining Permissions'

British Library Cataloguing in Publication Data
A catalogue record for this book is available from the British Library

Library of Congress Cataloguing in Publication Data
A catalogue record for this book is available from the Library of Congress

ISBN 0 7506 5367 1

For information on all Butterworth-Heinemann
publications visit our website at www.bh.com

Typeset by Integra Software Services Pvt. Ltd, Pondicherry, India
www.integra-india.com
Printed and bound in Great Britain by Biddles Ltd, www.biddles.co.uk

Contents

Foreword

Subject to specified exemptions, all building work in England and Wales (a separate system of building control applies to Scotland and Northern Ireland) is governed by Building Regulations. This is a statutory instrument, which sets out the minimum requirements and performance standards for the design and construction of buildings, and extensions to buildings.

The current regulations are the Building Regulations 2000. These take into consideration some major changes in technical requirement (such as conservation of fuel and power) and some procedural changes allowing local authorities to regularize unauthorized development.

Although the 2000 regulations are comparatively short, they rely on their technical detail being available in a series of Approved Documents and a vast number of British, European and international standards, codes of practice, drafts for development, published documents and other non-statuary guidance documents.

The main problem, from the point of view of the average builder and DIY enthusiast, is that the Building Regulations are too professional for their purposes. They cover every aspect of building, are far too detailed and contain too many options. All the builder or DIY person really requires is sufficient information to enable them to comply with the regulations in the simplest and most cost-effective manner possible.

Building inspectors, acting on behalf of local authorities, are primarily concerned with whether a building complies with the requirements of the Building Regulations and to do this, they need to 'see the calculations'. But how do the DIY enthusiast and/or builder obtain these calculations? Where can they find, for instance, the policy and requirements for load bearing elements of a structure?!

Builders, through experience, are normally aware of the overall requirements for foundations, drains, walls, central heating, air conditioning, safety, security, glazing, electricity, plumbing, roofing, floors, etc., but they still need a reminder when they come across a different situation for the first time (e.g. what if they are going to construct a building on soft soil, how deep should the foundations have to be?).

On the other hand, the DIY enthusiast, keen on building his own extension, conservatory, garage or workshop etc. usually has no past experience and needs the relevant information – but in a form that he can easily understand without having the advantage of many years experience. In fact, what he really needs is a rule of thumb guide to the basic requirements.

From a recent survey it has transpired that the majority of builders and virtually all DIY enthusiasts are self taught and most of their knowledge is gained through experience. When they hit a problem, it is usually discussed over a pint in the local pub with friends in the building trade as opposed to seeking professional help. What they really need is a reference book to enable them to understand (or remind themselves of) the official requirements.

The aim of my book, therefore, is to provide the reader with an in-brief guide that can act as an aide memoire to the current requirements of the Building Regulations. Intended readers are primarily builders and the DIY fraternity (who need to know the regulations but do not require the detail), but the book, with its ready reference and no-nonsense approach, will be equally useful to architects, designers, building surveyors and inspectors, etc.

Preface

The Great Fire of London in 1666 was probably the single most significant event to shape today's legislation! The rapid growth of fire through co-joined timber buildings highlighted the need to consider the possible spread of fire between properties and this consideration resulted in the publication of the first building construction legislation in 1667 requiring all buildings to have some form of fire resistance.

Two hundred years later, the Industrial Revolution had meant poor living and working conditions in ever expanding, densely populated urban areas. Outbreaks of cholera and other serious diseases, through poor sanitation, damp conditions and lack of ventilation, forced the government to take action and building control took on the greater role of health and safety through the first Public Health Act of 1875. This Act had two major revisions in 1936 and 1961, leading to the first set of national building standards (i.e. the Building Regulations 1965). Over the years these regulations have been amended and updated and the current document is the Building Regulations 2000.

The Building Regulations are approved by the Secretary of State and are intended to provide guidance to some of the more common building situations as well as providing a practical guide to meeting the requirements of Regulation 7 of the Building Act 1984, which states:

Materials and workmanship
7.   Building work shall be carried out –

(a)   with adequate and proper materials which –

   (i)   are appropriate for the circumstances in which they are used,
   (ii)   are adequately mixed or prepared, and
   (iii) are applied, used or fixed so as adequately to perform the
          functions for which they are designed; and

(b)   in a workmanlike manner.

What are the current regulations?

The current legislation is the Building Regulations 2000 (Statutory Instrument No 2531) which is made by the Secretary of State for the Environment under powers delegated by parliament under the Building Act 1984. Since then, the Building Regulations have received two Building Amendment Regulations (SI 2001 No 3335 and SI 2002 No 440).

Table P.1 Statutory instruments currently in place

The Building Regulations 2000 (SI 2000 No 2531)

Made	*13th September 2000*
Laid before Parliament	*22nd September 2000*
Coming into force	*1st January 2001*

Now includes the following amending statutory instruments: **SI 2001 No 3335**

Made	*4th October 2001*
Laid before Parliament	*11th October 2001*
Coming into force	*1st April 2002*

and **SI 2002 No 440**

Made	*28th February 2002*
Laid before Parliament	*5th March 2002*
Coming into force	*1st April 2002*

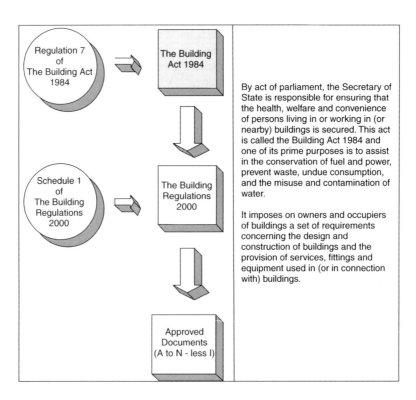

Figure P.1 The Building Act 1984

The Building Act 1984 consists of five parts:

Part 1   The Building Regulations
Part 2   Supervision of Building Work etc. other than by a Local Authority
Part 3   Other provisions about buildings
Part 4   General
Part 5   Supplementary

Part 5 then contains seven schedules whose prime function is to list the principal areas requiring regulation and to show how the Building Regulations are to be controlled by local authorities. These schedules are:

Schedule 1 – Building Regulations;
Schedule 2 – Relaxation of building regulations;
Schedule 3 – Inner London;
Schedule 4 – Provisions consequential upon public body's notice;
Schedule 5 – Transitional provisions;
Schedule 6 – Consequential amendments;
Schedule 7 – Repeals.

Schedule 1 is the most important (from the point of view of builders) as it shows, in general terms, how the Building Regulations are to be administered by local authorities, the approved methods of construction and the approved types of materials that are to be used in (or in connection with) buildings.

  The Building Act 1984 does not apply to Scotland or to Northern Ireland.

The Building Regulations describe the mandatory requirements for completing **all** building work including:

- accommodation for specific purposes (e.g. for disabled persons);
- air pressure plants;
- cesspools (and other methods for treating and disposing of foul matter);
- dimensions of rooms and other spaces (inside buildings);
- drainage (including waste disposal units);
- emission of smoke, gases, fumes, grit or dust (or other noxious or offensive substances);
- fire precautions (services, fittings and equipment, means of escape);
- lifts (escalators, hoists, conveyors and moving footways);
- materials and components (suitability, durability and use);
- means of access to and egress from;
- natural lighting and ventilation of buildings;
- open spaces around buildings;
- prevention of infestation;
- provision of power outlets;

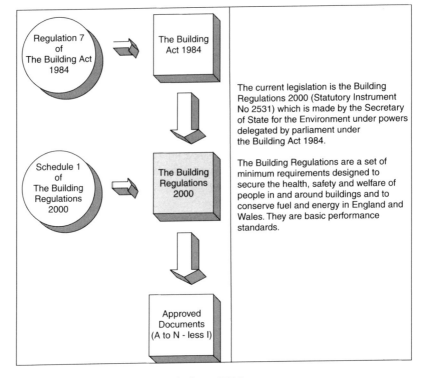

Figure P.2 The Building Regulations 2000

- resistance to moisture and decay;
- site preparation;
- solid fuel, oil, gas, electricity installations (including appliances, storage tanks, heat exchangers, ducts, fans and other equipment);
- standards of heating, artificial lighting, mechanical ventilation and air-conditioning;
- structural strength and stability (overloading, impact and explosion, under-pinning, safeguarding of adjacent buildings);
- telecommunications services (wiring installations for telephones, radio and television);
- third party liability (danger and obstruction to persons working or passing by building work);
- transmission of heat;
- transmission of sound;
- waste (storage, treatment and removal);
- water services, fittings and fixed equipment (including wells and bore-holes for supplying water) and
- matters connected with (or ancillary to) any of the foregoing matters.

The Building Regulations are legal requirements laid down by parliament and based on the Building Act 1984. The Building Regulations:

- are approved by parliament;
- deal with the minimum standards of design and building work for the construction of domestic, commercial and industrial buildings;
- set out the procedure for ensuring that building work meets the standards laid down;
- are designed to ensure structural stability;
- promote the use of suitable materials to provide adequate durability, fire and weather resistance, and the prevention of damp;
- stipulate the minimum amount of ventilation and natural light to be provided for habitable rooms;
- ensure the health and safety of people in and around buildings (by providing functional requirements for building design and construction);
- promote energy efficiency in buildings;
- contribute to meeting the needs of disabled people.

The level of safety and standards acceptable are set out as guidance in the approved documents. Compliance with the detailed guidance of the approved

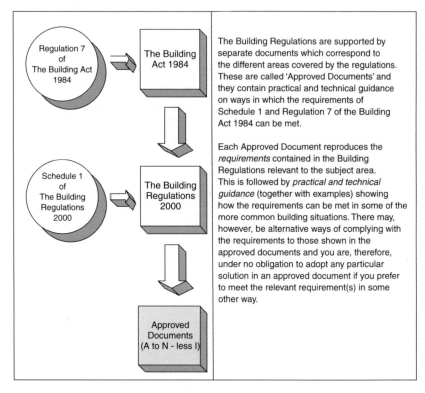

Figure P.3 Approved documents

documents is usually considered as evidence that the Building Regulations themselves have been complied with.

The approved documents are in 13 parts, A to N (less 'I') and consist of:

A   Structural
B   Fire safety
C   Site preparation and resistance to moisture
D   Toxic substances
E   Resistance to the passage of sound
F   Ventilation
G   Hygiene
H   Drainage and waste disposal
J   Combustion appliances and fuel storage systems
K   Protection from falling, collision and impact
L   Conservation of fuel and power
M   Access and facilities for disabled people
N   Glazing – safety in relation to impact, opening and cleaning

Approved Documents A to K and N (except for paragraphs H2 and J6) do not require anything to be done except for the purpose of securing reasonable standards of health and safety for persons in or about buildings (and any others who may be affected by buildings, or matters connected with buildings).

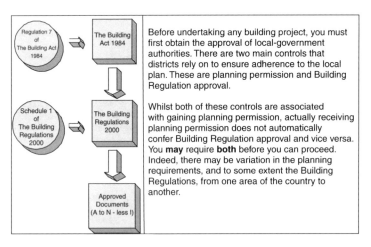

Figure P.4 Planning permission

Planning permission is the single biggest hurdle for anyone who has acquired land on which to build a house, or wants to extend or carry out other building work on property. There is never a guarantee that permission will be given and without it no project can start. Yet the system is not at all user-friendly.

There is a bewildering array of formalities to go through and ever more stringent requirements to satisfy. Planning permission has never been more difficult to get, nor so sought after. Every year over half-a-million applications are made and the number is rising.

The purpose of the planning system is to protect the environment as well as public amenities and facilities. It is **not** designed to protect the interests of one person over another. Within the framework of legislation approved by parliament, councils are tasked to ensure that development is allowed where it is needed, while ensuring that the character and amenity of the area are not adversely affected by new buildings or changes in the use of existing buildings and/or land.

Provided, however, that the work you are completing does not affect the external appearance of the building, you are allowed to make certain changes to your home without having to apply to the local council for permission. These are called 'Permitted Development Rights', but the majority of building work, that you are likely to complete will, however probably require you to have planning permission – so be warned!

The actual details of planning requirements are complex but for most domestic developments, the planning authority is only really concerned with construction work such as an extension to the house or the provision of a new garage or new outbuildings that is being carried out. Structures like walls and fences also need to be considered because their height or siting might well infringe the rights of their neighbours and other members of the community. The planning authority will also want to approve any change of use, such as converting a house into flats or running a business from premises previously occupied as a dwelling only.

Aim of this book

The prime aim of this book is to provide builders and DIY people with an aide memoire and a quick reference to the requirements of the Building Regulations. This book provides a user-friendly background to the Building Act 1984 and its associated Building Regulations. It explains the meaning of the Building Regulations, their current status, requirements, associated documentation and how local authorities and councils view their importance. It goes on to describe the content of the guidance documents (i.e. the 'Approved Documents') published by the Secretary of State and, in a series of 'what ifs', provides answers to the most common questions that DIY enthusiasts and builders might ask concerning building projects.

The book is structured as follows:

Chapter 1 – The Building Act 1984
Chapter 2 – The Building Regulations 2000
Chapter 3 – The requirements of the Building Regulations
Chapter 4 – Planning permission

Chapter 5 – How to comply with the requirements of the Building Regulations

Chapter 6 – Meeting the requirements of the Building Regulations

These chapters are then supported by the following annexes:

Annex A   Access and facilities for disabled people
Annex B   Conservation of fuel and power

and concludes with a list of all the main acronyms and abbreviations associated with the Building Regulations, together with a detailed glossary of terms and a full index.

The following symbols will help you get the most out of this book:

   an important requirement or point

   a good idea or suggestion

   further amplification or information

1

The Building Act 1984

1.1 Aim of the Building Act 1984 (Building Act 1984 Section 1)

By Act of Parliament, the Secretary of State is responsible for ensuring that the health, welfare and convenience of persons living in or working in (or nearby) buildings is secured.

This Act is called the Building Act 1984 and one of its prime purposes is to assist in the conservation of fuel and power, prevent waste, undue consumption, misuse and contamination of water. It imposes on owners and occupiers of buildings a set of requirements concerning the design and construction of buildings and the provision of services, fittings and equipment used in (or in connection with) buildings. These involve, and cover:

- a method of controlling (inspecting and reporting) buildings;
- how services, fittings and equipment may be used;
- the inspection and maintenance of any service, fitting or equipment used.

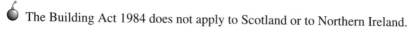 The Building Act 1984 does not apply to Scotland or to Northern Ireland.

1.2 What happens if I contravene any of these requirements? (Building Act 1984 Sections 2, 7, 35, 36 and 112)

If you contravene the Building Regulations or wilfully obstruct a person acting *in the execution of the Building Act 1984 or of its associated Building Regulations*, then on summary conviction, you could be liable to a fine or, in exceptional circumstances, even a short holiday in HM Prisons!

1.3 Who polices the Act?

Under the terms of the Building Act 1984, local authorities are responsible for ensuring that any building work being completed conforms to the requirements of the associated Building Regulations. They have the authority to:

- make you take down and remove or rebuild anything that contravenes a regulation;
- make you complete alterations so that your work complies with the Building Regulations;
- employ a third party (and then send you the bill!) to take down and rebuild non-conforming buildings or parts of buildings.

They can, in certain circumstances, even take you to court and have you fined – especially if you fail to complete the removal or rebuilding of the non-conforming work.

The above authority to prosecute and order remedial work to be completed applies equally whether you are the actual owner or merely the occupier – so be warned!

1.4 Are there any exemptions from Building Regulations? (Building Act 1984 Sections 3, 4 and 5)

The following are exempt from the Building Regulations:

- A 'public body' (i.e. local authorities, county councils and any other body 'that acts under an enactment for public purposes and not for its own profit'). This can be rather a grey area and it is best to seek advice if you think that you come under this category;
- buildings belonging to 'statutory undertakers' (e.g. a water board);
- schools or other educational establishments (provided that they are, or have been erected according to plans authorized and approved by the Secretary of State for Education and Science (or the Secretary of State for Wales) according to the Education Act 1980).

1.4.1 What about Crown buildings? (Building Act 1984 Section 44a d 87)

Although the majority of the requirements of the Building Regulations are applicable to Crown buildings (i.e. a building in which there is a Crown or Duchy of Lancaster or Duchy of Cornwall interest) or government buildings (held in trust for Her Majesty) there are occasional deviations and before submitting plans for work on a Crown building you should seek the advice of the Treasury.

1.4.2 What about buildings in Inner London? (Building Act 1984 Sections 44, 46 and 88)

You will find that the majority of the requirements found in the Building Regulations are also applicable to buildings in Inner London boroughs (i.e. Inner Temple and Middle Temple). There are, however, some important deviations (see Section 1.7.3) and before submitting plans you should seek the advice of the local authority concerned.

1.4.3 What about the UK Atomic Energy Authority? (Building Act 1984 Section 45)

The Building Regulations do not apply to buildings belonging to or occupied by the United Kingdom Atomic Energy Authority (UKAEA) unless they are dwelling houses and offices.

1.4.4 What about the British Airports Authority? (Building Act 1984 Section 45)

The Building Regulations do not apply to buildings belonging to or occupied by the British Airports Authority, unless it is a house, hotel or building used as offices or showrooms.

1.4.5 What about the Civil Aviation Authority? (Building Act 1984 Section 45)

The Building Regulations do not apply to buildings belonging to or occupied by the Civil Aviation Authority, unless it is a house, hotel or building used as offices or showrooms.

1.5 What about civil liability? (Building Act 1984 Section 38)

It is an aim of the Building Act 1984 that all building work is completed safely and without risk to people employed on the site or visiting the site etc. Any contravention of the Building Regulations that causes injury (or death) to any person is liable to prosecution in the normal way.

1.6 What does the Building Act 1984 contain?

The Building Act 1984 consists of five parts:

Part 1   The Building Regulations
Part 2   Supervision of building work etc. other than by a local authority

Part 3 Other provisions about buildings
Part 4 General
Part 5 Supplementary

These parts are then sub-divided into a number of sections as shown below:

Part 1	The Building Regulations	<ul><li>Power to make Building Regulations</li><li>Exemption from Building Regulations</li><li>Approved documents</li><li>Relaxation of Building Regulations</li><li>Type approval of building matter</li><li>Consultation</li><li>Passing of plans</li><li>Determination of questions</li><li>Proposed departure from plans</li><li>Lapse of deposit of plans</li><li>Tests for conformity with Building Regulations</li><li>Classification of buildings</li><li>Breach of Building Regulations</li><li>Appeals in certain cases</li><li>Application of Building Regulations to Crown etc.</li><li>Inner London</li></ul>
Part 2	Supervision of building work etc. otherwise than by a local authority	<ul><li>Supervision of plans and work by approved inspectors</li><li>Supervision of their own work by public bodies</li><li>Supplementary</li></ul>
Part 3	Other provisions about buildings	<ul><li>Drainage</li><li>Provision of sanitary conveniences</li><li>Buildings</li><li>Defective premises, demolition etc.</li><li>Yards and passages</li><li>Appeal to crown court</li><li>Application of provisions to crown property</li><li>Inner London</li><li>Miscellaneous</li></ul>
Part 4	General	<ul><li>Duties of local authorities</li><li>Documents</li><li>Entry on premises</li><li>Execution of works</li><li>Appeal against notice requiring works</li></ul>

- General provisions about appeals and applications
- Compensation, and recovery of sums
- Obstruction
- Prosecutions
- Protection of members etc. of authorities
- Default powers
- Local inquiries
- Orders
- Interpretation
- Savings

Part 5　Supplementary

- Schedule 1 – Building Regulations
- Schedule 2 – Relaxation of Building Regulations
- Schedule 3 – Inner London
- Schedule 4 – Provisions consequential upon public body's notice
- Schedule 5 – Transitional provisions
- Schedule 6 – Consequential amendments
- Schedule 7 – Repeals

Each section is then further sub-divided into a number of sub-sections describing the provisions of these sections (see Appendix A to Chapter 1).

1.7 What are the Supplementary Regulations?

Part 5 of the Building Act contains seven schedules whose function is to list the principal areas requiring regulation and to show how the Building Regulations are to be controlled by the local authority. These schedules are:

- Schedule 1 – Building Regulations;
- Schedule 2 – Relaxation of Building Regulations;
- Schedule 3 – Inner London;
- Schedule 4 – Provisions consequential upon public body's notice;
- Schedule 5 – Transitional provisions;
- Schedule 6 – Consequential amendments;
- Schedule 7 – Repeals.

1.7.1 What is Schedule 1 of Part 5 of the Building Act 1984?

The Building Regulations are a statutory instrument, authorized by parliament, which details how the generic requirements of the Building Act are to be met. Compliance with Building Regulations is required for **all**:

- alterations and extensions of buildings (including services, fixtures and fittings);
- provision of new services, fittings or equipment;

unless (in most circumstances) the increased area of the alteration or extension is less than $30\,m^2$ and the Building Regulations provide the generic and specific requirements for this work.

Building Regulations **also** apply to alterations and extensions being completed on buildings erected **before** the date on which the regulations came into force.

Schedule 1 of the Building Act 1984 shows, in general terms, how the Building Regulations are to be administered by local authorities, the approved methods of construction and the approved types of materials that are to be used in (or in connection with) buildings.

How are the Building Regulations controlled?

To assist local authorities, Section 1 shows:

- how notices are given;
- how plans of proposed work (or work already executed) are deposited;
- how copies of deposited plans are administered and retained;
- how documents are to be controlled;
- how work is tested;
- how samples are taken;
- how local authorities can seek external expertise to assist them in their duties;
- how certificates signifying compliance with the Building Regulations are to be issued;
- how local authorities can accept certificates from a person (or persons) nominated to act on their behalf;
- how proposed work can be prohibited;
- when a dispute arises, how local authorities can refer the matter to the Secretary of State;
- what fees (and what level of fees) local authorities can charge.

What are the requirements of the Building Regulations?

Schedule 1 describes the mandatory requirements for completing **all** building work. These include:

- accommodation for specific purposes (e.g. for disabled persons);
- air pressure plants;
- cesspools (and other methods for treating and disposing of foul matter);
- emission of smoke, gases, fumes, grit or dust (or other noxious and/or offensive substances);

- dimensions of rooms and other spaces (inside buildings);
- drainage (including waste disposal units);
- fire precautions (services, fittings and equipment, means of escape);
- lifts (escalators, hoists, conveyors and moving footways);
- materials and components (suitability, durability and use);
- means of access to and egress from;
- natural lighting and ventilation of buildings;
- open spaces around buildings;
- prevention of infestation;
- provision of power outlets;
- resistance to moisture and decay;
- site preparation;
- solid fuel, oil, gas and electricity installations (including appliances, storage tanks, heat exchangers, ducts, fans and other equipment);
- standards of heating, artificial lighting, mechanical ventilation and air-conditioning;
- structural strength and stability (overloading, impact and explosion, under-pinning, safeguarding of adjacent buildings);
- third party liability (danger and obstruction to persons working or passing by building work);
- transmission of heat;
- transmission of sound;
- waste (storage, treatment and removal);
- water services, fittings and fixed equipment (including wells a   ore-holes for supplying water)

and matters connected with (or ancillary to) any of the foregoing matters.

1.7.2 What is Schedule 2 of the Building Act 1984?

This Schedule provides guidance in connection with work that has been carried out prior to a local authority (under the Building Act 1984 Section 36) dispensing with or relaxing some of the requirements contained in the Building Regulations.

This Schedule is quite difficult to understand and if it affects you, then I would strongly advise that you discuss it with the local authority before proceeding any further.

1.7.3 What is Schedule 3 of the Building Act 1984?

Schedule 3 applies to how Building Regulations are to be used in Inner London and, as well as ruling which sections of the Act may be omitted, also details the requirements for drainage to Inner London buildings and shows how by-laws concerning the relation to the demolition of buildings (in Inner London) may be made.

What sections of the Building Act 1984 are not applied to Inner London?

In Inner London, because of its existing and changed circumstances (compared to other cities in England and Wales), certain sections of the Building Act (such as drainage to buildings) are inappropriate (see Tables 1.1 and 1.2) and additional requirements – which are applicable to Inner London **only** – have been approved instead.

What about the buildings and drainage to buildings in Inner London?

Under the terms of the Building Act 1984, it is not lawful in an Inner London borough to erect a house/other building, or to rebuild a house/other building that has been pulled down to (or below) floor level, **unless** that house/building is provided with drains in conformance with the borough council's requirements. These drains must be suitable for the drainage of the whole building and all works, apparatus and materials used in connection with these drains must satisfy the council's requirements.

It is not lawful to occupy a house or other building in Inner London that has been erected or rebuilt in contravention of the above restriction.

Table 1.1 Sections inapplicable to Inner London

Part	Description	Section	Sub-section
		Buildings	• Provision of food storage accommodation in house. • Entrances, exits etc. to be required in certain cases. • Means of escape from fire. • Raising of chimney. • Cellars and rooms below subsoil water level. • Consents under Section 74.
		Defective premises, demolition etc.	• Dangerous building. • Dangerous building – emergency measures. • Ruinous and dilapidated buildings and neglected sites. • Notice to local authority of intended demolition. • Local authority's power to serve notice about demolition. • Notices under Section 81. • Appeal against notice under Section 81.

Table 1.2 Sections inapplicable to Temples

Part	Description	Section	Sub-section
Part 3	Other provisions about buildings	Drainage	• Drainage of building. • Use and ventilation of soil pipes. • Repair etc. of drain.
		Buildings	• Provision of food storage accommodation in house. • Entrances, exits etc. to be required in certain cases. • Means of escape from fire. • Raising of chimney. • Cellars and rooms below subsoil water level. • Consents under Section 74.
		Defective premises, demolition etc.	• Dangerous building. • Dangerous building – emergency measures. • Ruinous and dilapidated buildings and neglected sites. • Notice to local authority of intended demolition. • Local authority's power to serve notice about demolition. • Notices under Section 81. • Appeal against notice under Section 81.

The basic requirements of all Inner London borough councils are that:

- the drains must be connected into a sewer that is (or is intended to be constructed) nearby;
- if a suitable sewer is not available then a covered cesspool or other place should be used, provided that it is not under any house or other building;
- the drains must provide efficient gravitational drainage at all times and under all circumstances and conditions.

If it is impossible or unfeasible to provide gravitational drainage to all parts of the building, then (but depending on the circumstances) the council may allow pumping and/or some other form of lifting apparatus to be used.

In **all** circumstances the council have the authority (under this Schedule of the Act) to order the owner/occupier:

- to construct a covered drain from the house or building into the sewer;
- to provide proper paved or water-resistant sloping surfaces for carrying surface water into the drain;
- to provide proper sinks, inlets and outlets (siphoned or otherwise trapped), for preventing the emission of effluvia from the drain – or any connection to it;

- to provide a proper water supply and water-supplying pipes, cisterns and apparatus for scouring the drain;
- to provide proper sand traps, expanding inlets and other apparatus for preventing the entry of improper substances into the drain.

You are not allowed to commence any work on drains, dig out the foundations of a house or to rebuild a house in Inner London unless, at least seven days previously, you have provided a notice of intent to the borough council.

Where a house or building in an Inner London borough, **whenever erected**, is without sufficient drainage and there is no proper sewer within 200 feet of any part of the house or building, the borough council may serve on the owner written notice requiring that person:

- to construct a covered watertight cesspool or tank or other suitable receptacle (provided that it is not under the house); and
- to construct and lay a covered drain leading from the house or building into that cesspool, tank or receptacle.

The Inner London Borough Council have the authority to carry out irregular inspections of drains and cesspools constructed by the owner and, if they are unsuitable, to require the owner to alter, repair or abandon them if they are unsuitable or contravene council regulations.

What about Inner London's by-laws?

By authority of the Building Act 1984, the Greater London Authority (GLA) may make by-laws in relation to the demolition of buildings in the Inner London boroughs and regulate and (in certain circumstances) mandate, concerning:

- the fixing of floor level fans on buildings undergoing demolition;
- the hoarding up of windows in a building where all the sashes and glass have been removed;
- the demolition of internal parts of buildings before any external walls are taken down;
- using screens and mats as a precaution against dust;
- the hours during which ceilings may be broken down and mortar may be shot, or be allowed to fall, into any lower floor.

The GLA may also make by-laws with respect to closets, sanitary conveniences, ashpits, cesspools and receptacles for dung (and their accessories) for buildings being erected or altered in Inner London.

1.7.4 What is Schedule 4 of the Building Act 1984?

Schedule 4 of the Building Act 1984 concerns the authority and ruling of public bodies' notices and certificates.

What is a public body's plans certificate?

When a public body (i.e. local authorities, county councils and any other body *'that acts under an enactment for public purposes and not for its own profit'*) is satisfied that the work specified in their (as well as another) public body's notices has been completed as detailed, and in full accordance with the Building Regulations, then that public body will give that local authority a certificate of completion.

This certificate is called a 'Public Bodies Plans Certificate' and can relate either to the whole, or to part of, the work specified in the public body's notice. Acceptance by the local authority signifies satisfactory completion of the planned work and the public body's notice ceases to apply to that work.

What is a public body's final certificate?

When a public body is satisfied that all work specified in their (or another's) public body's notice has been completed in compliance with the Building Regulations, then that public body will give the local authority a certificate of completion. This is referred to as a 'Final Certificate'.

How long is the duration of a public body's notice?

A public body's notice comes into force when it is accepted by the local authority and continues in force until the expiry of an agreed period of time.

Local authorities are authorized by the Building Regulations to extend the notice in certain circumstances.

1.7.5 What is Schedule 5 of the Building Act 1984?

Schedule 5 lists the transitional effect of the Building Act 1984 concerning:

- The Public Health Act 1936;
- The Clean Air Act 1956;
- The Housing Act 1957;
- The Public Health Act 1961;
- The London Government Act 1963;
- The Local Government Act 1972;
- The Health and Safety at Work etc. Act 1974;
- The Local Government (Miscellaneous Provisions) Act 1982.

1.7.6 What is Schedule 6 of the Building Act 1984?

Schedule 6 lists the consequential amendments that will have to be made to existing Acts of Parliament owing to the acceptance of the Building Act 1984. These amendments concern:

- The Restriction of Ribbon Development Act 1935;
- The Public Health Act 1936;
- The Atomic Energy Authority Act 1954;
- The Clean Air Act 1956;
- The Housing Act 1957;
- The Radioactive Substances Act 1960;
- The Public Health Act 1961;
- The London Government Act 1963;
- The Offices, Shops and Railway Premises Act 1963;
- The Faculty Jurisdiction Measure 1964;
- The Fire Precautions Act 1971;
- The Local Government Act 1972;
- The Safety of Sports Grounds Act 1975;
- The Local Land Charges Act 1975;
- The Development of Rural Wales Act 1976;
- The Local Government (Miscellaneous Provisions) 1976;
- The Interpretation Act 1978;
- The Highways Act 1980;
- New Towns Act 1981;
- The Local Government (Miscellaneous Provisions) 1982;
- The Public Health (Control of Disease) Act 1984.

1.7.7 What is Schedule 7 of the Building Act 1984?

Schedule 7 lists the cancellation (repeal) of some sections of existing Acts of Parliament, owing to acceptance of the Building Act 1984. These cancellations concern:

- The Public Health Act 1936;
- The Education Act 1944;
- The Water Act 1945;
- The Town and Country Planning Act 1947;
- The Atomic Energy Authority Act 1954;
- The Radioactive Substances Act 1960;
- The Public Health Act 1961;
- The London Government Act 1963;
- The Greater London Council (General Powers) Act 1967;
- The Fire Precautions Act 1971;
- The Local Government Act 1972;
- The Water Act 1973;
- The Health and Safety at Work etc. Act 1974;
- The Control of Pollution Act 1974;
- The Airports Authority Act 1975;
- The Local Government (Miscellaneous Provisions) Act 1976;
- The Criminal Law Act 1977;
- The City of London (Various Powers) Act 1977;

- The Education Act 1980;
- The Highways Act 1980;
- The Water Act 1981;
- The Civil Aviation Act 1982;
- The Local Government (Miscellaneous Provisions) Act 1982.

1.8 What are 'Approved Documents'? (Building Act 1984 Section 6)

The Secretary of State makes available a series of documents (called 'Approved Documents') which are intended to provide practical guidance with respect to the requirements of the Building Regulations (for details see Chapter 3).

1.9 What is the 'Building Regulations Advisory Committee'? (Building Act 1984 Section 14)

The Building Act allows the Secretary of State to appoint a committee (known as the Building Regulations Advisory Committees) to review, amend, improve and produce new Building Regulations and associated documentation (e.g. such as Approved Documents – see above).

1.10 What is 'type approval'? (Building Act 1984 Sections 12 and 13)

Type approval is where the Secretary of State is empowered to approve a particular type of building matter as complying, either generally or specifically, with a particular requirement of the Building Regulations. This power of approval is normally delegated by the Secretary of State to the local council or other nominated public body.

1.11 Does the Fire Authority have any say in Building Regulations? (Building Act 1984 Section 15)

When a requirement 'encroaches' on something that is normally handled by the Fire Authority under the Fire Precautions Act 1971 (e.g. provision of means of escape, structural fire precautions etc.) then the local authority **must** consult the fire authority before making any decision.

1.12 How are buildings classified? (Building Act 1984 Section 35)

For the purpose of the Building Act, buildings are normally classified:

- by reference to size;
- by description;
- by design;
- by purpose;
- by location;

or (to quote the Building Act of 1984) '*any other suitable characteristic*'.

1.13 What are the duties of the local authority? (Building Act 1984 Section 91)

It is the duty of local authorities to ensure that requirements of the Building Act 1984 are carried out (and that the appropriate associated Building Regulations are enforced) subject to:

- the provisions of Part I of the Public Health Act 1936 (relating to united districts and joint boards);
- section 151 of the Local Government, Planning and Land Act 1980 (relating to urban development areas);
- section 1(3) of the Public Health (Control of Disease) Act 1984 (relating to port health authorities).

1.13.1 What document controls must local authorities have in place? (Building Act 1984 Sections 92 and 93)

All notices, applications, orders, consents, demands and other documents, authorized, required by or given to, that are required by this Act or by a local authority (or an officer of a local authority), need to be in writing and in the format laid down by the Secretary of State.

All documents that a local authority is required to provide under the Building Act 1984 shall be signed by:

- the proper officer for this authority;
- the district surveyor (for documents relating to matters within his province);
- an officer authorized by the authority to sign documents (of a particular kind).

A document bearing the signature (including a facsimile of a signature by whatever process chosen) of an officer is deemed (for the purposes of the

Building Act 1984 and any of its associated Building Regulations and orders made under it) to have been given, made or issued by the local authority, unless otherwise proved.

1.13.2 How do local authorities 'serve' notices and documents?
(Building Act 1984 Section 94)

Any notice, order, consent, demand or other document that is authorized or required by the Building Act 1984 can be given or served to a person:

- by delivering it to the person concerned;
- by leaving it, or sending it in a prepaid letter addressed to him, at his usual or last known residence.

Or if it is not possible to ascertain the name and address of the person to or on whom it should be given or served (or if the premises are unoccupied) then the notice, order, consent, demand or other document can be addressed to the 'owner' or 'occupier' of the premises (naming them) and delivering it to *some person on the premises*' or, if there isn't anyone at the premises to whom it can be delivered, then a copy of the document can be fixed to a conspicuous part of the premises.

1.14 What are the powers of the local authority?
(Building Act 1984 Sections 97–101)

The powers of the local authority, as given by the Building Act 1984 and its associated Building Regulations, include:

- overall responsibility for the construction and maintenance of sewers and drains and the laying and maintenance of water mains and pipes;
- the authority to make the owner or occupier of any premises complete essential and remedial work in connection with the Building Act 1984 (particularly with respect to the construction, laying, alteration or repair of a sewer or drain);
- the authority to complete remedial and essential work themselves (on repayment of expenses) if the owner or occupier refuses to do this work himself;
- the ability to sell any materials that have been removed, by them, from any premises when executing works under this Act (paying all proceeds, less expenses, from this sale to the owner or occupier).

This does not apply to any refuse that is, or has been, removed by the local authority.

1.14.1 Have the local authority any power to enter premises?
(Building Act 1984 Section 95)

An authorized officer of a local authority has a right to enter any premises, at all reasonable hours (except for a factory or workplace in which 24 hours' notice has to be given) for the purpose of:

- ascertaining whether there is (or has been) a contravention of this Act (or of any Building Regulations) that it is the duty of the local authority to enforce;
- ascertaining whether or not any circumstances exist that would require local authority action or for them having to complete any work;
- taking any action, or executing any work, authorized or required by this Act, or by Building Regulations;
- carrying out their functions as a local authority.

If the local authority is refused admission to any premises (or the premises are unoccupied) then the local authority can apply to a Justice of the Peace for a warrant authorizing entry.

1.15 Who are approved inspectors? (Building Act 1984 Section 49)

An approved inspector is a person who is approved by the Secretary of State (or a body such as a local authority or county council designated by the Secretary of State) to inspect, supervise and to authorize building work. Lists of approved inspectors are available from all local authorities.

 Building Act 1984 Section 57
If an approved inspector gives a notice or certificate that falsely claims to comply with the Building Regulations and/or the Building Act of 1984, then he is liable to prosecution.

1.15.1 What is an initial notice? (Building Act 1984 Section 47)

An approved inspector will have to present an initial notice and plan of work to the local authority. Once accepted, the approved inspector is authorized to inspect and supervise all work being completed and to provide certificates and notices. Acceptance of an initial notice by a local authority is treated (for the purposes of conformance with section 13 of the Fire Precautions Act 1971 regarding suitable means of escape) as 'depositing plans of work'.

Under section 47 of the Building Act, the local authority is required to accept all certificates and notices, unless the initial notice and plans contravene a local ruling. Whilst the initial notice continues in force, the local authority are not allowed to give a notice in relation to any of the work being carried out or take any action for a contravention of Building Regulations.

If the local authority rejects the initial notice for any reason, then the Approved Inspector can appeal to a magistrates' court for a ruling. If still dissatisfied, he can appeal to the crown court.

Cancellation of initial notice (Building Act 1984 Sections 52 and 53)

If work has not commenced within three years (beginning the date on which the certificate was accepted), the local authority can cancel the initial notice.

If an approved inspector is unable to carry out or complete his functions, or is of the opinion that there is a contravention of the Building Regulations, then he can cancel the initial notice lodged with the local authority.

Equally, if the person carrying out the work has good reason to consider that the approved inspector is unable (or unwilling) to carry out his functions, then that person can cancel the initial notice given to the local authority.

💡 The fact that the initial notice has ceased to be in force does not affect the right of an approved inspector, however, to give a new initial notice relating to any of the work that was previously specified in the original notice.

1.15.2 What are plans certificates? (Building Act 1984 Section 50)

When an approved inspector has inspected and is satisfied himself that the plans of work specified in the initial notice do not contravene the Building Regulations in any way, he will provide a certificate (referred to as a 'plans certificate') to the local authority. This plans certificate:

- can relate to the whole or part of the work specified in the initial notice;
- does not have any effect unless the local authority accepts it;
- may only be rejected by the local authority 'on prescribed grounds'.

If, however, work has not commenced within three years (beginning the date on which the certificate was accepted), the local authority may rescind their acceptance, by notice to the approved inspector or the person shown on the initial notice, giving their grounds for cancellation.

1.15.3 What are final certificates? (Building Act 1984 Section 51)

Once the approved inspector is satisfied that all work has been completed in accordance with the work specified in the initial notice, he will provide a certificate (referred to as a 'Final Certificate') to the local authority and the person who carried out the work. This certificate will detail his acceptance of the work and, once acknowledged by the local authority, the approved inspector's job will have been completed and, from the point of view of local authority, he will have been considered '*to have discharged his duties*'.

1.15.4 Who retains all these records? (Building Act 1984 Section 56)

Local authorities are required to keep a register of all initial notices and certificates given by approved inspectors and to retain all relevant and associated documents concerning those notices and certificates. The local authority is further required to make this register available for public inspection during normal working hours.

1.15.5 Can public bodies supervise their own work? (Building Act 1984 Sections 54 and 55)

If a public body (e.g. local authority or county council) is of the opinion that building work that is to be completed on one of its own buildings can be adequately supervised by one of its employees or agents, then they can provide the local authority with a notice (referred to as a 'public bodies notice') together with their plan of work.

Once accepted by the local authority, the public body is authorized to inspect and supervise all work being completed and to provide certificates and notices. Acceptance by a local authority of public bodies notice is treated (for the purposes of conformance with section 13 of the Fire Precautions Act 1971 regarding suitable means of escape) as 'depositing plans of work'.

If the local authority rejects the public bodies notice for any reason, then they can appeal to a magistrates' court for a ruling. If still dissatisfied, they can appeal to the crown court.

1.16 What causes some plans for building work to be rejected? (Building Act 1984 Sections 16 and 17)

The local authority will reject all plans for building work that:

- are defective;
- contravene any of the Building Regulations.

In all cases the local authority will advise the person putting forward the plans why they have been rejected (giving details of the relevant regulation or section) and, where possible, indicate what amendments and/or modifications will have to be made in order to get them approved. The person who initially put forward the plans is then responsible for making amendments/alterations and resubmitting them for approval.

If, a plan for proposed building work is accompanied by a certificate from a person or persons approved by the Secretary of State (or someone designated by him), then only in extreme circumstances can the local authority reject the plans.

1.17 Can I apply for a relaxation in certain circumstances? (Building Act 1984 Sections 7–11, 30 and 39)

The Building Act allows the local authority to dispense with, or relax, a Building Regulation if they believe that that requirement is unreasonable in relation to a particular type of work being carried out.

Schedule 2 of the Building Act 1984 provides guidance and rules for the application of Building Regulations to work that has been carried out **prior to** the local authority (under the Building Act 1984 Section 36) dispensing with, or relaxing, some of the requirements contained in the Building Regulations. This schedule is quite difficult to understand and if it affects you, then I would strongly advise that you discuss it with the local authority before proceeding any further.

For the majority of cases, applications for dispensing with, or relaxing Building Regulations, can be settled locally. In more complicated cases, however, the local authority can seek guidance from the Secretary of State who will give a direction as to whether the requirement may be relaxed or dispensed with (unconditionally or subject to certain conditions).

If a question arises between the local authority and the person who has executed (or has proposed to execute any work) regarding:

- the application of Building Regulations;
- whether the plans are in conformity with the Building Regulations;
- whether the work has been executed in conformance with these plans;

then the question can be referred to the Secretary of State for determination. In these cases, the Secretary of State's decision will be deemed final.

The Building Act allows the local authority to charge a fee for reviewing and deciding on these matters with different fees for different cases.

1.18 Can I change a plan of work once it has been approved? (Building Act 1984 Section 31)

If the person intending to carry out building work has had their plan (or plans) passed by the local authority, but then wants to change them, that person will have to submit (to the local authority) a set of revised plans showing precisely how they want to deviate from the approved plan and ask for their approval. If the deviation or change is a small one this can usually be achieved by talking to the local planning officer, but if it is a major change, then it could result in the resubmission of a complete plan of the revised building work.

1.19 Must I complete the approved work in a certain time? (Building Act 1984 Section 32)

Once a building plan has been passed by the local authority, then '**work must commence**' within three years from the date that it was approved. Failure to do so could result in the local authority cancelling the approved plans and you will have to resubmit them if you want to carry on with your project.

 The phrase 'work must commence' can vary from local authority to local authority. Normally this will mean physically laying the foundations of the building but in other cases it could mean that far more work has to be completed in the three year time span. It is always best to check with the local authority and ask for clarification about this restriction, when your plans are first approved.

1.20 How is my building work evaluated for conformance with the Building Regulations? (Building Act 1984 Section 32)

Part of the local authority's duty is to make regular checks that all building work being completed is in conformance with the approved plan and the Building Regulations. These checks would normally be completed at certain stages of the work (e.g. the excavation of foundations) and tests will include:

- tests of the soil or subsoil of the site of the building;
- tests of any material, component or combination of components that has been, is being or is proposed to be used in the construction of a building;
- tests of any service, fitting or equipment that has been, is being or is proposed to be provided in or in connection with a building.

The cost of carrying out these tests will normally be charged to the owner or occupier of the building.

💣 The local authority has the power to ask the person responsible for the building work to complete some of these tests on their behalf.

1.20.1 Can I build on a site that contains offensive material?
(Building Act 1984 Section 29)

If the site you are intending to erect a building or extension on is:

- ground that has been filled up with material impregnated with faecal or offensive animal or offensive vegetable mater;
- ground upon which any such material has been deposited;

then that material must be removed or rendered innocuous before work can commence.

💡 This requirement normally rests with the current owner/occupier of the building, but in certain circumstances (for example, if the site was previously used as a chicken farm or similar) then the previous owner might be held responsible. If the requirements of this particular section are applicable to you, then it is recommended that you seek guidance from the local authority before committing yourself!

1.21 What about dangerous buildings? (Building Act 1984 Sections 77 and 78)

If a building, or part of a building or structure, is in such a dangerous condition (or is used to carry loads which would make it dangerous) then the local authority may apply to a magistrates' court to make an order requiring the owner:

- to carry out work to avert the danger;
- to demolish the building or structure, or any dangerous part of it, and remove any rubbish resulting from the demolition.

💣 The local authority can also make an order restricting its use until such time as a magistrates' court is satisfied that all necessary works have been completed.

1.21.1 Emergency measures

In emergencies, the local authority can make the owner take immediate action to remove the danger or they can complete the necessary action themselves. In these cases, the local authority is entitled to recover from the owner such expenses reasonably incurred by them. For example:

- fencing off the building or structure;
- arranging for the building/structure to be monitored.

1.21.2 Can I demolish a dangerous building? (Building Act 1984 Section 80)

You must have good reasons for knocking down a building, such as making way for rebuilding or improvement (which in most cases would be incorporated in the same planning application).

 Be careful! Penalties are severe for demolishing something illegally.

You are not allowed to begin any demolition work (even on a dangerous building) unless you have given the local authority notice of your intention and this has either been acknowledged by the local authority or the relevant notification period has expired. In this notice you will have to:

- specify the building to be demolished;
- state the reason(s) for wanting to demolish it;
- show how you intend to demolish it.

Copies of this notice will have to be sent to:

- the local authority;
- the occupier of any building adjacent to the building in question;
- British Gas;
- the area electricity board in whose area the building is situated.

This regulation does not apply to the demolition of an internal part of an occupied building, or a greenhouse, conservatory, shed or prefabricated garage (that forms part of that building) or an agricultural building defined in section 26 of the General Rate Act 1967.

1.21.3 Can I be made to demolish a dangerous building? (Building Act 1984 Sections 81, 82 and 83)

If the local authority considers that a building is so dangerous that it should be demolished, they are entitled to issue a notice to the owner requiring him:

- to shore up any building adjacent to the building to which the notice relates;
- to weatherproof any surfaces of an adjacent building that are exposed by the demolition;
- to repair and make good any damage to an adjacent building caused by the demolition or by the negligent act or omission of any person engaged in it;
- to remove material or rubbish resulting from the demolition and clear the site;
- to disconnect, seal and remove any sewer or drain in or under the building;

- to make good the surface of the ground that has been disturbed in connection with this removal of drains etc.;
- (in accordance with the Water Act 1945 (interference with valves and other apparatus) and the Gas Act 1972 (public safety)), to arrange with the relevant statutory undertakers (e.g. the water authority, British Gas and the electricity supplier) for the disconnection of gas, electricity and water supplies to the building;
- to leave the site in a satisfactory condition following completion of all demolition work.

Before complying with this notice, the owner must give the local authority 48 hours' notice of commencement.

In certain circumstances, the owner of an adjacent building may be liable to assist in the cost of shoring up their part of the building and waterproofing the surfaces. It could be worthwhile checking this point with the local authority!

1.22 What about defective buildings? (Building Act 1984 Sections 76, 79 and 80)

If a building or structure is, because of its ruinous or dilapidated condition, liable to cause damage to (or be a nuisance to) the amenities of the neighbourhood, then the local authority can require the owner:

- to carry out necessary repairs and/or restoration, or
- to demolish the building or structure (or any part of it) and to remove all of the rubbish or other material resulting from this demolition.

If, however, the building or structure is in a defective state and remedial action envisaged under sections 93 to 96 of the Public Health Act would cause an unreasonable delay, then the local authority can serve an abatement notice stating that within nine days **they** intend to complete such works as they deem necessary to remedy the defective state and recover the 'expenses reasonably incurred in so doing' from the person on whom the notice was sent.

If appropriate, the owner can (within seven days) after the local authority's notice has been served, serve a counter-notice stating that he intends to remedy the defects specified in the first-mentioned notice himself.

A local authority is **not** entitled to serve a notice, or commence any work in accordance with a notice that they have served, if the execution of the works would (to their knowledge) be in contravention of a building preservation order that has been made under section 29 of the Town and Country Planning Act.

1.23 What are the rights of the owner or occupier of the premises? (Building Act 1984 Sections 102–7)

When a person has been given a notice by a local authority to complete work, he has the right to appeal to a magistrates' court on any of the following grounds:

- that the notice or requirement is not justified by the terms of the provision under which it purports to have been given;
- that there has been some informality, defect or error in (or in connection with) the notice;
- that the authority have refused (unreasonably) to approve completion of alternative works, or that the works required by the notice to be executed are unreasonable or unnecessary;
- that the time limit set to complete the work is insufficient;
- that the notice should lawfully have been served on the occupier of the premises in question instead of on the owner (or vice versa);
- that some other person (who is likely to benefit from completion of the work) should share in the expense of the works.

1.24 Can I appeal against a local authority's ruling? (Building Act 1984 Sections 40 and 41)

If you have grounds for disagreeing with the local authority's ruling to remove or renew 'offending work', then you are entitled to appeal to the local magistrates' court and they will rule whether the local authority were correct and entitled to give you this ruling, or whether they should withdraw the notice.

If you then disagree with the magistrates' ruling, you have the right to appeal to the Crown Court.

💡 Where the Secretary of State has given a ruling, however, this ruling shall be considered as being final.

1.24.1 What about compensation? (Building Act 1984 Sections 103–110)

If an owner or occupier considers that a ruling obtained from the local authority is incorrect, he can appeal (in the first case) to the local magistrates' court. If on appeal, the magistrates rule against the local authority, then the owner/occupier of the building concerned is entitled to compensation from the local authority. If, on the other hand the magistrates rule in favour of the local

authority, then the local authority is entitled to recover any expenses that they
have incurred.

 Be sure of your facts before you ask a magistrates' court for a ruling!

1.24.2 What happens if the plans mean building over an existing sewer etc.? (Building Act 1984 Section 18)

Before the local authority can approve a plan for building work involving first
erecting a building or extension over an existing sewer or drain, they must
notify and seek the advice of the water authority.

As part of the Public Health Act 1936 and the Control of Pollution Act
1974, local authorities are required to keep maps of all sewers etc.

Appendix A Contents of the Building Regulations

Part	Description	Section	Sub-section
Part 1	The Building Regulations	Power to make building regulations	• Power to make building regulations. • Continuing requirements.
		Exemption from building regulations	• Exemption of particular classes of buildings etc. • Exemption of educational buildings and buildings of statutory undertakers. • Exemption of public bodies from procedural requirements of building regulations.
		Approved documents	• Approval of documents for purposes of building regulations. • Compliance or non-compliance with approved documents.
		Relaxation of building regulations	• Relaxation of building regulations. • Application for relaxation. • Advertisement of proposal for relaxation of building regulations. • Type relaxation of building regulations.
		Type approval of building matter	• Power of Secretary of State to approve type of building matter. • Delegation of power to approve.

Part	Description	Section	Sub-section
		Consultation	• Consultation with Building Regulations Advisory Committee and other bodies. • Consultation with fire authority.
		Passing of plans	• Passing or rejection of plans. • Approval of persons to give certificates etc. • Building over sewers etc. • Use of short-lived materials. • Use of materials unsuitable for permanent building. • Provision of drainage. • Drainage of buildings in combination. • Provision of facilities for refuse. • Provision of exits etc. • Provision of water supply. • Provision of closets. • Provision of bathrooms. • Provision for food storage. • Site containing offensive material.
		Determination of questions	
		Proposed departure from plans	
		Lapse of deposit of plans	
		Tests for conformity with building regulations	
		Classification of buildings	
		Breach of building regulations	• Penalty for contravening building regulations. • Removal or alteration of offending work. • Obtaining of report where Section 36 notice given. • Civil liability.
		Appeals in certain cases	• Appeal against refusal etc. to relax building regulations. • Appeal against Section 36 notice. • Appeal to crown court. • Appeal and statement of case to high court in certain cases. • Procedure on appeal to Secretary of State on certain matters.
		Application of building regulations to crown etc.	• Application to crown. • Application to United Kingdom Atomic Energy Authority.
		Inner London	

Part	Description	Section	Sub-section
Part 2	Supervision of Building Work etc. other than by a local authority	Supervision of plans and work by approved inspectors	• Giving and acceptance of initial notice. • Effect of initial notice. • Approved inspectors. • Plans certificates. • Final certificates. • Cancellation of initial notice. • Effect of initial notice ceasing to be in force.
		Supervision of their own work by public bodies	
		Supplementary	• Appeals. • Recording and furnishing of information. • Offences. • Construction of Part 11.
Part 3	Other provisions about buildings	Drainage	• Drainage of building. • Use and ventilation of soil pipes. • Repair etc. of drain. • Disconnection of drain. • Improper construction or repair of water-closet or drain.
		Provision of sanitary conveniences	• Provision of closets in building. • Provision of sanitary conveniences in workplace. • Replacement of earth-closets etc. • Loan of temporary sanitary conveniences. • Erection of public conveniences.
		Buildings	• Provision of water supply in occupied house. • Provision of food storage accommodation in house. • Entrances, exits etc. to be required in certain cases. • Means of escape from fire. • Raising of chimney. • Cellars and rooms below subsoil water level. • Consents under Section 74.
		Defective premises, demolition etc.	• Defective premises. • Dangerous building. • Dangerous building – emergency measures. • Ruinous and dilapidated buildings and neglected sites. • Notice to local authority of intended demolition. • Local authority's power to serve notice about demolition.

Part	Description	Section	Sub-section
			• Notices under Section 81. • Appeal against notice under Section 81.
		Yards and passages	• Paving and drainage of yards and passages. • Maintenance of entrances to courtyards.
		Appeal to crown court	
		Application of provisions to crown property	
		Inner London	
		Miscellaneous	• References in Acts to building by-laws. • Facilities for inspecting local Acts.
Part 4	General	Duties of local authorities	
		Documents	• Form of documents. • Authentication of documents. • Service of documents.
		Entry on Premises	• Power to enter premises. • Supplementary provisions as to entry.
		Execution of works	• Power to require occupier to permit work. • Content and enforcement of notice requiring works. • Sale of materials. • Breaking open of streets.
		Appeal against notice requiring works	
		General provisions about appeals and applications	• Procedure on appeal or application to magistrates' court. • Local authority to give effect to appeal. • Judge not disqualified by liability to rates.
		Compensation, and recovery of sums	• Compensation for damage. • Recovery of expenses etc. • Payments by instalments. • Inclusion of several sums in one complaint. • Liability of agent or trustee. • Arbitration.
		Obstruction	
		Prosecutions	• Prosecution of offences. • Continuing offences.
		Protection of members etc. of authorities	
		Default powers	• Default powers of Secretary of State.

Part	Description	Section	Sub-section
			• Expenses of Secretary of State.
			• Variation or revocation of order transferring powers.
		Local inquiries	
		Orders	
		Interpretation	• Meaning of 'building'.
			• Meaning of 'Building Regulations'.
			• Meaning of 'construct' and 'erect'.
			• Meaning of deposit of plans.
			• Construction and availability of sewers.
			• General interpretation.
			• Construction of certain references concerning Temples.
		Savings	• Protection for dock and railway undertakings.
			• Saving for Local Land Charges Act 1975.
			• Saving for other laws.
			• Restriction of application of Part IV to Schedule 3.
Part 5	Supplementary	Supplementary	• Transitional provisions.
			• Consequential amendments and repeals.
			• Commencement.
			• Short title and extent.
		Schedule 1 – Building regulations.	
		Schedule 2 – Relaxation of building regulations.	
		Schedule 3 – Inner London.	
		Schedule 4 – Provisions consequential upon public body's notice.	
		Schedule 5 – Transitional provisions.	
		Schedule 6 – Consequential amendments.	
		Schedule 7 – Repeals.	

2

The Building Regulations 2000

Even when planning permission is not required, most building works, including alterations to existing structures, are subject to minimum standards of construction to safeguard public health and safety.

2.1 What is the purpose of the Building Regulations?

The Building Regulations are legal requirements laid down by parliament, based on the Building Act 1984. They are approved by parliament and deal with the minimum standards of design and building work for the construction of domestic, commercial and industrial buildings.

Building Regulations ensure that new developments or alterations and/or extensions to buildings are all carried out to an agreed standard that protects the health and safety of people in and around the building.

Building standards are enforced by your local building control officer, but for matters concerning drainage or sanitary installations, you will need to consult their technical services department.

Builders and developers are required by law to obtain building control approval, which is an independent check that the Building Regulations have been complied with. There are two types of building control providers – the local authority and approved inspectors.

2.2 Why do we need the Building Regulations?

As mentioned in the Preface, the Great Fire of London in 1666 was the single most significant event to have shaped today's legislation. The rapid growth of the fire through timber buildings built next to each other highlighted the need for builders to consider the possible spread of fire between properties when rebuilding work commenced. This resulted in the first building construction legislation that required all buildings to have some form of fire resistance.

During the Industrial Revolution (200 years later) poor living and working conditions in ever expanding, densely populated urban areas, caused outbreaks of cholera and other serious diseases. Poor sanitation, damp conditions and lack of ventilation forced the government to take action and building control took on the greater role of health and safety through the first Public Health Act of 1875. This Act had two major revisions in 1936 and 1961 and led to the first set of national building standards – the Building Regulations 1965.

The current legislation is the Building Regulations 2000 (Statutory Instrument No 2531) which is made by the Secretary of State for the Environment under powers delegated by parliament under the Building Act of 1984.

The Building Regulations are a set of minimum requirements designed to secure the health, safety and welfare of people in and around buildings and to conserve fuel and energy in England and Wales. They are basic performance standards and the level of safety and acceptable standards are set out as guidance in the Approved Documents. Compliance with the detailed guidance of the Approved Documents is usually considered as evidence that the Regulations themselves have been complied with.

Alternate ways of achieving the same level of safety, or accessibility, are also acceptable.

2.3 What building work is covered by the Building Regulations?

The Building Regulations cover all new building work. This means that if you want to put up a new building, extend or alter an existing one, or provide fittings in a building such as drains or heat-producing appliances, washing and sanitary facilities and hot water storage (i.e. unvented hot water systems), the Building Regulations will probably apply. They may also apply to certain changes of use of an existing building (even though construction work may not be intended) as the change of use could involve the building having to meet different requirements of the Regulations.

It should be remembered, however, that although it may appear that the Regulations do not apply to some of the work you wish to undertake, the end result of doing that work could well lead to you contravening some of the Regulations. You should also recognize that some work – whether or not controlled – could have implications for an adjacent property. In such cases it would be advisable to take professional advice and consult the local authority or an approved inspector. Some examples are:

- removing a buttressed support to a party wall;
- underpinning part of a building;
- removing a tree close to a wall of an adjoining property;

- adding floor screed to a balcony which may reduce the height of a safety barrier;
- building parapets which may increase snow accumulation and lead to an excessive increase in loading on roofs.

2.4 What are the requirements associated with the Building Regulations?

The Building Regulations contain a list of requirements (referred to as 'Schedule 1') that are designed to ensure the health and safety of people in and around buildings; to promote energy conservation; and to provide access and facilities for disabled people. In total there are 13 parts (A–N less I) to these requirements and these cover subjects such as structure, fire safety, ventilation, drainage etc.

The requirements are expressed in broad, functional terms in order to give designers and builders the maximum flexibility in preparing their plans.

2.5 What are the Approved Documents?

Approved Documents contain practical and technical guidance on ways in which the requirements of each part of the Building Regulations can be met.

Each approved document reproduces the *requirements* contained in the Building Regulations relevant to the subject area. This is followed by *practical and technical guidance*, with examples, on how the requirements can be met in some of the more common building situations. There may, however, be alternative ways of complying with the requirements to those shown in the Approved Documents and you are, therefore, under no obligation to adopt any particular solution in an Approved Document if you prefer to meet the relevant requirement(s) in some other way.

The Building Regulations are constantly reviewed to meet the growing demand for better, safer and more accessible buildings. Any changes necessary are brought into operation after consultation with all interested parties. This has meant several amendments since 2000 with the emphasis in more recent years being on:

- increased thermal insulation to conserve energy and reduce global warming;
- providing better access and facilities for disabled people;
- a more comprehensive, one stop approach to fire safety requirements.

Just because an Approved Document has not been complied with, however, does not necessarily mean that the work is wrong. The circumstances of every particular case are considered when an application is made to make sure that adequate levels of safety will be achieved.

The approved documents are in 13 parts, A to N (less 'I') and consist of:

A   Structure
B   Fire safety
C   Site preparation and resistance to moisture
D   Toxic substances
E   Resistance to the passage of sound
F   Ventilation
G   Hygiene
H   Drainage and waste disposal
J   Combustion appliances and fuel storage systems
K   Protection from falling, collision and impact
L   Conservation of fuel and power
M   Access and facilities for disabled people
N   Glazing – safety in relation to impact, opening and cleaning

💡 Approved documents A to K and N (except for paragraphs H2 and J6) do not require anything to be done except for the purpose of securing reasonable standards of health and safety for persons in or about buildings (and any others who may be affected by buildings, or matters connected with buildings).

📠 You can buy a copy of the approved documents (and the Building Act 1984 if you wish) from The Stationery Office (TSO), PO Box 29, Duke Street, Norwich, NR3 1GN (Tel: 0870 600 5522, Fax: 0870 600 5533, www.tso. co.uk), or some book shops. Sometimes they are available from libraries. Alternatively, you can download .pdf copies of the Approved Documents from www.hsmo.gov.uk (then legislation/uk/acts) or www.safety.odpm.gov.uk.

2.5.1 Part A Structure

So that buildings do not collapse, requirements ensure that:

- all structural elements of a building can safely carry the loads expected to be placed on them;
- foundations are adequate for any movement of the ground (for example, movement caused by trees, etc);
- large buildings are strong enough to withstand an explosion without collapsing.

2.5.2 Part B Fire safety

The Regulations consider six aspects of fire safety in the construction of buildings. These are:

(1)   that the design of a building enables occupants to escape to a place of safety, by their own efforts, in the event of a fire;
(2)   that the internal linings of a building do not support a rapid spread of fire;

(3)   that the structure of the building does not collapse prematurely;
(4)   that the slow spread of fire through the building (as well as in unseen cavities and voids) is prevented by providing fire resisting walls and/or partitions where necessary;
(5)   that the spread of fire between buildings is limited by spacing them apart and controlling the number and size of openings on boundaries;
(6)   that the building is designed to enable the fire brigade to fight a fire and rescue any persons caught in a fire.

2.5.3 Part C Site preparation and resistance to moisture

There are four requirements to this part:

(1)   that before any Building Works commence, all vegetation and topsoil is removed;
(2)   that any contaminated ground is either treated, neutralized or removed before a building is erected;
(3)   that subsoil drainage is provided to waterlogged sites;
(4)   that the roof, walls, and floor are weatherproofed against damp and rain penetration.

2.5.4 Part D Toxic substances

This part requires walls to be constructed in such a way that any fumes filling a cavity are prevented from penetrating the building.

2.5.5 Part E Resistance to passage of sound

This part has three main requirements:

(1)   that walls of dwellings have reasonable resistance to the passage of air-borne sound;
(2)   that floors and stairs in flats have reasonable resistance to the passage of airborne sound from below;
(3)   that floors in flats have reasonable resistance to the passage of impact sound from above.

2.5.6 Part F Ventilation

There are two aspects considered by this part:

(1)   adequate ventilation must be provided to kitchens, bath and shower rooms, sanitary accommodation and to other habitable rooms (both domestic and non-domestic);
(2)   roofs need to be well vented, or designed, to prevent moist air causing condensation damage.

2.5.7 Part G Hygiene

There are three aspects included in this part:

(1) buildings are required to have satisfactory sanitary conveniences and washing facilities;
(2) all dwellings are required to have a fixed bath or shower with hot and cold water;
(3) unvented hot water systems over a certain size are required to have safety provisions to prevent explosion.

2.5.8 Part H Drainage and waste disposal

There are four aspects of this part:

(1) new drains taking foul water from buildings are required to discharge to a foul water sewer (or other suitable outfall), be watertight, and be accessible for cleaning;
(2) where no public sewer is available, holding tanks or sewage treatment plants should be made available;
(3) new drains taking rainwater from roofs of buildings need to be watertight, accessible for cleaning and (if there is no sewer available) discharge to a suitable surface water sewer or ditch, soakaway, or watercourse;
(4) storage facilities, reasonably close to the building, need to be provided for refuse collection.

2.5.9 Part J Combustion appliances and fuel storage systems

There are three main aspects to this part:

(1) heat producing appliances must be provided with a supply of fresh air to prevent carbon monoxide poisoning to the building's occupants;
(2) chimneys and flues need to be adequately designed so that smoke and other products of combustion are safely discharged to the outside air;
(3) fireplaces and heat producing appliances should be designed and positioned so as to avoid the building's structure from igniting.

2.5.10 Part K Protection from falling, collision and impact

There are five main aspects to this part:

(1) to avoid accidents on stairs, ladders and ramps, the physical dimensions need to be suitable for the use of the building;
(2) to avoid persons falling off stairwells, balconies, floors, some roofs, light wells and basement areas or similar sunken areas connected to a building need to be suitably guarded according to the building's use;

(3)   to avoid vehicles falling off buildings, car park floors, ramps and other raised areas need to be provided with vehicle barriers;
(4)   to avoid danger to people from colliding with an open window, skylight, or ventilator, some form of guarding may be needed;
(5)   measures need to be taken to avoid the opening and closing of powered sliding or open-upwards doors and gates falling onto any person or trapping them.

2.5.11 Part L Conservation of fuel and power

There are four main aspects to this part:

(1)   roofs, walls, windows, doors and floors need to have resistance to loss of heat (the amount will vary according to the size and use of building);
(2)   controls need to be available to enable occupants to turn off electric lighting;
(3)   controls need to be available to enable low energy lights to be used;
(4)   controls need to be provided for boilers so as to avoid inefficient usage and waste.

2.5.12 Part M Access and facilities for disabled people

There are five main aspects to this part:

(1)   buildings are to be designed so that people with a disability can use them safely and conveniently. Specifically:
       – dwellings should have wider corridors, passages, stairs, and internal doors in the entrance storey;
       – all switches and electrical sockets in a new dwelling should be positioned between 450 mm and 1200 mm from floor level;
       – common areas of new flats should be made easier for disabled people to access the flats;
       – lifts and stairs should be designed with wheelchair users in mind.
(2)   new buildings and some extensions to buildings (but not dwelling extensions) must allow access for people with disabilities into and within buildings;
(3)   sanitary conveniences need to be provided (of adequate size and fittings) for people with a wide range of disabilities;
(4)   new housing must have a WC that is accessible to wheelchair users in the entrance storey (if there are no habitable rooms in the entrance storey, the accessible WC could be in the principal storey.);
(5)   audience or spectator seating in theatres, audit, sports grounds etc. are required to have suitable facilities provided for wheelchair users.

2.5.13 Part N Glazing – safety in relation to impact, opening and cleaning

There are four main aspects in this part:

(1) glazing in locations where people might collide with the glass should either be robust enough not to break, or be constructed of safety glass, or be provided with suitable guarding;
(2) large sheets of glazing need to be made obvious so that people do not collide with that glazing;
(3) where non-dwelling windows, skylights and ventilators are openable by people, controls and/or limiters need to be provided to ensure safe operation and prevent persons falling through a window;
(4) safe access for cleaning both sides of non-dwelling windows, skylights etc. over 2.0 m above ground needs to be available.

2.6 Are there any exemptions?

The Building Regulations do not apply to:

* schools or other educational establishments (provided that they h?· been erected according to plans approved by the Secretary for Educa   and Science or the Secretary of State for Wales);
* a building belonging to statutory undertakers;
* a building belonging to the United Kingdom Atomic Energy Authority;
* a building belonging to the British Airports Authority;
* a building belonging to the Civil Aviation Authority;

unless it is a house, hotel or a building used as offices or showrooms not forming part of any of the above premises.

2.7 What happens if I do not comply with an Approved Document?

Not actually complying with an Approved Document doesn't mean that a person is liable to any civil or criminal prosecution. If, however, that person has contravened a Building Regulation then not having complied with the recommendations contained in the Approved Documents may be held against them.

2.8 Do I need Building Regulations approval?

If you are considering carrying out building work to your property then you may need to apply to your local authority for *Building Regulations approval*.

For most types of building work (e.g. extensions, alterations, conversions and drainage works), you will be required to submit a Building Regulations application prior to commencing any work.

Certain types of extensions and small detached buildings are exempt from Building Regulations control. However, you may still be required to apply for planning permission.

🎯 If you are in any doubt about whether you need to apply for permission, you should contact your Local Authority Building Control Department before commencing any work to your property (in all cases, you may require Planning Permission).

2.8.1 Building work needing formal approval

The Building Regulations apply to any building that involves:

- the erection of a new building or re-erection of an existing building;
- the extension of a building;
- the 'material alteration' of a building;
- the 'material change of use' of a building;
- the installation, alteration or extension of a controlled service or fitting to a building.

2.8.2 Typical examples of work needing approval:

- Home extensions such as for a kitchen, bedroom, lounge, etc.;
- Loft conversions;
- Internal structural alterations, such as the removal of a load-bearing wall or partition;
- Installation of baths, showers, WCs that involve new drainage or waste plumbing;
- Installation of new heating appliances (other than electric);
- New chimneys or flues;
- Underpinning of foundations;
- Altered openings for new windows in roofs or walls;
- Replacing roof coverings (unless exactly like for like repair);
- Installation of cavity insulation;
- Erection of new buildings that are not exempt.

2.8.3 Exempt buildings

There are certain buildings and work that are exempt from control. This is generally because they are buildings controlled by other legislation or because it would not be reasonable to control.

The following list, although not extensive, provides an indication of the main exemptions. These Regulations do not apply to:

- local authorities;
- county councils;
- public bodies;
- the Metropolitan Police Authority;
- any building constructed in accordance with the Explosives Acts 1875 and 1923;
- any building erected under the Nuclear Installations Act 1965;
- any building included in the schedule of monuments maintained under Section 1 of the Ancient Monuments and Archaeological Areas Act 1979;
- buildings not frequented by people;
- greenhouses and agricultural buildings – unless they are being used for retailing, packing or exhibiting;
- temporary buildings, i.e. a building which is not intended to remain erected for more than 28 days;
- ancillary buildings, e.g. an office on a building site;
- a detached building with a floor area less than $15\,m^2$ and containing no sleeping accommodation;
- a small detached building with a floor area leas than $30\,m^2$, which contains no sleeping accommodation, is less than 1 m from the boundar   d is constructed substantially of non-combustible material;
- a detached building designed and intended to shelter people from the effects of nuclear, chemical or conventional weapons;
- a conservatory whose floor area is less than $30\,m^2$ provided that it is wholly or partly glazed;
- a porch whose floor area is less than $30\,m^2$ provided that it is wholly or partly glazed;
- a covered yard or covered way whose floor area is less than $30\,m^2$;
- a carport open on at least two sides whose floor area is less than $30\,m^2$.

The power to dispense with or relax any requirement contained in these Regulations rests with the local authority. It is, therefore advisable to contact your local authority Building Control Officer with details of your particular exemption claim so that you obtain a written reply agreeing the exemption. This will aid any future sale of the property!

Are there any other exemptions from the requirement to give building notice or deposit full plans?

The installations listed in Table 2.1 are exempt from having to give building notice or deposit full plans, **provided** that the person carrying out the work is as indicated in the second column.

In addition, provided any associated building work required to ensure that the appliance (service or fitting detailed above) complies with the applicable

Table 2.1 Exemptions from the requirement to give building notice or to deposit full plans

Type of work	Person carrying out work
Installation of a **heat-producing gas appliance**.	A person, or an employee of a person, who is a member of a class of persons approved in accordance with Regulation 3 of the Gas Safety (Installation and Use) Regulations 1998.
Installation of an **oil-fired combustion appliance** which has a rated heat output of 45 kilowatts or less and which is installed in a building with no more than three storeys (excluding any basement).	An individual registered under the Oil Firing Registration Scheme by the Oil Firing Technical Association for the Petroleum Industry Ltd in respect of that type of work.
Installation of **oil storage tanks** and the pipes connecting them to combustion appliances.	An individual registered under the Oil Firing Registration Scheme by the Oil Firing Technical Association for the Petroleum Industry Ltd in respect of that type of work.
Installation of a **solid fuel burning combustion appliance** which has a rated heat output of 50 kilowatts or less and which is installed in a building with no more than three storeys (excluding any basement).	An individual registered under the Registration Scheme for Companies and Engineers involved in the Installation and Maintenance of Domestic Solid Fuel Fired Equipment by HETAS Ltd in respect of that type of work.
Installation of a **service or fitting** which is installed in or in connection with a building with no more than three storeys (excluding any basement) and which does not involve connection to a drainage system at a depth greater than 750 mm from the surface.	An individual registered under the Approved Contractor Person Scheme (Building Regulations) by the Institute of Plumbing in respect of that type of work.
Installation of a **foul water drainage system** which is installed in or in connection with a building with no more than three storeys (excluding any basement) and which does not involve connection to a drainage system at a depth greater than 750 mm from the surface.	An individual registered under the Approved Contractor Person Scheme (Building Regulations) by the Institute of Plumbing in respect of that type of work.
Installation of **a rainwater drainage system** in relation to which paragraph H3 of Schedule 1 imposes a requirement, which is installed in or in connection with a building with no more than three storeys (excluding any basement) and which does not involve connection to a drainage system at a depth greater than 750 mm from the surface.	An individual registered under the Approved Contractor Person Scheme (Building Regulations) by the Institute of Plumbing in respect of that type of work.
Installation of **a hot water vessel** which is installed in or in connection with a building with no more than three storeys (excluding any basement) and which does not involve connection to a drainage system at a depth greater than 750 mm from the surface.	An individual registered under the Approved Contractor Person Scheme (Building Regulations) by the Institute of Plumbing in respect of that type of work.
Installation, as a replacement, of a **window, rooflight, roof window or door in an existing building**.	A person registered under the Fenestration Self-Assessment Scheme by Fensa Ltd in respect of that type of work.

requirements contained in Schedule 1 – unless it is a heat producing gas appliance) which

(a) has a net rated heat input of 70 kilowatts or less; and
(b) is installed in a building with no more than 3 storeys (excluding any basement).

 'appliance' includes any fittings or services, other than a hot water storage vessel that does not incorporate a vent pipe to the atmosphere, which form part of the space heating or hot water system served by the combustion appliance; and

 'building work' does **not** include the provision of a masonry chimney.

2.8.4 Where can I obtain assistance in understanding the requirements?

Local councils can provide assistance with:

- applying for Building Regulations approval;
- what type of application is most appropriate for your proposal;
- how to prepare your application, and what information we need;
- advice about the use of materials;
- advice on fire safety measures including safe evacuation of buildings in the event of an emergency;
- advice about how to incorporate the most efficient energy safety measures into your scheme;
- how to provide adequate access for disabled people;
- at what stages local councils need to inspect your work;
- what your Building Regulation Completion Certificate means to you.

2.9 How do I obtain Building Regulations approval?

You, as the owner or builder, are required to fill in an application form and return it, along with basic drawings and relevant information, to the building control office at least two days before work commences. Alternatively, you may submit full detailed plans for approval. Whatever method you adopt, it may save time and trouble if you make an appointment to discuss your scheme with the building control officer well before you intend carrying out any work.

The building control officer will be happy to discuss your intentions, including proposed structural details and dimensions together with any lists of the materials you intend to use, so that he can point out any obvious contra-vention of the Building Regulations before you make an official application for approval. At the same time he can suggest whether it is necessary to approach other authorities to discuss planning, sanitation, fire escapes and so on.

The building control officer will ask you to inform the office when crucial stages of the work are ready for inspection (by a surveyor) in order to make sure the work is carried out according to your original specification. Should the surveyor be dissatisfied with any aspect of the work, he may suggest ways to remedy the situation.

When the building is finished you must notify the council, and it would be to your advantage to ask for written confirmation that the work was satisfactory as this will help to reassure a prospective buyer when you come to sell the property.

2.10 What are Building Control Bodies?

Your local authority has a general duty to see that all building work complies with the Building Regulations. To ensure that your particular building work complies with the Building Regulations you must use one of the two services available to check and approve plans, and to inspect your work as appropriate. The two services are the local authority building control service or the service provided by the private sector in the form of an approved inspector. Both building control bodies will charge for their services. Both may offer advice before work is started.

2.10.1 What will the local authority do?

This rather depends on whether you are submitting a full plans application or a building notice. There are two options for you to choose from if you decide to ask the building control office to assess your planning application:

(1)   full plans application submission, or
(2)   building notice application.

In both cases the building control office will carry out site inspections at various stages.

 The total fee is the same whichever method is chosen.

Full plans

If you use the full plans procedure, the local authority will check your plans and consult any appropriate authorities (such as fire and water authorities). If your plans comply with the Building Regulations you will receive a notice that they have been approved. If the local authority are not satisfied, then you may be asked to make amendments or provide more details. Alternatively, a conditional approval may be issued which will either specify modifications that must be made to the plans, or will specify further plans that must be deposited. A local authority may only apply conditions if you have either requested them to do so or have consented to them doing so. A request or consent must be made in writing. If your plans are rejected the reasons will be stated in the notice.

Building notice

If you use the building notice procedure, as with full plans applications, the work will normally be inspected as it proceeds; but you will not receive any notice indicating whether your proposal has been passed or rejected. Instead, you will be advised where the work itself is found (by the building control officer) not to comply with the Regulations.

Where a building notice has been given, the person carrying out building work or making a material change of use is required to provide plans showing how they intend conforming with the requirements of the Building Regulations. The local authority may also require further information such as structural design calculations of plans.

2.10.2 What will the approved inspector do?

If you use an approved inspector they will give you advice, check plans, issue a plans certificate, inspect the work, etc. as agreed between you both. You and the Inspector will jointly notify the local authority on what is termed an initial notice. Once that has been accepted by the local authority, the approved inspector will then be responsible for the supervision of building work. Although the local authority will have no further involvement, you may still have to supply them with limited information to enable them to be satisfied about certain aspects linked to Building Regulations (e.g. about the point of connection to an existing sewer).

If the approved inspector is not satisfied with your proposals you may alter your plans according to his advice; or you may seek a ruling from the Secretary of State regarding any disagreement between you. The approved inspector might also suggest an alternative form of construction, and, provided that the work has not been started, you can apply to the local authority for a relaxation or a dispensation from one (or more) of the Regulations' requirements and, in the event of a refusal by the local authority, appeal to the Secretary of State.

If, however, you do not exercise these options and you do not do what the approved inspector has advised to achieve compliance, the inspector will not be able to issue a final certificate. The inspector will also be obliged to notify the local authority so that they can consider whether to use their powers of enforcement.

2.10.3 What is the difference between a full plans application and the building notice procedure?

For a full plans application, plans need to be produced showing all constructional details, preferably well in advance of your intended commencement on site. For the building notice procedure less detailed plans are required. In both cases, your application or notice should be submitted to the local authority and should be accompanied by any relevant structural calculations, to demonstrate compliance with safety requirements on the structure of the building.

If the use of the building is a 'designated use' under the Fire Precautions Act 1971, the application method **must** be a 'full plans' submission. This is to allow the local building control office to consult the fire brigade to see if they have any comments on the adequacy of the building's proposed means of escape in the event of fire. Approved plans are valid for at least three years.

2.11 How do I apply for Building Control?

If your prospective work will involve any form of structure, you could need building control approval.

Some types of work may need both planning permission and building control approval; others may need only one or the other. The process of assessing a proposed building project is carried out through an evaluation of submitted information and plans, and the inspection of work as the building progresses.

Take advantage of the free advice that local authorities offer, and discuss your ideas well in advance.

A person who intends carrying out any building work or making a material change of use to a building, shall:

- either provide the local authority with a building notice, or
- deposit full plans with the local authority

subject to the exclusions listed in Section 2.11.1.

2.11.1 What applications do not require submission plans?

The following building works do not require the submission of plans:

- where a person intends to have installed (by a person, or an employee of a person approved in accordance with Regulation 3 of the Gas Safety (Installation and Use) Regulations 1998) a heat-producing gas appliance;

- where Regulation 20 of the Building Regulations 2000 (Approved Inspectors local authority powers in relation to partly completed work) applies;
- in respect of any work specified in an initial notice, an amendment notice or a public body's notice that is in force.

2.11.2 Other considerations

Depending on the type of work involved, you may need to get approval from several sources before starting. The list below provides a few examples:

- There may be legal objections to alterations being made to your property.
- A solicitor might need to be consulted to see if any covenants or other forms of restriction are listed in the title deeds to your property, and if any other person or party needs to be consulted before you carry out your work.
- You may need planning permission for a particular type of development work.
- If a building is listed or is within a Conservation Area or an Area of Outstanding Natural Beauty, special rules apply.

2.12 Full plans application

This type of application can be used for any type of building work, but it **must** be used where the proposed premises are to be used as a factory, office, shop, hotel, boarding house or railway premises.

A full plans application requires the submission of fully detailed plans, specifications, calculations and other supporting details to enable the building control officer to ascertain compliance with the Building Regulations. The amount of detail depends on the size and type of building works proposed, but as a minimum will have to consist of:

- a description of the proposed building work or material change of use;
- plan(s) showing what work will be completed; plus
- a location plan showing where the building is located relative to neighbouring streets.

The full plans application may be accompanied by a request (from the person carrying out such building work) that on completion of the work, he wishes the local authority to issue a completion certificate.

Two copies of the full plans application need to be sent to the local authority except in cases where the proposed building work relates to the erection, extension or material alteration of a building (other than a dwelling-house or flat) and where fire safety imposes an additional requirement, in which case five copies are required.

A full plans application will be thoroughly checked by the local authority who are required to pass or reject your plans within a certain time limit (usually eight weeks); or they may add conditions to an approval, with your written agreement. If they are satisfied that the work shown on the plans complies with the Regulations, you will be issued with an approval notice (within a period of five weeks or up to two months) showing that your plans were approved as complying with the Building Regulations.

If your plans are rejected, and you do not consider it is necessary to alter them, you will have two options available to you:

- you may seek a 'determination' from the Secretary of State if you believe your work complies with the Regulations (but you must apply before work starts);
- if you acknowledge that your proposals do not necessarily comply with a particular requirement in the Regulations and feel that it is too onerous in your particular circumstances, you may apply for a relaxation or dispensation of that particular requirement from the local authority. You can make this sort of application at any time you like but it is obviously sensible to do so as soon as possible and preferably before work starts. If the local authority refuses your application, you may then appeal to the Secretary of State within a month of the date of receipt of the rejection notice.

2.12.1 Consultation with sewerage undertaker

Where applicable, the local authority shall consult the sewerage undertaker as soon as practicable after the plans have been deposited, and before issuing any completion certificate in relation to the building work.

2.12.2 Advantages of submitting full plans application

The advantages of the full plans method of submission are:

- that the plans can be examined and approved in advance for an advance payment of (typically) 25% of the total fee;
- a formal notice of approval or rejection will be issued within five weeks (unless the applicant agrees to extend this to two months);
- only when work starts on site, and the building control officer has completed his initial visit, is the remaining part of the fee invoiced;
- that a (free) completion certificate will be issued on satisfactory completion of the work.

2.13 Building notice procedure

Under the Building Notice procedure no approval notice is given. There is also no procedure to seek a determination from the Secretary of State if there is a disagreement between you and the local authority – unless plans are subsequently deposited. However, the advantage of the building notice procedure is that it will allow you to carry out minor works without the need to prepare full plans. You must, however, feel confident that the work will comply with the Regulations or you risk having to correct any work you carry out at the request of the local authority.

A building notice is particularly suited to minor works (for example, a householder wishing to install another WC). For such building work, detailed plans are unnecessary and most matters can be agreed when the building control officer visits your property. You do not need to have detailed plans prepared, but in some cases you may be asked to supply extra information.

As no formal approval is given, good liaison between the builder and the building control officer is essential to ensure that work does not have to be re-done. This method is **not** allowed for any work on listed buildings or buildings in a Conservation Area.

The submission of a marked up sketch showing the location of the building, although not mandatory, is recommended.

This type of application may be used for all types of building work, so long as no part of the premises are used for any of the purposes mentioned above under the full plans application.

2.13.1 What do I have to include in a Building Notice?

A building notice shall:

- state the name and address of the person intending to carry out the work;
- be signed by that person or on that person's behalf;
- contain, or be accompanied by:
  - a description of the proposed building work or material change of use;
  - particulars of the location of the building;
  - the use or intended use of that building.

Extension of a building

When planning a building extension, the building notice needs to be accompanied by:

- a plan to a scale of not less than 1:1250 showing:
  - the size and position of the building, or the building as extended and its relationship to adjoining boundaries;
  - the boundaries of the curtilage of the building, or the building as extended, and the size, position and use of every other building or proposed building within that curtilage;
  - the width and position of any street on or within the boundaries of the curtilage of the building or the building as extended;

- a statement specifying the number of storeys (each basement level being counted as one storey), in the building to which the proposal relates;
- particulars of:
  - the provisions to be made for the drainage of the building or extension;
  - the steps to be taken to comply with any local enactment which applies.

Insertion of insulating material into the cavity walls of a building

For cavity wall insulations, the building notice needs to be accompanied by a statement which specifies:

- the name and type of insulating material to be used;
- the name of any European Technical Approval issuing body that has approved the insulating material;
- the requirements of Schedule 1 in relation to which the issuing body has approved the insulating material;
- any European Economic Area (EEA) national standard with which the insulating material conforms;
- the name of any body that has issued any current approval to the installer of the insulating material.

Provision of a hot water storage system

A building notice in respect of a proposed hot water system shall be accompanied by a statement which specifies:

- the name, make, model and type of hot water storage system to be installed;
- the name of the body (if any) that has approved or certified the system;
- the name of the body (if any) that has issued any current registered operative identity card to the installer or proposed installer of the system.

2.14 How long is a building notice valid?

A building notice shall cease to have effect on the expiry of three years from the date on which that notice was given to the local authority, unless before the expiry of that period:

- the building work to which the notice related has commenced; or
- the material change of use described in the notice was made.

The approved plans may be built to for at least three years, **even if the Building Regulations change during this time**.

2.15 What can I do if my plans are rejected?

If your plans were initially rejected, you can start work provided you give the necessary notice of commencement required under Regulation 14 of the Building Regulations and are satisfied that the building work itself now complies with the Regulations. However, it would **not** be advisable to follow this course if you are in any doubt and have not taken professional advice. Instead you should:

- resubmit your full plans application with amendments to ensure that they comply with Building Regulations; or
- if you think your plans comply and that the decision to reject is therefore not justified, you can refer the matter to the Secretary of State for the Environment, Transport and the Regions, or the Secretary of State for Wales (as appropriate) for their determination, but usually only before the work has started; or
- you could (in particular cases) ask the local authority to relax or dispense with the rejection. If the local authority refuse your application you could then appeal to the appropriate Secretary of State within one month of the refusal.

In the first two cases, the address to write to is the Department of the Environment, Transport and the Regions (DETR), 3/C1, Eland House, Bressenden Place, London SW1E 5DU. In Wales, you should refer the matter to the Secretary of State for Wales, Welsh Office, Crown Buildings, Cathays Park, Cardiff CF1 3NQ.

A fee is payable for determinations but not for appeals. The fee is half the plan fee (excluding VAT) subject to a minimum of £50 and a maximum of £500. The DETR or the Welsh Office will then seek comments from the local authority on your application (or appeal) which will be copied to you. You will then have a further opportunity to comment before a decision is issued by the Secretary of State.

2.15.1 Do my neighbours have the right to object to what is proposed in my Building Regulations application?

Basically – *no*! But whilst there is no requirement in the Building Regulations to consult neighbours, it would be prudent to do so. In any event, you should be careful that the work does not encroach on their property since this could well lead to bad feeling and possibly an application for an injunction for the removal of the work.

Objections may be raised under other legislation, particularly if your proposal is subject to approval under the Town and Country Planning legislation or the Party Wall etc. Act of 1996.

💡 A free explanatory booklet on the Party Wall etc. Act 1996 is available from the DETR Free Literature, PO Box No 236, Wetherby, LS23 7NB (Tel: 0870 1226 236, Fax: 0870 1226 237).

2.16 What happens if I wish to seek a determination but the work in question has started?

You will only need to seek a determination if you believe the proposals in your full plans application comply with the Regulations but the local authority disagrees. You may apply for a determination either before or after the local authority has formally rejected your full plans application. The legal procedure is intended to deal with compliance of '*proposed*' work only and, in general, applications relating to work which is substantially completed cannot be accepted. Exceptionally, however, applications for '*late*' determinations may be accepted – but it is in your best interest to always ensure that you apply for a determination well before you start work.

2.17 When can I start work?

Again, it depends on whether you are using the local authority or the approved inspector.

2.17.1 Using the local authority

Once you have given a Building Notice or submitted a full plans application, you can start work at any time. However, you must give the local authority a commencement notice at least two clear days (not including the day on which you give notice and any Saturday, Sunday, bank or public holiday) before you start; and if you start work before you receive a decision on your full plans application, you will prejudice your ability to seek a determination from the Secretary of State if there is a dispute.

2.17.2 Using an approved inspector

If you use an approved inspector you may, subject to any arrangements you may have agreed with the inspector, start work as soon as the initial notice is accepted by the local authority (or is deemed to have been accepted if nothing is heard from the local authority within five working days of the notice being given). Work may not start if the initial notice is rejected, however.

2.18 Planning officers

Before construction begins, planning officers determine whether the plans for the building or other structure comply with the Building Regulations and if they are suited to the engineering and environmental demands of the building site. Building inspectors are then responsible for inspecting the structural quality and general safety of buildings.

2.19 Building inspectors

Building inspectors examine the construction, alteration, or repair of buildings, highways and streets, sewer and water systems, dams, bridges, and other structures to ensure compliance with building codes and ordinances, zoning regulations, and contract specifications.

Building codes and standards are the primary means by which building construction is regulated in the UK to assure the health and safety of the general public. Inspectors make an initial inspection during the first phase of construction and then complete follow-up inspections throughout the construction project in order to monitor compliance with regulations.

The inspectors will visit the worksite before the foundation is poured to inspect the soil condition and positioning and depth of the footings. Later, they return to the site to inspect the foundation after it has been completed. The size and type of structure, as well as the rate of completion, determine the number of other site visits they must make. Upon completion of the project, they make a final comprehensive inspection.

2.20 Notice of commencement and completion of certain stages of work

A person who proposes carrying out building work shall not start work unless:

- he has given the local authority notice that he intends to commence work; and
- at least two days have elapsed since the end of the day on which he gave the notice.

2.20.1 Notice of completion of certain stages of work

The person responsible for completing the building work is also responsible for notifying the local authority a minimum of five days prior to commencing any work involving excavations for foundations, foundations themselves, any damp-proof course; any concrete or other material to be laid over a site and drains or sewers.

Upon completion of this work (especially work that will eventually be covered up by later work) the person responsible for the building work shall give five days' notice of intention to backfill.

A person who has laid, haunched or covered any drain or sewer shall (not more than five days after that work has been completed) give the local authority notice to that effect.

Where a building is being erected, and that building (or any part of it) is to be occupied before completion, the person carrying out that work shall give the local authority at least five days' notice before the building, or any part of it is, occupied.

The person carrying out the building work shall **not**:

- cover up any foundation (or excavation for a foundation), any damp-proof course or any concrete or other material laid over a site; or
- cover up (in any way) any drains or sewers **unless** he has given the local authority notice that he intends to commence that work, and at least one day has elapsed since the end of the day on which he gave the notice.

Where a person fails to comply with the above, then the local authority can insist that he shall cut into, lay open or pull down so much of the work as to enable the authority to ascertain whether these Regulations have been complied with, or not. If the local authority then notifies the owner/builder that certain work contravenes the requirements in these Regulations, then the owner/builder shall, after completing the remedial work, notify the local authority of its completion.

This requirement does not apply in respect of any work specified in an initial notice, an amendment notice or a public body's notice that is in force.

2.20.2 What kind of tests are the local authorities likely to make?

To establish whether building work has been carried out in conformance with the Building Regulations, local authorities will test to ensure that all work has been carried out:

- in a workmanlike manner;
- with adequate and proper materials which:
  - are appropriate for the circumstances in which they are used,
  - are adequately mixed or prepared, and
  - are applied, used or fixed so as adequately to perform the functions for which they are designed;
- complies with the requirements of Part H of Schedule 1 (drainage and waste disposal);
- so as to enable them to ascertain whether the materials used comply with the provisions of these Regulations.

2.20.3 Energy rating

Where a new dwelling is being created, the person carrying out the building work shall calculate (and inform the local authority of) the dwelling's energy rating not later than five days after the work has been completed and, where a new dwelling is created, at least five days before intended occupation of the dwelling.

If the building is not to be immediately occupied, then the person carrying out the building work shall affix (not later than five days after the work has been completed) in a conspicuous place in the dwelling, a notice stating the energy rating of the dwelling.

💡 Details of the correct procedures for calculating the energy rating are available from local authorities.

2.21 What are the requirements relating to building work?

In all cases, building work shall be carried out so that it:

(a) *it complies with the applicable requirements contained in Schedule 1; and*
(b) *in complying with any such requirement there is no failure to comply with any other such requirement.*

*Building work shall be carried out so that, **after** it has been completed:*

(a) *any building which is extended or to which a material alteration is made; or*
(b) *any building in, or in connection with which, a controlled service or fitting is provided, extended or materially altered; or*
(c) *any controlled service or fitting, complies with the applicable requirements of Schedule 1 or, where it did not comply with any such requirement, is no more unsatisfactory in relation to that requirement than before the work was carried out.*

2.22 Do I need to employ a professional builder?

Unless you have a reasonable working knowledge of building construction it would be advisable before you start work to get some professional advice (e.g. from an architect, or a structural engineer, or a building surveyor) and/or choose a recognized builder to carry out the work. It is also advisable to consult the local authority building control officer or an approved inspector in advance.

2.23 Unauthorized building work

If, for any reason, building work has been done without a building notice or full plans of the work being deposited with the local authority; or a notice of commencement of work being given, then the applicant may apply, in writing, to the local authority for a regularization certificate.

This application will need to include:

- a description of the unauthorized work,
- a plan of the unauthorized work, and
- a plan showing any additional work that is required for compliance with the requirements relating to building work in the Building Regulations.

Local authorities may then '*require the applicant to take such reasonable steps, including laying open the unauthorised work for inspection by the authority, making tests and taking samples, as the authority think appropriate to ascertain what work, if any, is required to secure that the relevant requirements are met*'.

When the applicant has taken any such steps required by the local authority, the local authority shall notify the applicant:

- of the work which is required to comply with the relevant requirements;
- of the requirements which can be dispensed with or relaxed; or
- that no work is required to secure compliance with the relevant requirements.

2.23.1 What happens if I do work without approval?

The local authority has a general duty to see that all building work complies with the Regulations – except where it is formally under the control of an approved inspector. Where a local authority is controlling the work and finds after its completion that it does not comply, then the local authority may require you to alter or remove it. If you fail to do this the local authority may serve a notice requiring you to do so and you will be liable for the costs.

2.23.2 What are the penalties for contravening the Building Regulations?

If you contravene the Building Regulations by building without notifying the local authority or by carrying out work which does not comply, the local authority can prosecute. If you are convicted, you are liable to a penalty not exceeding £5000 (at the date of publication) plus £50 for each day on which each individual contravention is not put right after you have been convicted. If you do not put the work right when asked to do so, the local authority have power to do it themselves and recover costs from you.

2.24 Why do I need a completion certificate?

A completion certificate certifies that the local authority are satisfied that the work complies with the relevant requirements of Schedule 1 of the Building Regulations, '*in so far as they have been able to ascertain after taking all reasonable steps*'.

A completion certificate is a valuable document that should be kept in a safe place!

2.25 How do I get a completion certificate when the work is finished?

The local authority shall give a completion certificate only when they have received the completion notice and have been able to ascertain that the relevant requirements of Schedule 1 (specified in the certificate) have been satisfied.

Where full plans are submitted for work that is also subject to the Fire Precautions Act 1971, the local authority must issue you with a completion certificate concerning compliance with the fire safety requirements of the Building Regulations once work has finished. In other circumstances, you may ask to be given one when the work is finished, but you must make your request **when you first submit your plans**.

If you use an approved inspector, they must issue a final certificate to the local authority when the work is completed.

2.26 Where can I find out more?

You can find out more from:

- the local authority's building control department;
- an approved inspector; or
- other sources.

2.26.1 Local authority

Each local authority in England and Wales (i.e. unitary, district and London boroughs in England and county and county borough councils in Wales) has a building control section whose general duty is to see that work complies with the Building Regulations – except where it is formally under the control of an approved inspector. Most local authorities have their own website and these usually contain a wealth of useful information, the majority of which is downloadable as read-only .pdf files.

Individual local authorities co-ordinate their services regionally and nationally (and provide a range of national approval schemes) via LABC Services. You can find out more about LABC Services through its website at www.labc-services.co.uk but your local authority building control department will be pleased to give you information and advice. They may offer to let you see their copies of the Building Act 1984, the Building Regulations 2000 and their associated Approved Documents that provide additional guidance.

The *Fire and Building Regulations Procedural Guide* which deals with procedures for building work to which the Fire Precautions Act 1971 applies, and the Department of the Environment, Transport and the Regions (DETR) leaflet on safety of garden walls, are amongst the documentation and advice that is available, free of charge, from your local authority.

The DETR's and the Welsh Office's separate booklets on planning permission for small businesses and householders are also available free of charge from your local authorities.

2.26.2 Approved inspectors

Approved inspectors are companies or individuals authorized under the Building Act 1984 to carry out building control work in England and Wales.

The Construction Industry Council (CIC) is responsible for deciding all applications for approved inspector status. You can find out more about the CIC's role (including how to apply to become an approved inspector) through its website at www.cic.org.uk/cicair/cicair.htm.

A list of approved inspectors can be viewed at the Association of Corporate Approved Inspectors (ACAI) website at www.acai.org.uk.

A DTLR 'circular letter' lists the inspectors approved by the Secretary of State (before the CIC took over this responsibility).

2.26.3 Other sources

Most of the documents, including Approved Documents A to N and Regulation 7, can be purchased from The Stationery Office (TSO – formally HMSO), 29 Duke Street, Norwich, NR3 1GN or from any main bookshop. Orders to TSO can be telephoned to 0870 600 5522 or faxed to 0870 600 5533 and their website is www.tso.co.uk. Amendments and new Approved Documents are issued from time to time. Copies should also be available in public reference libraries, from www.hmso.gov.uk or from www.safety.odm.gov.uk

Appendix A Typical planning permission form

TORRIDGE DISTRICT COUNCIL	Planning and Technical Services Department Building Act 1984 – Building Regulations 2000 **DEPOSIT OF FULL PLANS**	 *Local Authority* **BUILDING CONTROL**

Submit **Two** Copies of **Forms** To:	Complete in BLOCK Capitals and BLACK Ink.
Director of Planning and Technical Services Torridge District Council Riverbank House Bideford EX39 2QG Tel (01237) 428700 Fax (01237) 478849 E-Mail building.control@torridge.gov.uk	APPLICATION NUMBER:................... EXPIRY DATE: .... PLAN FEE INSPECTION FEE Cheque/PO/Cash: .... VAT: .... Net Amount .... Total Fee: .... VAT .... Accepted: .... Total Fee .... *For Office Use Only*

Please Read The Notes Before Completing This Form

1. **Applicants Details (See Note 1):**   Owners Details (if different)
   Name:
   Address:

   Post Code: Tel. No:
   Fax No: E-Mail:

2. **Agents Details** (if any) to whom correspondence should be sent
   Name:
   Address:

   Post Code: Tel. No:
   Fax No: E-Mail:

3. Location of building to which work relates
   Address:

   Post Code: Tel. No:

4. **Proposed Work**
   Description:
   Date of commencement:

5. **Use of building**
   a. If new building or extension state proposed use:
   b. If existing building state present use:

6. **Method of Drainage**
   a. Foul water: ☐
   b. Surface water: ☐

7. **Means of water supply:**

8. Have you received Planning Consent for this work?   YES ☐ / NO ☐
   If yes, please give Consent No. and Date of Decision:

9. **Conservation of Fuel & Power (See Note 5.3)** Please indicate the method used to show compliance with Part L

For Dwellings	For Other Buildings
☐ Elemental Method Estimated SAP;	☐ Elemental Method
☐ Target U-Value ☐60 or less ☐Over 60	☐ Calculation Method
☐ SAP Energy Rating (Calculations to be enclosed)	☐ Energy Use Method
☐ Material Alteration / Change of Use	☐ Material Alteration / Change of Use

PLEASE TURN OVER	**JANUARY 2001**

10.	**Fire Safety Requirements**

Is the building proposed to be put, or is currently put, to a use as a Workplace to which the Fire Precautions (Workplace) Regulations 1997 applies and/or is Designated as defined in the fire Precautions Act 1971) ?
Yes ☐ / No ☐

If Yes, please indicate the design method you have used to show compliance with the Requirements of Part B of Schedule 1 to the Building Regulations.

An additional copy of all plans/specifications are to be submitted if you have answered Yes to the above.

11.	**Conditions**

Do you consent to the plans being passed subject to conditions where appropriate Yes ☐ / No ☐

12.	**Extension of time**

Do you consent to the prescribed period being extended to two months Yes ☐ / No ☐

13.	**Fees** (See Guidance Note on Fees for information)

   a. If Table 1 work - please state number of small domestic buildings:
                  - please state number of different types of buildings:
   b. If Table 2 work - please state floor area : m^2
   c. If Table 3 work - please state estimated cost of work excluding VAT £:

14.	**Statement**

This notice is given in relation to the building work as described, and is submitted in accordance with Regulation 11(1)(b) and is accompanied by the appropriate fee. I understand that further fees will be payable following the first inspection by the Local Authority.

Name: Signature: Date:

Should you have difficulty in filling out these forms please contact this office.

NOTES

1 The applicant is the person proposing to carry out the building work, eg the building's owner.

2 Two copies of this notice should be completed and submitted with two copies of plans and particulars in accordance with the provisions of Building Regulation 13.If a sewer connection is proposed an additional set of plans will be required.

3 Subject to certain exceptions a Full Plans Submission attracts fees payable by the person by whom or on whose behalf the work is to be carried out. Fees are payable in two stages. The first must accompany the deposit of plans and the second fee is payable after the first site inspection of work in progress. This second fee is a single payment in respect of each individual building, to cover all site visits and consultations which may be necessary until the work is satisfactorily completed.

3.1 schedule 1 prescribes the plan and inspection fees payable for small domestic buildings. Schedule 2 prescribes the fees payable for small alterations and extensions to a dwelling home, and the addition of a small garage or carport. Schedule 3 prescribes the fees payable for all other cases.

3.2 the appropriate fee is dependent upon the type of work proposed. Fee scales and methods of calculation are set out in the Guidance Notes on Fees which is available on request.

4 Subject to certain provisions of the Public Health Act 1936 owners and occupiers of premises are entitled to have their private foul and surface water drains and sewers connected to the public sewers, where available. Special arrangements apply to trade effluent discharge. Persons wishing to make such connections must give not less than 21 days notice to the appropriate authority.

5 The requirements of Part L (Conservation of Fuel and Power) must be considered for all dwellings and other buildings whose floor area exceeds 30m2. The requirement also applies to a material Change of Use.

5.1 extensions to dwellings not exceeding 10m^2 in floor area need only be as energy efficient as the existing dwelling. Therefore question 8 is not applicable.

5.2 for other buildings with a floor area exceeding 100m^2 floor area, details of artificial lighting systems will be required with the application.

5.3 Regulation 14a requires the provision of Energy Ratings calculated by the Governments Standard Assessment Procedure (SAP) for new dwellings and dwellings created as a result of material changes of use. The SAP calculation must be provided to the Authority within a minimum of five days prior to the occupation of the dwelling or five days after the completion, whichever comes first. Dwellings includes Flats. The person carrying out the building work must display in the dwelling, as soon as practicable after the energy rating has been calculated, a notice of that rating. The notice must be displayed in a conspicuous place in the dwelling. Should occupation of the dwelling take place before physical completion, then the energy rating notice should be given to the occupier.
The requirement to display an energy rating notice in the dwelling, or give a notice to the occupier, is not necessary where the person carrying out the building work intends to occupy, or occupies, the dwelling as his/her residence.

6 Section 16 of the Building Act 1984 provides for the passing of plans subject to conditions. The conditions may specify modifications to the deposited plans and/or that further plans shall be deposited.

7 These notes are for general guidance only, particulars regarding the deposit of plans are contained in Regulation 13 of the Building Regulations 2000 and, in respect of fees, in the Building (Prescribed Fees, etc) Regulations 1991 as amended.

8 Persons proposing to carry out building work or make a material change of use of a building are reminded that permission may be required under the Town and Country Planning Acts.

9 Further information concerning the Building Regulations and Planning matters may be obtained from your Local Authority.

Appendix B Application for listed building consent

TORRIDGE DISTRICT COUNCIL

Planning and Technical Services Department

Application For
Listed Building Consent

Local Authority
BUILDING CONTROL

Planning (Listed Buildings and Conservation Areas) Act 1990

Submit Three Copies of Plans and Forms To: (See Note 1.)	Complete in BLOCK Capitals and BLACK Ink.
Director of Planning and Technical Services Torridge District Council Riverbank House Bideford EX39 2QG	APPLICATION No: .... DATE RECEIVED: .... *For Office Use Only*

Please Read The Notes Before Completing This Form

1. **Applicants Details**
   Name:
   Address:

   Post Code: Tel. No.:

   Are you, or your partner related to or connected with an elected member or officer of the Torridge District Council? If yes give details:

2. **Agents Details** (if any) to whom correspondence should be sent
   Name:
   Address:

   Post Code: Tel. No.:

3. Full address or location of the building to which this application relates:
   Address:

   Post Code: Tel. No.:

4. Applicants interest in building (e.g. owner, lease, prospective purchaser, etc.)

5. Proposed work e.g. demolition, alteration, extension. Description:

6. Justification why the works are considered desirable or necessary.

7. Drawings and plans submitted with application.
   Details:

Note: The plans should be sufficient to identify the buildings and all the alterations and extensions should be shown in detail; the works should also be shown in relation to any adjacent buildings.

PLEASE TURN OVER

8.   * I/We hereby apply for listed building consent to execute the works described in the application and all the accompanying plans and drawings and in accordance therewith.

DATE: SIGNED: ....

On behalf of:
                                          (insert applicants name if signed by agent)
* Delete where appropriate

Planning (Listed Buildings and Conservation Areas) Act 1990 provide that an application for listed building consent shall not be entertained unless it is accompanied by one of four certificates. If you are the sole owner (see(a) below) of all the land, Certificate A, which is printed below, is appropriate: only one copy need be completed. If you can not complete certificate A you will have to give notice to the other owners and complete certificate B. Certificate C and D are appropriate only if you have made efforts to trace the other owners and have failed.
The forms of these notices and certificates are prescribed in the regulations.

9.   Planning (Listed Buildings and Conservation Areas) Act 1990.

CERTIFICATE A

I hereby certify that no person other than *myself / the applicant was an owner (a) of the building to which the application relates at the beginning of the period of 20 days before the date of the accompanying application.

SIGNED: ....   Date:
*On behalf of:
  • Delete where appropriate   (insert applicants name if signed by agent)

NOTE: (a) "owner" means a person having freehold interest or a leasehold interest the unexpired term of which was not less than 7 years.

NOTES:

1. Planning Policy Guidance Note 15 (para. 3.4) requires that the application must make clear the justification for the works (see question 6).

2. In support of the application, please provide 3no. plans / drawings showing , in red, the location of the property and the work to be carried out. The drawings should preferably be of a scale of no less than 1:100 and should clearly indicate existing and new work

3. Any object or structure fixed to the listed building or forming part of the land and comprised within the curtilage of the building is treated as part of the listed building.

4. If an appeal is made to the Secretary of State concerning this application, the regulations require that a copy of the following documents shall be furnished to the Secretary of State by the applicant:

(a) the application made to the Local Authority Planning Department together with all the relevant plans, drawings, particulars and documents (including a copy of the certificate) submitted with it.
(b) the notice of decision (if any) and all other relevant correspondence with the Local Planning Authority.

5. If consent is granted for the demolition of a listed building, the effect of section 8(2)(b) of the 1990 Act is that demolition may not be undertaken until notice of the proposal has been given to the Royal Commission and the Commission subsequently have either been given reasonable access to the building for at least one month following the grant of consent and before works commenced or have stated that they have completed their record of the building or that they do not wish to record it.

Appendix C Application for agricultural/forestry determination

Planning and Technical Services Department

TORRIDGE DISTRICT COUNCIL

Application For
Agricultural/Forestry Determination

Local Authority
BUILDING CONTROL

Town and Country Planning (General Permitted Development) Order 1995 Part 6 & 7, Class A, Schedule 2

Submit Two Copies of Plans and Forms To:	Complete in BLOCK Capitals and BLACK Ink.
Director of Planning and Technical Services Torridge District Council Riverbank House Bideford EX39 2QG	APPLICATION No: .... DATE RECEIVED: .... *For Office Use Only*

Please Read The Notes Before Completing This Form

1. **Applicants Details**
   Name:
   Address:
                   Post Code: Tel. No.:

2. **Agents Details** (if any) to whom correspondence should be sent
   Name:
   Address:
                   Post Code: Tel. No.:

3. **Full Postal address & Location of Works / Building(s)**

4. **Description of proposed works / building(s) [see note 4]**
   Details:

5. Please state:
   (a) Does this holding consist of more that one separate block of land?   Yes ☐ / No ☐
   If yes, please give the size of each block of land and indicate which block the proposal relates to :-

   (b) Size of holding   [see note 4]

   (c) Size of building(s) in metres
   (d) Height (e.g. to ridge) in metres
   (e) Materials to be used
   Purpose to which building(s) works will be put

<table>
<tr><td>6.</td><td>Please give details on any Agricultural Building(s)/extensions erected within the previous two years</td></tr>
</table>

* I / We enclose the appropriate fee of **£35.00**

DATE: **SIGNED:**

* Delete where appropriate   **On behalf of:**

(insert applicants name if signed by agent)

In accordance with the scale of charges, I enclose a remittance of £

NOTES:

1. An application for Agricultural Determination is required for development consisting of the erection of a building or the significant alteration of a private way. (Note: "Significant extension and significant alteration" mean any extension or alteration of the building where the cubic content of the original building will be exceeded by more than 10%, or the height of the building as extended or altered would exceed the height of the original building.)

2. The accompanying plans must show the siting, design and external appearance of the building, or as the case may be, siting and means of construction of the private way.

3. Please ensure that any dimensions shown on any drawings/plans accompanying this application are given in metric measurement.

4. If you intend to erect a new building as opposed to altering or extending one, it will be necessary to provide a plan demonstrating that the size of the holding exceeds 5 hectares.

5. If the holding is less than 5 hectares, planning permission will always be required. In these cases this application form should NOT be used.

6. The application must be accompanied by a fee of £35.00

Appendix D Application for consent to display advertisements

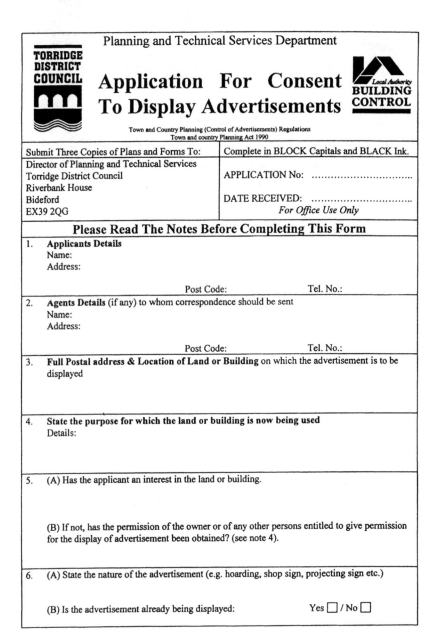

Planning and Technical Services Department

TORRIDGE DISTRICT COUNCIL

Application For Consent To Display Advertisements

Local Authority **BUILDING CONTROL**

Town and Country Planning (Control of Advertisements) Regulations
Town and country Planning Act 1990

Submit Three Copies of Plans and Forms To:	Complete in BLOCK Capitals and BLACK Ink.
Director of Planning and Technical Services	
Torridge District Council	APPLICATION No: ....
Riverbank House	
Bideford	DATE RECEIVED: ....
EX39 2QG	*For Office Use Only*

Please Read The Notes Before Completing This Form

1. **Applicants Details**
    Name:
    Address:

    Post Code: Tel. No.:

2. **Agents Details** (if any) to whom correspondence should be sent
    Name:
    Address:

    Post Code: Tel. No.:

3. **Full Postal address & Location of Land or Building** on which the advertisement is to be displayed

4. **State the purpose for which the land or building is now being used**
    Details:

5. (A) Has the applicant an interest in the land or building.

    (B) If not, has the permission of the owner or of any other persons entitled to give permission for the display of advertisement been obtained? (see note 4).

6. (A) State the nature of the advertisement (e.g. hoarding, shop sign, projecting sign etc.)

    (B) Is the advertisement already being displayed: Yes ☐ / No ☐

Planning and Technical Services Department

TORRIDGE DISTRICT COUNCIL

Building Act 1984 – Building Regulations 2000

REGULARISATION

Local Authority
**BUILDING
CONTROL**

Submit **Two** Copies of **Forms** To:	Complete in BLOCK Capitals and BLACK Ink.
Director of Planning and Technical Services Torridge District Council Riverbank House Bideford EX39 2QG Tel (01237) 428700 Fax (01237) 478849 E-Mail building.control@torridge.gov.uk	Application No: .... FEE: .... Cheque/PO/Cash: .... VAT: .... Total Fee: .... Accepted: .... *For Office Use Only*

Please Read The Notes Before Completing This Form

1. **Applicants Details** (See Note 1):   Owners Details (if different)
   Name:
   Address:

   Post Code: Tel. No.:
   Fax No: E-Mail:

2. **Agents Details** (if any) to whom correspondence should be sent
   Name:
   Address:

   Post Code: Tel. No.:
   Fax No: E-Mail:

3. **Location** of building to which work relates
   Address:

   Post Code: Tel. No.:

4. **Work requiring Regularisation certificate** (See Note 3):
   Description:

5. **Use of building**
   Previous use:
   Present use:

6. **Services**
   a.   Foul water: ☐
   b.   Surface water: ☐
   c.   Means of water supply: ☐

7. Have you received Planning Consent for this work?   Yes ☐ / No ☐
   If yes, please give Consent No. & Date of Decision:

8. **Conservation of Fuel & Power** (See Note 3.3) Please indicate the method used to show compliance with Part L
   For Dwellings   For Other Buildings
   ☐ Elemental Method   Estimated SAP;   ☐ Elemental Method
   ☐ Target U-Value   ☐60 or less ☐Over 60   ☐ Calculation Method
   ☐ SAP Energy Rating (Calculations to be enclosed)   ☐ Energy Use Method
   ☐ Material Alteration / Change of Use   ☐ Material Alteration / Change of Use

9. **Particulars of building work**
   Date work commenced: Date work completed:

   Name of Builder: Address:

   Postcode: Tel. No.:

PLEASE TURN OVER   **JANUARY 2001**

10. **Fire Safety Requirements**
     Is the Building proposed to be put, or is currently put, to a use as a Workplace to which the Fire Precautions (Workplace) Regulations 1997 applies and/or is Designated as defined in the Fire Precautions Act 1971? Yes ☐ / No ☐

     If Yes, please indicate the method you have used to show compliance with the Requirements of Part B of Schedule 1 to the Building Regulations.

     An additional copy of all plans/specifications are to be submitted if you have answered Yes to the above.

11. **Fees** (See Guidance Note on Fees for information)
     a. If Schedule 1 work - please state number of small domestic buildings:
                              - Please state number of different types of buildings:
     b. If Schedule 2 work - please state floor area m^2 :
     c. If Schedule 3 work - please state estimated cost of work excluding VAT £:

12. **Statement**
     This notice is given in relation to the building work as described, and is submitted in accordance with Regulation 13(a) and is accompanied by the appropriate fee.

     Name: Signature: Date:

     Should you have difficulty in filling out these forms please contact this office.

NOTES

1 The applicant is the building's owner.

2 Two copies of this notice should be completed and submitted.

3 Where the work includes the erection of a new building, material alteration, material change of use or extension, this notice shall be accompanied, so far as is reasonably practicable, by the following:

3.1 a block plan to a scale of not less than 1:1250 showing:

3.1.1 a plan of the unauthorised work;

3.1.2 the provision made for the drainage of the building or extension;

3.1.3 a plan showing any additional work required to be carried out to secure the unauthorised work complies with the requirements relating to building work in the building regulations which were applicable to that work when it was carried out.

3.2 where the building or extension has been erected over a sewer or drain shown on the relative map of public sewers, the precautions taken in building over a sewer or drain.

3.3 the requirements of Part L (Conservation of Fuel and Power) must be considered for all dwellings and all other buildings whose floor area exceeds 30m^2. The requirements also apply to material changes of use.

3.3.1 extensions to dwellings not exceeding 10m^2 in floor area need only be as energy efficient as the existing dwelling. Therefore Question 9 is not applicable.

3.3.2 for other buildings with a floor area exceeding 100m^2 floor area, details of artificial lighting systems will be required with the application.

3.3.3 Regulation 14a requires the provision of Energy Ratings calculated by the Governments Standard Assessment Procedure (SAP) for new dwellings and dwellings created as the result of material changes of use. The SAP calculations must be provided to the Authority as soon as possible. Dwellings includes Flats. The Person who carried out the building work must display in the dwelling, as soon as practicable after the energy rating has been calculated, a notice of that rating. The notice must be displayed in a conspicuous place in the dwelling. Should occupation take place before physical completion, then the energy rating notice should be given to the occupier.
The requirement to display an energy rating notice in the dwelling, or give a notice to the occupier, is not necessary where the person who carried out the building work intends to occupy, or occupies, the dwelling as his/her residence.

4 Where the work involved the insertion of insulating material into the cavity walls of a building this application shall be accompanied by a statement as to:

4.1 the name and type of insulating material used;

4.2 whether or not the insulating material is approved by the British Board of Agreement or conforms to a British Standard specification;

4.3 whether or not the installer was a person who is subject of a British Standards Institution Certification of Registration or has been approved by the British Board of Agreement for the insertion of that material.

5 Where the work involved the provision of an unvented hot water storage system, this application shall be accompanied by a statement as to:

5.1 the name and type of system to be provided;

5.2 whether or not the system is approved by the British Board of Agreement;

5.3 whether or not the installer has been approved by the British Board of Agreement for the provision of that system.

6 In accordance with Regulation 13(a), the Council may require an applicant to take such reasonable steps, including laying open the unauthorised work for inspection, making tests and taking samples as the Authority think appropriate to ascertain what work, if any, is required to secure compliance with the relevant Regulations.

7 The Regularisation Fee is payable at the time the submission is made. A Guidance Note on Fees is available on request.

8 These notes are for general guidance only, particulars regarding the submission of Regularisation applications are contained in Regulation 13(a) of the Building Regulations 2000 and, in respect of fees, in the Building (Prescribed Fees etc) Regulations 1991 - as amended.

9 Further information and advice concerning the Building Regulations and planning matters may be obtained from your Local Authority.

3

The requirements of the Building Regulations

The Building Regulations 2000 (SI 2000 No 2531)

Made   *13th September 2000*
Laid before Parliament   *22nd September 2000*
Came into force   *1st January 2001*

Including the following amending statutory instruments:

SI 2001 No 3335

Made   *4th October 2001*
Laid before Parliament   *11th October 2001*
Came into force   *1st April 2002*

and **SI 2002 No 440**

Made   *28th February 2002*
Laid before Parliament   *5th March 2002*
Came into force   *1st April 2002*

3.1 Part A – Structure

Number	Title	Regulation	Requirement (in a nutshell)
A1	Loading	(1) The building shall be constructed so that the combined dead, imposed and wind loads are sustained and transmitted by it to the ground – (a) safely; and (b) without causing such deflection or deformation of any part of the building, or such movement of the ground, as will impair the stability of any part of another building. (2) In assessing whether a building complies with sub-paragraph (1) regard shall be had to the imposed and wind loads to which it is likely to be subjected in the ordinary course of its use for the purpose for which it is intended.	The safety of a structure depends on: • the loading (see BS 6399, Parts 1 and 3); • properties of materials; • design analysis; • details of construction; • safety factors; • workmanship.
A2	Ground movement	The building shall be constructed so that ground movement caused by – (a) swelling, shrinkage or freezing of the subsoil; or (b) land-slip or subsidence (other than subsidence arising from shrinkage), in so far as the risk can be reasonably foreseen, will not impair the stability of any part of the building.	• Horizontal and vertical ties should be provided.
A3	Disproportionate collapse	The building shall be constructed so that in the event of an accident the building will not suffer collapse to an extent disproportionate to the cause.	Requirement A3 applies only to a building having five or more storeys (each basement level being counted as one storey) excluding a storey within the roof space where the slope of the roof does not exceed 70° to the horizontal.

3.2 Part B – Fire safety

Number	Title	Regulation	Requirement (in a nutshell)
B1	**Means of warning and escape**	*The building shall be designed and constructed so that there are appropriate provisions for the early warning of fire, and appropriate means of escape in case of fire from the building to a place of safety outside the building capable of being safely and effectively used at all material times.* Requirement B1 does not apply to any prison provided under Section 33 of the Prisons Act 1952 (power to provide prisons etc.).	For a typical one or two storey dwelling, the requirement is limited to the provision of smoke alarms and to the provision of openable windows for emergency exit. For all other types of buildings, in case of fire, escape routes should be provided that: • are sufficient in number and capacity according to the size and use of the building; • are suitably located to enable persons to escape to a place of safety in the event of fire; • are sufficiently protected from the effects of fire (by enclosure where necessary); • are adequately lit; • are suitably signed; • either limit the ingress of smoke to the escape route(s) or restrict the fire and remove smoke.
B2	**Internal fire spread (linings)**	*(1) To inhibit the spread of fire within the building the internal linings shall –* *(a) adequately resist the spread of flame over their surfaces; and* *(b) have, if ignited, a rate of heat release which is reasonable in the circumstances.* *(2) In this paragraph 'internal linings' mean the materials lining any partition, wall, ceiling or other internal structure.*	As a fire precaution, all materials used for internal linings of a building should have a low rate of surface flame spread and (in some cases) a low rate of heat release.

B3   Internal fire spread (structure)

(1)   The building shall be designed and constructed so that, in the event of fire, its stability will be maintained for a reasonable period.

(2)   A wall common to two or more buildings shall be designed and constructed so that it adequately resists the spread of fire between those buildings. For the purposes of this sub-paragraph a house in a terrace and a semi-detached house are each to be treated as a separate building.

(3)   To inhibit the spread of fire within the building, it shall be sub-divided with fire-resisting construction to an extent appropriate to the size and intended use of the building.

(4)   The building shall be designed and constructed so that the unseen spread of fire and smoke within concealed spaces in its structure and fabric is inhibited.

🔦 Requirement B3(3) does not apply to material alterations to any prison provided under Section 33 of the Prisons Act 1952.

- All structural, loadbearing elements of a building shall be capable of withstanding the effects of fire for an appropriate period without loss of stability.
- Ideally the building should be sub-divided by elements of fire-resisting construction into compartments.
- All openings in fire-separating elements shall be suitably protected in order to maintain the integrity of the continuity of the fire separation.
- Any hidden voids in the construction shall be sealed and sub-divided to inhibit the unseen spread of fire and products of combustion, in order to reduce the risk of structural failure, and the spread of fire.

B4   External fire spread

(1)   The external walls of the building shall adequately resist the spread of fire over the walls and from one building to another, having regard to the height, use and position of the building.

(2)   The roof of the building shall adequately resist the spread of fire over the roof and from one building to another, having regard to the use and position of the building.

- External walls shall be constructed so that the risk of ignition from an external source, and the spread of fire over their surfaces, is restricted.
- The amount of unprotected area in the side of the building shall be restricted so as to limit the amount of thermal radiation that can pass through the wall.

Number	Title	Regulation	Requirement (in a nutshell)
			• The roof shall be constructed so that the risk of spread of flame and/or fire penetration from an external fire source is restricted. • The risk of a fire spreading from the building to a building beyond the boundary, or vice versa shall be limited.
B5	Access and facilities for the fire service	*(1) The building shall be designed and constructed so as to provide reasonable facilities to assist fire fighters in the protection of life.* *(2) Reasonable provision shall be made within the site of the building to enable fire appliances to gain access to the building.*	For dwellings and other small buildings, it is usually only necessary to ensure that the building is sufficiently close to a point accessible to fire brigade vehicles. In more detail this includes: • vehicle access for fire appliances; • access for fire-fighting personnel; • the provision of fire mains within the building (for non-domestic buildings); • venting for heat and smoke from basement areas.

3.3 Part C – Site preparation and resistance to moisture

Number	Title	Regulation	Requirement (in a nutshell)
C1	**Preparation of site**	*The ground to be covered by the building shall be reasonably free from vegetable matter.*	Buildings should be safeguarded from the adverse effects of: • vegetable matter; • contaminants on or in the ground to be covered by the building; • ground water.
C2	**Dangerous and offensive substances**	*Precautions shall be taken to avoid danger to health and safety caused by substances found on or in the ground to be covered by the building.* This requirement does not apply to a building or space within a building – • into which people do not normally go; or • which is used solely for storage; or • which is a garage used solely in connection with a single dwelling.	Buildings should be safeguarded from the adverse effects of: • vegetable matter; • contaminants on or in the ground to be covered by the building; • ground water.
C3	**Subsoil drainage**	*Subsoil drainage shall be provided if it is needed to avoid –* *(a) the passage of the ground moisture to the interior of the building;* *(b) damage to the fabric of the building.* This requirement does not apply to a building or space within a building – • into which people do not normally go; or • which is used solely for storage; or • which is a garage used solely in nection with a single dwelling.	Buildings should be safeguarded from the adverse effects of: • vegetable matter; • contaminants on or in the ground to be covered by the building; • ground water.

Number	Title	Requirement (in a nutshell)
		Regulation
C4	**Resistance to weather and ground moisture**	*The walls, floors and roof of the building shall resist the passage of moisture to the inside of the building.*

- A solid or suspended floor shall be built next to the ground to prevent undue moisture from reaching the upper surface of the floor.
- A wall shall be erected to prevent undue moisture from the ground reaching the inside of the building, and (if it is an outside wall) adequately resisting the penetration of rain and snow to the inside of the building.
- The roof of the building shall be resistant to the penetration of moisture from rain or snow to the inside of the building.
- All floors next to the ground, walls and roof shall not be damaged by moisture from the ground, rain or snow and shall not carry that moisture to any part of the building which it would damage.

3.4 Part D – Toxic substances

Number	Title	Requirement (in a nutshell)
		Regulation
D1	**Cavity insulation**	*If insulating material is inserted into a cavity in a cavity wall reasonable precautions shall be taken to prevent the subsequent permeation of any toxic fumes from that material into any part of the building occupied by people.*

Fumes given off by insulating materials such as by urea formaldehyde (UF) foams should not be allowed to penetrate occupied parts of buildings to an extent where they could become a health risk to persons in the building by reaching an irritant concentration.

3.5 Part E – Resistance to the passage of sound

Number	Title	Regulation	Requirement (in a nutshell)
E1	**Airborne sound (walls)**	*A wall which –* *(a) separates a dwelling from another building or from another dwelling, or* *(b) separates a habitable room or kitchen within a dwelling from another part of the same building which is not used exclusively as part of the dwelling,* *shall have reasonable resistance to the transmission of airborne sound.*	Dwellings shall be designed and built so that the noise from normal domestic activity in an adjoining dwelling (or other building) is kept down to a level that: • does not affect the health of the occupants of the dwelling; • will allow them to sleep, rest and engage in their normal domestic activities in satisfactory conditions.
E2	**Airborne sound (floors and stairs)**	*A floor or a stair which separates a dwelling from another dwelling, or from another part of the same building which is not used exclusively as part of the dwelling, shall have reasonable resistance to the transmission of airborne sound.*	Dwellings shall be designed and built so that the noise from normal domestic activity in an adjoining dwelling (or other building) is kept down to a level that: • does not affect the health of the occupants of the dwelling; • will allow them to sleep, rest and engage in their normal domestic activities in satisfactory conditions.
E3	**Impact sound (floors and walls)**	*A floor or a stair above a dwelling which separates it from another dwelling, or from another part of the same building which is not used exclusively as part of the dwelling, shall have reasonable resistance to the transmission of impact sound.*	Dwellings shall be designed and built so that the noise from normal domestic activity in an adjoining dwelling (or other building) is kept down to a level that: • does not affect the health of the occupants of the dwelling; • will allow them to sleep, rest and engage in their normal domestic activities in satisfactory conditions.

3.6 Part F – Ventilation

Number	Title	Regulation	Requirement (in a nutshell)
F1	**Means of ventilation**	*There shall be adequate means of ventilation provided for people in the building* Requirement F1 does not apply to a building or space within a building – (a) into which people do not normally go; or (b) which is used solely for storage; or (c) which is a garage used solely in connection with a single dwelling.	• Ventilation (mechanical and/or air-conditioning systems designed for domestic buildings) shall be capable of restricting the accumulation of moisture and pollutants originating within a building.
F2	**Condensation in roofs**	*Adequate provision shall be made to prevent excessive condensation –* (a) *in a roof; or* (b) *in a roof void above an insulated ceiling.*	Condensation in a roof and in spaces above insulated ceilings shall be limited (by the ventilation of cold deck roofs) so that, under normal conditions: • the thermal performance of the insulating materials; and • the structural performance of the roof construction will not be substantially and permanently reduced.

3.7 Part G – Hygiene

Number	Title	Regulation	Requirement (in a nutshell)
G1	Sanitary conveniences and washing facilities	*(1) Adequate sanitary conveniences shall be provided in rooms provided for that purpose, or in bathrooms.* Any such room or bathroom shall be separated from places where food is prepared. *(2) Adequate washbasins shall be provided in –* *(a) rooms containing water closets; or* *(b) rooms or spaces adjacent to rooms containing water closets.* Any such room or space shall be separated from places where food is prepared. *(3) There shall be a suitable installation for the provision of hot and cold water to washbasins provided in accordance with paragraph (2).* *(4) Sanitary conveniences and washbasins to which this paragraph applies shall be designed and installed so as to allow effective cleaning.*	**All** dwellings (house, flat or maisonette) should have at least one closet and one washbasin: • closets (and/or urinals) should be separated by a door from any space used for food preparation or where washing-up is done; • washbasins should, ideally, be located in the room containing the closet; • the surfaces of a closet, urinal or washbasin should be smooth, non-absorbent and capable of being easily cleaned; • closets (and/or urinals) should be capable of being flushed effectively; • closets (and/or urinals) should only be connected to a flush pipe or discharge pipe; • washbasins should have a supply of hot and cold water; • closets fitted with flushing apparatus should discharge through a trap and discharge pipe into a discharge stack or a drain.
G2	Bathrooms	*A bathroom shall be provided containing either a fixed bath or shower bath, and there shall be a suitable installation for the provision of hot and cold water to the bath or shower bath.* ☀ Requirement G2 applies only to dwellings.	All dwellings (house, flat or maisonette) should have at least one bathroom with a fixed bath or shower and the bath or shower should: • have a supply of hot and cold water; • discharge through a grating, a trap and branch discharge pipe to a discharge stack or (if on a ground floor);

Number	Title	Regulation	Requirement (in a nutshell)
			• discharge into a gully or directly to a foul drain; • be connected to a macerator and pump (of an approved type) if there is no suitable water supply or means of disposing foul water.
G3	**Hot water storage**	*A hot water storage system that has a hot water storage vessel which does not incorporate a vent pipe to the atmosphere shall be installed by a person competent to do so, and there shall be precautions:* *(a) to prevent the temperature of stored water at any time exceeding 100°C; and* *(b) to ensure that the hot water discharged from safety devices is safely conveyed to where it is visible but will not cause danger **to** persons in or about the building.* Requirement G3 does not apply to: (a) a hot water storage system that has a storage vessel with a capacity of 15 litres or less; (b) a system providing space heating only; (c) a system that heats or stores water for the purposes only of an industrial process.	A hot water storage system shall: • be installed by a competent person; • not exceed 100°C; • discharge safely; • not cause danger to persons in or about the building.

3.8 Part H – Drainage and waste disposal

Number	Title	Regulation	Requirement (in a nutshell)
H1	Foul water drainage	*(1) An adequate system of drainage shall be provided to carry foul water from appliances within the building to one of the following, listed in order of priority –* *(a) a public sewer; or, where that is not reasonably practicable,* *(b) a private sewer communicating with a public sewer; or, where that is not reasonably practicable,* *(c) either a septic tank which has an appropriate form of secondary treatment or another wastewater treatment system; or, where that is not reasonably practicable,* *(d) a cesspool.* *(2) In this Part 'foul water' means waste water which comprises or includes* *(a) waste from a sanitary convenience, bidet or appliance used for washing receptacles for foul waste; or* *(b) water which has been used for food preparation, cooking or washing.* Requirement H1 does not apply to the diversion of water which has been used for personal washing or for the washing of clothes, linen or other articles to collection systems for reuse.	The foul water drainage system shall: • convey the flow of foul water to a foul water outfall (i.e. sewer, cesspool, septic tank or settlement (i.e. holding) tank); • minimize the risk of blockage or leakage; • prevent foul air from the drainage system from entering the building under working conditions; • be ventilated; • be accessible for clearing blockages; • not increase the vulnerability of the building to flooding. H1 is applicable to domestic buildings and small non-domestic buildings. Further guidance on larger buildings is provided in Appendix A to Approved Document H. Complex systems in larger buildings should be designed in accordance with BS EN 12056.

Number	Title	Regulation	Requirement (in a nutshell)
H2	**Wastewater treatment systems and cesspools**	*(1) Any septic tank and its form of secondary treatment, other wastewater treatment system or cesspool, shall be so sited and constructed that –* *(a) it is not prejudicial to the health of any person;* *(b) it will not contaminate any watercourse, underground water or water supply;* *(c) there are adequate means of access for emptying and maintenance; and* *(d) where relevant, it will function to a sufficient standard for the protection of health in the event of a power failure.* *(2) Any septic tank, holding tank which is part of a wastewater treatment system or cesspool shall be –* *(a) of adequate capacity;* *(b) so constructed that it is impermeable to liquids; and* *(c) adequately ventilated.* *(3) Where a foul water drainage system from a building discharges to a septic tank, wastewater treatment system or cesspool, a durable notice shall be affixed in a suitable place in the building containing information on any continuing maintenance required to avoid risks to health.*	Wastewater treatment systems shall: • have sufficient capacity to enable breakdown and settlement of solid matter in the wastewater from the buildings; • be sited and constructed so as to prevent overloading of the receiving water. Cesspools shall have sufficient capacity to store the foul water from the building until they are emptied. Wastewater treatment systems and cesspools shall be sited and constructed so as not to: • be prejudicial to health or a nuisance; • adversely affect water sources or resources; • pollute controlled waters; • be in an area where there is a risk of flooding. Septic tanks and wastewater treatment systems and cesspools shall be constructed and sited so as to: • have adequate ventilation; • prevent leakage of the contents and ingress of subsoil water; • having regard to water table levels at any time of the year and rising groundwater levels. Drainage fields shall be sited and constructed so as to: • avoid overloading of the soakage capacity; and • provide adequately for the availability of an aerated layer in the soil at all times.

H3   **Rainwater drainage**

(1)   Adequate provision shall be made for rainwater to be carried from the roof of the building.

(2)   Paved areas around the building shall be so constructed as to be adequately drained.

(3)   Rainwater from a system provided pursuant to sub-paragraphs (1) or (2) shall discharge to one of the following, listed in order of priority –

(a)   an adequate soakaway or some other adequate infiltration system; or, where that is not reasonably practicable,

(b)   a watercourse; or, where that is not reasonably practicable,

(c)   a sewer.

Requirement H3(2) applies only to paved areas –

(a)   which provide access to the building pursuant to paragraph M2 of Schedule 1 (access for disabled people);

(b)   which provide access to or from a place of storage pursuant to paragraph H6(2) of Schedule 1 (solid waste storage); or

(c)   in any passage giving access to the building, where this is intended to be used in common by the occupiers of one or more other buildings.

Requirement H3(3) does not apply to the gathering of rainwater for reuse.

Rainwater drainage systems shall:

- minimize the risk of blockage or leakage;
- be accessible for clearing blockages;
- ensure that rainwater soaking into the ground is distributed sufficiently so that it does not damage foundations of the proposed building or any adjacent structure;
- ensure that rainwater from roofs and paved areas is carried away from the surface either by a drainage system or by other means;
- ensure that the rainwater drainage system carries the flow of rainwater from the roof to an outfall (e.g. a soakaway, a watercourse, a surface water or a combined sewer).

Number	Title	Regulation	Requirement (in a nutshell)
H4	**Building over sewers**	*(1) The erection or extension of a building or work involving the underpinning of a building shall be carried out in a way that is not detrimental to the building or building extension or to the continued maintenance of the drain, sewer or disposal main.* *(2) In this paragraph 'disposal main' means any pipe, tunnel or conduit used for the conveyance of effluent to or from a sewage disposal works, which is not a public sewer.* *(3) In this paragraph and paragraph H5 'map of sewers' means any records kept by a sewerage undertaker under Section 199 of the Water Industry Act 1991.* *Requirement H4 applies only to work carried out –* *(a) over a drain, sewer or disposal main which is shown on any map of sewers; or* *(b) on any site or in such a manner as may result in interference with the use of, or obstruction of the access of any person to, any drain, sewer or disposal main which is shown on any map of sewers.*	Building or extension or work involving underpinning shall: • be constructed or carried out in a manner which will not overload or otherwise cause damage to the drain, sewer or disposal main either during or after the construction; • not obstruct reasonable access to any manhole or inspection chamber on the drain, sewer or disposal main; • in the event of the drain, sewer or disposal main requiring replacement, not unduly obstruct work to replace the drain, sewer or disposal main, on its present alignment; • reduce the risk of damage to the building as a result of failure of the drain, sewer or disposal main.

H5 **Separate systems of drainage**

Any system for discharging water to a sewer which is provided pursuant to paragraph H3 shall be separate from that provided for the conveyance of foul water from the building.

Requirement H5 applies only to a system provided in connection with the erection or extension of a building where it is reasonably practicable for the system to discharge directly or indirectly to a sewer for the separate conveyance of surface water which is –

(a) shown on a map of sewers; or

(b) under construction either by the sewerage undertaker or by some other person (where the sewer is the subject of an agreement to make a declaration of vesting pursuant to Section 104 of the Water Industry Act 1991).

Separate systems of drains and sewers shall be provided for foul water and rainwater where:

(a) the rainwater is not contaminated; and

(b) the drainage is to be connected either directly or indirectly to the public sewer system and either –

  (i) the public sewer system in the area comprises separate systems for foul water and surface water; or

  (ii) a system of sewers which provides for the separate conveyance of surface water is under construction either by the sewerage undertaker or by some other person (where the sewer is the subject of an agreement to make a declaration of vesting pursuant to Section 104 of the Water Industry Act 1991).

H6 **Solid waste storage**

(1) Adequate provision shall be made for storage of solid waste.

(2) Adequate means of access shall be provided –

(a) for people in the building to the place of storage; and

(b) from the place of storage to a collection point (where one has been specified by the waste collection authority under Section 46 (household waste) or Section 47 (commercial waste) of the Environmental Protection Act 1990 or to a street (where no collection point has been specified)).

Solid waste storage shall be:

- designed and sited so as not to be prejudicial to health;
- of sufficient capacity having regard to the quantity of solid waste to be removed and the frequency of removal;
- sited so as to be accessible for use by people in the building and of ready access from a street for emptying and removal.

3.9 Part J – Combustion appliances and fuel storage systems

Number	Title	Regulation	Requirement (in a nutshell)

J1	**Air supply**	*Combustion appliances shall be so installed that there is an adequate supply of air to them for combustion, to prevent over-heating and for the efficient working of any flue*
		
		 Requirements J1 only applies to fixed combustion appliances (including incinerators).
			The building shall:
			 • enable the admission of sufficient air for:
			  – the proper combustion of fuel and the operation of flues; and
			  – the cooling of appliances where necessary;
			 • enable normal operation of appliances without the products of combustion becoming a hazard to health;

J2	**Discharge of products of combustion**	*Combustion appliances shall have adequate provision for the discharge of products of combustion to the outside air.* Requirements J2 only applies to fixed combustion appliances (including incinerators).	• enable normal operation of appliances without their causing danger through damage by heat or fire to the fabric of the building; • have been inspected and tested to establish suitability for the purpose intended; • have been labelled to indicate performance capabilities.
J3	**Protection of building**	*Combustion appliances and flue-pipes shall be so installed, and fireplaces and chimneys shall be so constructed and installed, as to reduce to a reasonable level the risk of people suffering burns or the building catching fire in consequence of their use* Requirements J3 only applies to fixed combustion appliances (including incinerators).	Oil and LPG fuel storage installations shall be located and constructed so that they are reasonably protected from fires that may occur in buildings or beyond boundaries Oil storage tanks used wholly or mainly for private dwellings shall be: • reasonably resistant to physical damage and corrosion; • designed and installed so as to minimize the risk of oil escaping during the filling or maintenance of the tank;

J4

Provision of information

Where a hearth, fireplace, flue or chimney is provided or extended, a durable notice containing information on the performance capabilities of the hearth, fireplace, flue or chimney shall be affixed in a suitable place in the building for the purpose of enabling combustion appliances to be safely installed.

- incorporate secondary containment when there is a significant risk of pollution;
- be labelled with information on how to respond to a leak.

J5

Protection of liquid fuel storage systems

Liquid fuel storage systems and the pipes connecting them to combustion appliances shall be so constructed and separated from buildings and the boundary of the premises as to reduce to a reasonable level the risk of the fuel igniting in the event of fire in adjacent buildings or premises.

Requirement J5 applies only to –
(a) fixed oil storage tanks with capacities greater than 90 litres and connecting pipes; and
(b) fixed liquefied petroleum gas storage installations with capacities which are located outside the building and which serve fixed combustion appliances (including incinerators) in the building.

Number	Title	Regulation	Requirement (in a nutshell)
J6	**Protection against pollution**	*Oil storage tanks and the pipes connecting them to combustion appliances shall –* *(a) be so constructed and protected as to reduce to a reasonable level the risk of the oil escaping and causing pollution; and* *(b) have affixed in a prominent position a durable notice containing information on how to respond to an oil escape so as to reduce to a reasonable level the risk of pollution.* Requirement J6 applies only to fixed oil storage tanks with capacities of 3500 litres or less, and connecting pipes, which are – (a) located outside the building; and (b) serve fixed combustion appliances (including incinerators) in a building used wholly or mainly as a private dwelling, but does not apply to buried systems.	

3.10 Part K – Protection from falling, collision and impact

Number	Title	Regulation	Requirement (in a nutshell)
K1	**Stairs, ladders and ramps**	*Stairs, ladders and ramps shall be so designed, constructed and installed as to be safe for people moving between different levels in or about the building.* Requirement K1 applies only to stairs, ladders and ramps which form part of the building.	• All stairs, steps and ladders shall provide reasonable safety between levels in a building. • In a public building the standard of stair, ladder or ramp may be higher than in a dwelling, to reflect the lesser familiarity and greater number of users.
K2	**Protection from falling**	*(a) Any stairs, ramps, floors and balconies and any roof to which people have access, and* *(b) any light well, basement area or similar sunken area connected to a building,* *shall be provided with barriers where it is necessary to protect people in or about the building from falling.* Requirement K2 (a) applies only to stairs and ramps which form part of the building.	• Pedestrian guarding should be provided for any part of a floor, gallery, balcony, roof, or any other place to which people have access and any light well, basement area or similar sunken area next to a building.

Number	Title	Regulation	Requirement (in a nutshell)
K3	**Vehicle barriers and loading bays**	*(1) Vehicle ramps and any levels in a building to which vehicles have access, shall be provided with barriers where it is necessary to protect people in or about the building.* *(2) Vehicle loading bays shall be constructed in such a way, or be provided with such features, as may be necessary to protect people in them from collision with vehicles.*	• Vehicle barriers should be provided that are capable of resisting or deflecting the impact of vehicles. • Loading bays shall be provided with an adequate number of exits (or refuges) to enable people to avoid being crushed by vehicles.
K4	**Protection from collision with open windows, etc.**	*Provision shall be made to prevent people moving in or about the building from colliding with open windows, skylights or ventilators.* Requirement K4 does not apply to dwellings.	• All windows, skylights, and ventilators shall be capable of being left open without danger of people colliding with them.
K5	**Protection against impact from and trapping by doors**	*(1) Provision shall be made to prevent any door or gate –* *(a) which slides or opens upwards, from falling onto any person; and* *(b) which is powered, from trapping any person.* *(2) Provision shall be made for powered doors and gates to be opened in the event of a power failure.* *(3) Provision shall be made to ensure a clear view of the space on either side of a swing door or gate.*	Requirement K5 does not apply to – (a) dwellings, or (b) any door or gate that is part of a lift.

3.11 Part L1 – Conservation of fuel and power in dwellings

Number	Title	Regulation	Requirement (in a nutshell)

L1	**Dwellings**	*Reasonable provision shall be made for the conservation of fuel and power in dwellings by –*
		(a) limiting the heat loss:
		  (i) through the fabric of the building;
		  (ii) from hot water pipes and hot air ducts used for space heating;
		  (iii) from hot water vessels.
		(b) providing space heating and hot water systems which are energy-efficient;
		(c) providing lighting systems with appropriate lamps and sufficient controls so that energy can be used efficiently;
		(d) providing sufficient information with the heating and hot water services so that building occupiers can operate and maintain the services in such a manner as to use no more energy than is reasonable in the circumstances.
		
		The requirement for sufficient controls in paragraph L1(c) applies only to external lighting systems fixed to the building.
			• Energy efficiency measures shall be provided.

3.12 Part L2 – Conservation of fuel and power in buildings other than dwellings

Number	Title	Regulation	Requirement (in a nutshell)

L2	**Buildings other than dwellings**	*Reasonable provision shall be made for the conservation of fuel and power in buildings other than dwellings by –*
		(a) limiting the heat losses and gains through the fabric of the building;
		(b) limiting the heat loss:
		  (i) from hot water pipes and hot air ducts used for space heating;
		  (ii) from hot water vessels and hot water service pipes.
		(c) providing space heating and hot water systems that are energy-efficient;
		(d) limiting exposure to solar overheating;
		(e) making provision where air conditioning and mechanical ventilation systems are installed, so that no more energy needs to be used than is reasonable in the circumstances;
		(f) limiting the heat gains by chilled water and refrigerant vessels and pipes and air ducts that serve air conditioning systems;
		(g) providing lighting systems that are energy-efficient;
		(h) providing sufficient information with the relevant services so that the building can be operated and maintained in such a manner as to use no more energy than is reasonable in the circumstances.
			Requirements L2(e) and (f) apply only within buildings and parts of buildings where more than 200 m^2 of floor area is to be served by air conditioning or mechanical ventilation systems.
			
			Requirement L2(g) applies only within buildings and parts of buildings where more than 100 m^2 of floor area is to be served by artificial lighting.

3.13 Part M – Access and facilities for disabled people

Number	Title	Regulation	Requirement (in a nutshell)
M1	Interpretation	In this part 'disabled people' means people who have – (a) an impairment which limits their ability to walk or which requires them to use a wheelchair for mobility, or (b) impaired hearing or sight. The requirements of this Part do not apply to – (a) a material alteration; (b) an extension to a dwelling, or any other extension that does not include a ground storey; (c) any part of a building that is used solely to enable the building or any service or fitting in the building to be inspected, repaired or maintained.	All precautions shall be taken to ensure that all new dwellings, other buildings and student living accommodation shall be reasonably safe and convenient for disabled people to: • gain access to and within, buildings other than dwellings and to use them; • visit new dwellings and to use the principal storey.
M2	Access and use	Reasonable provision shall be made for disabled people to gain access to and to use the building.	
M3	Sanitary conveniences	(1) Reasonable provision shall be made in the entrance storey of a dwelling for sanitary conveniences, or where the entrance storey contains no habitable rooms, reasonable provision for sanitary conveniences shall be made in either the entrance storey or principal storey.	

Number	Title	Regulation	Requirement (in a nutshell)
		(2) In this paragraph 'entrance storey' means the storey which contains the principal entrance to the dwelling, and 'principal storey' means the storey nearest to the entrance storey which contains a habitable room, or if there are two such storeys equally near, either such storey.	
		(3) If sanitary conveniences are provided in any building which is not a dwelling, reasonable provision shall be made for disabled people.	
M4	**Audience or spectator seating**	*If the building contains audience or spectator seating, reasonable provision shall be made to accommodate disabled people.*	Requirement M4 does not apply to dwellings.

3.14 Part N – Glazing – safety in relation to impact, opening and cleaning

Number	Title	Regulation	Requirement (in a nutshell)
N1	**Protection against impact**	*Glazing with which people are likely to come into contact whilst moving in or about the building shall – (a) if broken on impact, break in a way which is unlikely to cause injury; or (b) resist impact without breaking; or (c) be shielded or protected from impact.*	All glazing installed in buildings shall be: • sufficiently robust to withstand impact from a falling or passing person; or • protected from a falling or passing person.
N2	**Manifestation of glazing**	*Transparent glazing with which people are likely to come into contact while moving in or about the building, shall incorporate features which make it apparent.*	Requirement N2 does not apply to dwellings.
N3	**Safe opening and closing of windows, etc.**	*Windows, skylights and ventilators which can be opened by people in or about the building shall be so constructed or equipped that they may be opened, closed or adjusted safely.*	Requirement N3 does not apply to dwellings.
N4	**Safe access for cleaning windows, etc.**	*Provision shall be made for any windows, skylights, or any transparent or translucent walls, ceilings or roofs to be safely accessible for cleaning.*	Requirement N4 does not apply to – (a) dwellings; or (b) any transparent or translucent elements whose surface are not intended to be cleaned.

Notes

1 Statutory Instrument 2001 No 3335

In addition to minor manuscript amendments, Statutory Instrument 2001 No 3335 amended the Building Regulations 2000 (Statutory Instrument 2000 No 2531) as follows.

(a) The definition of 'controlled service or fitting' in Regulation 2(1) is extended to include services or fittings in relation to which Part L (conservation of fuel and power) imposes a requirement.

(b) A new regulation 3(1A) is inserted to limit the definition of building work in relation to the provision of certain controlled services and fittings in existing dwellings.

(c) The applicable requirements relating to material changes of use contained in Regulation 6 are extended to include paragraphs H1 (foul water drainage) and L2 (conservation of fuel and power in buildings and parts of buildings other than dwellings) of Schedule 1.

(d) Paragraphs H2 and J6 of Schedule 1 are excluded from the limitation on requirements contained in Regulation 8.

(e) A new Regulation 12(4A) is inserted to extend the cases in which a person is required to deposit full plans.

(f) A new regulation 14A is inserted to require the local authority to consult with the sewerage undertaker in certain cases.

(g) The powers of the local authority to test drains and private sewers contained in Regulation 18 are extended to cover tests of any building work.

(h) A new Part H extends the existing requirements (by requiring, in H1 and H3, the provision of adequate foul water and rainwater drainage and by adding a requirement in H6 to provide a means of access to a collection point) and adds new requirements on building over sewers and separate systems of sewers and for the provision of information on wastewater treatment systems.

(i) A new Part J extends the existing requirements (by extending J1 to cover the prevention of overheating and by extending J3 to cover the risk of burns to people) and adds new requirements on the protection of liquid fuel storage systems from fire, protection against pollution and for the provision of information.

(j) A new Part L amends the existing requirements by separating the requirements that apply to dwellings and other buildings and adds new requirements on lighting systems, solar overheating, mechanical ventilation systems and for the provision of information on services.

2 Statutory Instrument 2002 No 440

In addition to minor manuscript amendments, Statutory Instrument 2002 No 440 amended the Building Regulations 2000 (Statutory Instrument 2000

No 2531 as previously amended by Statutory Instrument 2001 No 3335) as follows:

(a)   A new Regulation 16A makes special provision for building work consisting of the installation of replacement windows, rooflights, roof windows and doors in existing buildings. It also authorizes a local authority to accept, as evidence that the work complies with Regulations 4 and 7, a certificate to that effect by a person registered under the Fenestration Self-Assessment Scheme. Regulation 16A also provides for the notification of the completion of work where no certificate is to be given.

3.15 Useful addresses for further information

Details of the schemes referred to in these Regulations can be obtained from the following addresses:

* Fenestration Self-Assessment Scheme – Fensa Ltd, 44–48 Borough High Street, London SE1 1XB (Tel: 020 7207 5874; Fax 020 7357 7458).
* Oil Firing Registration Scheme – OFTEC, Century House, 100 High Street, Banstead, Surrey SM7 2NN (Tel: 01737 373311; Fax: 01737 373553).
* Registration Scheme for Companies and Engineers involved in the Installation and Maintenance of Domestic Solid Fuel Fired Equipment – HETAS Ltd, 12 Kestrel Walk, Letchworth, Hertfordshire SG6 2TB (Tel: 01462 634721; Fax: 01462 674329).
* Approved Contractor Person Scheme (Building Regulations) – Institute of Plumbing, 64 Station Lane, Hornchurch, Essex RM12 6NH (Tel: 01708 472791; Fax: 01708 448987).

4

Planning permission

Before undertaking any building project, you must first obtain the approval of local-government authorities. Many people (particularly householders) are initially reluctant to approach local authorities because, according to local gossip, they are 'likely to be obstructive'. In fact the reality of it is quite the reverse as their purpose is to protect all of us from irresponsible builders and developers and they are normally most sympathetic and helpful to any builder and/or DIY person who wants to comply with the statutory requirements and has asked for their advice.

There are two main controls that districts rely on to ensure that adherence to the local plan is ensured, namely planning permission and building regulation approval. Quite a lot of people are confused as to their exact use and whilst both of these controls are associated with gaining planning permission, actually receiving planning permission does not automatically confer Building Regulation approval and vice versa. You **may** require **both** before you can proceed. Indeed, there may be a variation in the planning requirements (and to some extent the Building Regulations) from one area of the country to another. Consequently, the information given on the following pages should be considered as a guide only and not as an authoritative statement of the law.

You are allowed to make certain changes to your home without having to apply to the local council for permission **provided** that it does not affect the external appearance of the building. These are called permitted development rights. The majority of building work that you are likely to complete will, however, probably require you to have planning permission and it is the nation's planning system that plays an important role in today's society by helping to protect the environment in our towns, cities and the countryside.

For example, if you are thinking about carrying out work on a listed building or work that requires the pruning or felling of a tree protected by a tree preservation order, then you will need to contact your local authority planning department before carrying out any work. You never know, you might even require listed building consent or be required to follow certain procedures if carrying out work to trees.

You do **not** require planning permission to carry out any internal alterations to your home, house, flat or maisonette, provided that it does not affect the external appearance of the building.

4.1 Planning controls

Planning controls exist primarily to regulate the use and siting of buildings and other constructions – as well as their appearance. What might seem to be a minor development in itself, could have far-reaching implications that you had not previously considered (for example, erecting a structure that would ultimately obscure vision at a busy junction and thereby constitute a danger to traffic). Equally, the local authority might refuse permission on the grounds that the planned scheme would not blend sympathetically with its surroundings. Your property could also be affected by legal restrictions such as a right of way, which could prejudice planning permission.

The actual details of planning requirements are complex but in respect of domestic developments, the planning authority is concerned primarily with the construction work such as an extension to the house or the provision of a new garage or new outbuildings that is being carried out. Structures like walls and fences also need to be considered because their height or siting might well infringe the rights of neighbours and other members of the community. The planning authority will also want to approve any change of use, such as converting a house into flats or running a business from premises previously occupied as a dwelling only.

4.1.1 Why are planning controls needed?

The purpose of the planning system is to protect the environment as well as public amenities and facilities. It is not designed to protect the interests of one person over another. Within the framework of legislation approved by parliament, councils are tasked to ensure that development is allowed where it is needed, while ensuring that the character and amenity of the area are not adversely affected by new buildings or changes in the use of existing buildings or land.

Some people think the planning system should be used to prevent any change in their local environment, while others may think that planning controls are an unnecessary interference on their individual rights. The present position is that **all** major works need planning permission from the council but many minor works do not. Parliament thinks this is the right balance as it enables councils to protect the character and amenity of their area, while individuals have a reasonable degree of freedom to alter their property.

If you live in a listed building of historical or architectural interest or your house is in a Conservation Area, you should seek advice before considering any alterations.

4.2 Who requires planning permission?

Although the rules and requirements vary according to whether you actually own a house or a flat/maisonette, generally speaking, the principles and procedures for making planning applications are exactly the same for owners of houses and for freeholders (or leaseholders) of flats and maisonettes. Planning regulations, however, have to cover many different situations and so even the provisions that affect the average householder are quite detailed.

💡 You will not need to apply for planning permission to paint your flat or maisonette but, if you are a leaseholder, you may first need to get permission from your landlord or management company.

4.3 Who controls planning permission?

National policy is mainly set out in Planning Policy Guidance Notes (PPGs). Each region then has its own Regional Planning Guidance Notes (RPGs) and Structure Plans will set out county planning policy. Local plans then set out the planning policy for the district and these must be in broad conformity with PPGs and RPGs.

Through the Building Act 1984, parliament has given the main responsibility for planning to local planning authorities. The structure plan, together with the local plan, forms most of the development plan for the areas of the district.

4.3.1 County structure plan

The structure plan sets the broad planning policies for the area and is prepared by the county council.

4.3.2 District local plan

Local plans are prepared by district councils for their areas (except local plans concerning waste and minerals, which are prepared by the county council), and they set out the planning policies for the whole of the district and are used as the basis for assessing **all** planning applications. The district council is responsible for keeping their plan under constant review and for making it available to everybody (usually via their council website).

The published local plan is used as a guide to the location of development over a ten-year period. For example, they:

- will identify where new homes, jobs and other types of development may be built;
- may require related development to be provided, such as children's play areas, parking facilities and road improvements;
- will outline restrictions where certain types of development are unacceptable.

In preparing local plans, districts are responsible for consulting local people and for ensuring that their views are taken into account, thereby giving them a chance to influence the way in which their area is affected.

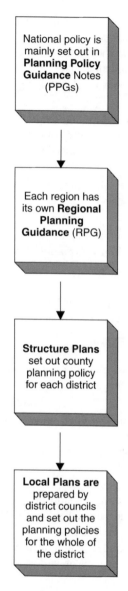

Figure 4.1 Planning responsibilities

👆 At all times, local plans must also take account of national, regional and county planning policy.

A local plan consists of a written statement (which sets out and explains the policies and proposals) and the proposals map, which shows where they apply. Together, these elements of the plan:

- allow local people to clearly see if their homes, businesses or other property would be affected by what is proposed;
- give guidance to anyone who wants to build on a piece of land or change the use of a building in the area; and
- provide a basis for decisions on planning applications.

Planning permission is needed for most building works, engineering works and use of land, and the following are some common examples of when you would need to apply for planning permission:

- you want to make additions or extensions to a flat or maisonette (including those converted from houses);

💡 But you do not need planning permission to carry out internal alterations or work, provided that it does not affect the external appearance of the building.

- you want to divide off part of your house for use as a separate home (for example, a self-contained flat or bed-sit) or use a caravan in your garden as a home for someone else;

💡 But you do not need planning permission to let one or two of your rooms to lodgers.

- you want to divide off part of your home for business or commercial use (for example, a workshop);
- you want to build a parking place for a commercial vehicle;
- you want to build something which goes against the terms of the original planning permission for your house – for example, your house may have been built with a restriction to stop people putting up fences in front gardens because it is on an 'open plan' estate. Your council has a record of all planning permissions in its area;
- the work you want to complete might obstruct the view of road users;
- the work would involve a new or wider access to a trunk or classified road.

💡 If you have any queries about a particular case, the first thing to do is to ask the planning department of your local council. You may also be able to find out more about planning law in your local library. If you are concerned about a legal problem involving planning, you may need to get professional advice or ask your local Citizens Advice Bureau.

💡 A DETR booklet (*Planning Permission, A Guide for Business*) giving advice about working from home and whether Planning Permission is likely to be required is available from Councils.

4.4 What is planning permission?

The planning control process is administered by your local authority and the system '*exists to control the development and use of land and buildings for the best interests of the community*'.

The process is intended to make the environment better for everyone and acts as a service to manage the types of constructions, modifications of premises, uses of land, and ensures the right mix of premises in any one vicinity (that individuals may plan to make) is maintained. The key feature of the process is to allow a party to propose a plan, and for other parties to object if they wish to, or are qualified to.

4.5 What types of planning permission are available?

There are three types of planning permission available: outline, reserved and full.

4.5.1 Outline

This is an application for a development 'in principle' without giving too much detail on the actual building or construction. It basically lets you know, in advance, whether the development is likely to be approved. Assuming permission is granted under these circumstances you will then have to submit a further application in greater detail. In the main, this applies to large-scale developments only and you will be better off making a full application in the first place.

4.5.2 Reserved matters

This is the follow-up stage to an outline application to give more substance and more detail.

4.5.3 Full planning permission

Is the most widely used and is for erection or alteration of buildings or changes of use. There are no preliminary or outline stages and when consent is granted it is for a specific period of time. If this is due to lapse, a renewal of limited permission can be applied for.

4.6 How do I apply for planning permission?

There are fees to pay for each application for planning permission and your local planning office can provide you with the relevant details. You must make sure you have paid the correct fee – as permission can be refused if there is a discrepancy on fees paid.

When your forms and plans are ready, they need to be submitted to the planning office. The planning office will arrange for them to be listed in the local newspaper under 'latest planning applications' and will write to each neighbouring property and (normally) give them 21 days in which to raise any objections.

At the planning office, officials will produce a file after the 21 days has expired, with any objections or supporting information, and will make a recommendation on the application, ready for presenting it at the next planning committee or subcommittee meeting. At this meeting, they will discuss the case, reject it, ask for modifications or accept it. Whichever the decision the planning officer will feed back the decision to the applicant.

💡 There is an appeal procedure, which your local authority planning officer can advise you about.

4.7 Do I really need planning permission?

Most alterations and extensions to property and changes of use of land need to have some form of Planning Permission, which is achieved by submitting a Planning Application to the local authority. The purpose of this control is to protect and enhance our surroundings, to preserve important buildings and natural areas and strengthen the local economy.

However, not all extensions and alterations to dwelling houses require planning permission. Certain types of development are permitted without the need to make an official request, and it is always wise to contact the local authority before commencing any work.

Whether or not Planning Permission is required, good design is always important. Extensions and alterations should be in scale and in harmony with the remainder of the house. The builder should ensure that details such as window openings and matching materials are taken into account.

💡 Householders are encouraged (by councils) to employ a skilled designer when preparing plans for extensions and alterations. Alternatively, the authority's planning officers are able to offer general design guidance prior to the submission of your scheme.

Table 4.1 provides an indication of the basic requirements for planning permission and building regulation approval.

💡 Table 4.1 is only meant as guidance. A more complete description of the above synopsis is contained in Chapter 5. In all circumstances it is recommended that you talk to your local planning officer before contemplating any work. The cost of a local phonecall could save you a lot of money (and stress) in the long term!

Table 4.1 Basic requirements for planning permission and building regulation approval

Type of work	Planning permission		Building Regulation approval	
Decoration and repair inside and outside a building	No	Unless it is a listed building or within a Conservation Area.	No	Unless it is a listed building or within a Conservation Area.
		Consult your local authority.		Consult your local authority.
Structural alterations inside	No	As long as the use of the house is not altered.	Possibly	Consult your local authority.
	Yes	If the alterations are major such as removal or part removal of a load bearing wall or altering the drainage system.	Yes	
	Yes	If they are to an office or shop.	Yes	
Replacing windows and doors	No	Unless:	Possibly	Consult your local authority.
		• they project beyond the foremost wall of the house facing the highway		
		• the building is a listed building		
		• the building is in a Conservation Area.		
	Yes	To replace shop windows.	Yes	
Electrical work	No		No	But it must comply with IEE Regulations.
Plumbing	No		No	For replacements (but you will need to consult the technical services department for any installation that alters present internal or external drainage).
			Yes	For an unvented hot water system.
Central heating	No		No	If electric.
			Yes	If gas, solid fuel or oil.

Table 4.1 Basic requirements for planning permission and building regulation approval (*Continued*)

Type of work	Planning permission		Building Regulation approval	
Oil-storage tank	No	Provided that it is in the garden and has a capacity of not more than 3500 litres (778 gallons) and **no** point is more than 3 m (9 ft 9″) high and **no** part projects beyond the foremost wall of the house facing the highway.	No	
Planting a hedge	No	Unless it obscures view of traffic at a junction or access to a main road.	No	
Building a garden wall or fence	Yes	If it is more than 1 m (3 ft 3″) high and is a boundary enclosure adjoining a highway.	No	
	Yes	If it is more than 2 m (6 ft 6″) high elsewhere.		
Felling or lopping trees	No	Unless the trees are protected by a tree preservation order or you live in a Conservation Area.	No	
Laying a path or a driveway	No	Unless it provides access to a main road.	No	
Building a hard standing for a car	No	Provided that it is within your boundary and is not used for a commercial vehicle.	No	
Installing a swimming pool	Possibly	Consult your local planning officer.	Yes	For an indoor pool.
Erecting aerials, satellite dishes and flagpoles	No	Unless it is a standalone antenna or mast greater than 3 m in height.	No	
	Possibly	If erecting a satellite dish, especially in a Conservation Area or if it is a listed building (consult your local planning officer).		
Advertising	No	If the advertisement is less than 0.3 m² and not illuminated.	Possibly	Consult your local planning officer.

Building a porch	No	Unless: • the floor area exceeds 3 m^2 (3.6 y^2) • any part is more than 3 m (9 ft 9") high • any part is less than 2 m (6 ft 6") from a boundary adjoining a highway or public footpath.	Yes	If area exceeds 30 m^2 (35.9 y^2).
Constructing a small outbuilding	Possibly	Provided the building is less than 10 m^3 (13.08 y^2) in volume, not within 5 m (16 ft 3") of the house or an existing extension. Erecting outbuildings can be a potential minefield and it is best to consult the local planning officer before commencing work.	Yes	If area exceeds 30 m^2 (35.9 y^2). 🔔 If it is within 1 m (3 ft 3") of a boundary, it must be built from incombustible materials.
Building a garage	Possibly	You can build a garage up to 10 m^3 (13.08 y^3) in volume without planning permission, if it is within 5 m (16 ft 3") of the house or an existing extension. Further away than this, it can be up to half the area of the garden, but the height must not exceed 4 m (13 ft)		Yes
Building a conservatory	Possibly	You can extend your house by building a conservatory, provided that the total of both previous and new extensions does not exceed the permitted volume.	Yes	If area exceeds 30 m^2 (35.9 y^2)
Loft conversions and roof extensions	No	Provided the volume of the house is unchanged and the highest part of the roof is not raised.	Yes	
	Yes	For front elevation dormer windows or rear ones over a certain size.		
Building an extension	Possibly	You can extend your house by building an extension, provided that the total of both previous and new extensions does not exceed the permitted volume. Building extensions can be a potential minefield and it is best to consult the local planning officer before contemplating any work.	Yes	If area exceeds 30 m^2 (35.9 y^2)

Table 4.1 Basic requirements for planning permission and building regulation approval (*Continued*)

Type of work	Planning permission		Building Regulation approval	
Converting a house to business premises (including bedsitters)	Yes	Even where construction works may not be intended.	Yes	Unless you are not proposing any building work to make the change.
Converting an old building	Yes		Yes	
Material change of use	Possibly	Even if no building or engineering work is proposed.	Yes	
Building a new house	Yes		Yes	
Infilling	Possibly	Consult your local planning officer.	Yes	If a new development.
Demolition	Yes	If it is a listed building or in a Conservation Area.	No	For a complete detached house.
		If the whole house is to be demolished.	Yes	For a partial demolition to ensure that the remaining part of the house (or adjoining buildings/extensions) are structurally sound.
	Possibly	For partial demolition (seek advice from your local planning officer before proceeding).		

4.8 How should I set about gaining planning permission?

If you are in the planning stages for your work, and you know planning permission will be required, it is wise to get the plans passed before you go to any expense or make any decisions that you may find hard to reverse – such as signing a contract for work. If your plans are rejected, you will still have to pay your architect or whoever prepared your plans for submission but you won't have to pay any penalty clauses to the building contractor.

It is always best to submit an application in the early stages – if you try to be clever by submitting plans at the last minute, in the hope that neighbours will not have time to react, then you could be in for an expensive mistake! It's much better to do things properly and up-front.

An architect (surveyor or general contractor) can be asked to prepare and submit your plans on your behalf if you like, but as the owner and person requiring the development, it will be your name that goes on the application, even if all the correspondence goes between your architect and the planning department.

You don't have to own the land to make a planning application for work upon it, but you will need to disclose your interest in the property. This might happen if you plan to buy land, with the intention of developing it, subject to planning approval. You would obtain the consent before the purchase proceeds.

To submit you application you will need to use the official forms, available from the local authority planning department. It's a good idea to collect these personally, as you may get the opportunity to talk through your ideas with a planning officer, and get some useful feedback. You will also need to include detailed plans of the present and proposed layout as well as the property's position in relation to other properties and roads or other features.

New work requires details of materials used, dimensions and all related installations, similar to that required for Building Regulations.

4.9 What sort of plans will I have to submit?

There are three types of plans (namely site, block and building) that can accompany your application and, as indicated above, the choice will depend on the work proposed.

4.9.1 Site plan

A site plan indicates the development location and relationship to neighbouring property and roads etc. Minimum scale is 1:2500 (or 1:1250 in a built-up area). The land to which the application refers is outlined in red ink. Adjacent land, if owned by the applicant, is outlined in blue ink.

Block plan

A block plan is a detailed plan of a construction or structural alteration that shows the existing and proposed building, all trees, waterways, ways of access, pipes and drainage and any other important features. Minimum scales are 1:1500.

Building plans

Building plans are the detailed drawings of the proposed building works and would show plans, elevations and cross-sections to accurately describe every feature of the proposal. These plans are normally very thorough and include types of material, colour and texture, the layers of foundations, floor constructions, and roof constructions etc.

4.10 What is meant by 'building works'?

In the context of the Building Regulations, 'building works' means:

(a)   the erection or extension of a building;
(b)   the provision or extension of a controlled service or fitting:

- in or in connection with an existing dwelling; and
- which is a service or fitting in relation to which paragraph L1 (but not Parts G, H or J) of Schedule 1 imposes a requirement; and
- where there shall only be building work where that work consists of the provision of a window, rooflight, roof window, door (being a door which together with its frame has more than 50% of its internal face area glazed), a space heating or hot water service boiler, or a hot water vessel.

(c)   the material alteration of a building, or a controlled service or fitting;
(d)   work required by Regulation 6 (requirements relating to material change of use);
(e)   the insertion of insulating material into the cavity wall of a building;
(f)   work involving the underpinning of a building.

4.11 What important areas should I take into consideration?

The following are some of the most important areas that should be considered before you submit a planning application.

4.11.1 Advertisement applications

If your proposal is to display an advertisement, you will need to make a separate application on a special set of forms. Three copies of the forms and the relevant drawings must be supplied. These must include a location plan and sufficient detail to show the size, materials and colour of the sign and its position. No certificate of ownership is needed, but it is illegal to display signs on the property without the consent of the owner.

4.11.2 Listed building consent

You will need to apply for listed building consent if either of the following cases apply:

- you want to demolish a listed building;
- you want to alter or extend a listed building in a manner which would affect its character as a building of special architectural or historic interest.

You may also need listed building consent for any works to separate buildings within the grounds of a listed building. Check the position carefully with the council – it is a criminal offence to carry out work which needs listed building consent without obtaining it beforehand.

4.11.3 Conservation Area consent

If you live in a Conservation Area, you will need Conservation Area consent to do the following:

- demolish a building with a volume of more than $115 \, m^3$ (there are a few exceptions and further information will be available from your council);
- demolish a gate, fence, wall or railing over 1 m high if it is next to a highway (including a public footpath or bridleway) or public open space; or over 2 m high elsewhere.

4.11.4 Trees

Many trees are protected by Tree Preservation Orders (TPOs), which mean that, in general, you need the council's consent to prune or fell them. In addition, there are controls over many other trees in Conservation Areas.

💡 Ask the council for a copy of the department's free leaflet *Protected Trees: a guide to tree preservation procedures*.

4.12 What are the government's restrictions on planning applications?

All applications for planning permission will have to take into account the following Acts and regulations.

Planning (Listed Buildings and Conservation Areas) Act 1990

Under the terms of the Planning (Listed Buildings and Conservation Areas) Act 1990, local councils must maintain a list of buildings within their boroughs, which have been classified as being of special architectural or historic interest. Councils are also required to keep maps showing which properties are within Conservation Areas.

Town and Country Planning (Control of Advertisements) Regulations 1992

In accordance with the Town and Country Planning (Control of Advertisements) Regulations 1992, councils need to maintain a publicly available register of applications and decisions for consent to display advertisements.

The Local Government (Access to Information) (Variation) Order 1992

The Local Government (Access to Information) (Variation) Order 1992 ensures that information relating to proposed development by councils cannot be treated as exempt when the planning decision is made.

Town and Country Planning (General Development Procedure) Order 1995

Every council must keep the following registers available for public inspection in accordance with the Town and Country Planning (General Development Procedure) Order 1995:

- planning applications, including accompanying plans and drawings;
- applications for a certificate of lawfulness of existing or proposed use or development;
- Enforcement Notices and any related stop notices.

All applications for planning permission must receive publicity.

Other areas

As well as the legal requirement to make the planning register available for public inspection, councils will also allow the public to have access to all other relevant information such as letters of objection/support for an application or correspondence about considerations. Three clear days before any committee meeting, the file will normally be made available for public inspection and this file will remain available (i.e. for further public inspection) after the committee meeting. Although commercial confidentiality could well be a valid consideration, the council will not use it so as to prevent important information about materials and facilities also being available.

4.13 How do I apply for planning permission?

Once you have established that planning permission is required, you will need to submit a planning application. Remember, it may take up to eight weeks, or even longer, to get planning permission, so apply early.

You will have to prepare a plan showing the position of the site in question (i.e. the site plan) so that the authority can determine exactly where the building is located. You must also submit another, larger-scale, plan to show the relationship of the building to other premises and highways (i.e. the block plan). In addition, it would help the council if you also supplied drawings to give a clear idea of what the new proposal will look like, together with details of both the colour and the kind of materials you intend using. You may prepare the drawings yourself, provided you are able to make them accurate.

Under normal circumstances you will have to pay a fee in order to seek planning permission, but there are exceptions. The planning department will advise you.

4.13.1 Application forms and plans

It is important to make sure that you make your planning application correctly. The following checklist may help:

- Obtain the application forms from the planning department or from the local council's website.
- Read the 'Notes for Applicants' carefully – again available from the planning department, or their website.
- Fill in the relevant parts of the forms and remember to sign and date them.
- Submit the correct number and type of supporting plans. Each application should be accompanied by a site plan of not less than 1:2500 scale and detailed plans, sections and elevations, where relevant.
- Fill in and sign the relevant certificate relating to land ownership.

It is in your own interest to provide plans of good quality and clarity and so it is probably advisable to get help from an architect, surveyor, or similarly qualified person to prepare the plans and carry out the necessary technical work for you. You can obtain the necessary application form from the planning department of your local council and you will find that this is laid out simply, with guidance notes to help you fill it in. Alternatively, you can ask a builder or architect to make the application on your behalf. This is sensible if the development you are planning is in any way complicated, because you will have to include measured drawings with the application form.

4.14 What is the planning permission process?

If you think you might need to apply for planning permission, then this is the process to follow:

Step 1

Contact the planning department of your council. Tell the planning staff what you want to do and ask for their advice.

Step 2

If they think you need to apply for planning permission, ask them for an application form. They will tell you how many copies of the form you will need to send back and how much the application fee will be. Ask if they foresee any difficulties which could be overcome by amending your proposal. It can save time or trouble later if the proposals you want to carry out also reflect what the council would like to see. The planning department will also be able to tell you if Building Regulations approval will also be required.

Step 3

Decide what type of application you need to make. In most cases this will be a full application but there are a few circumstances when you may want to make an outline application – for example, if you want to see what the council thinks of the building work you intend to carry out before you go to the trouble of making detailed drawings (but you will still need to submit details at a later stage).

Step 4

Send the completed application forms, forms and supporting documents to your council, together with the correct fee. Each form must be accompanied by a plan of the site and a copy of the drawings showing the work you propose to carry out. (The council will advise you on what drawings are needed.)

   Extracts from Ordnance Survey maps can be supplied for planning applications submitted by private individuals and for school/college use. There is usually a charge for this service.

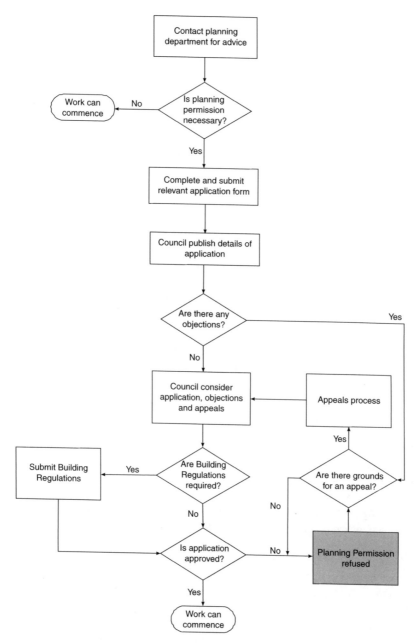

Figure 4.2 Planning permission

Step 5

The planning department will acknowledge receipt of your application, and publicly announce it – via letters to the neighbourhood parish council and anyone directly affected by the proposal, by publishing details of the application in the local press, notifying your neighbours and/or putting up a notice on or near the site. The council may also consult other organizations, such as the highway authority or the parish council (or community council in Wales).

A copy of the application will also be placed on the planning register at the council offices so that it can be inspected by any interested member of the public. Anyone can object to the proposal, but there is a limited period of time in which to do this and they must specify the grounds for objection.

💡 Under the Local Government Act 1972 (as amended), the public have the right to inspect and copy the following documents:

- the agenda for a council committee or sub-committee meeting; reports for the public part of the meeting;
- the minutes of such meetings and any background papers, including planning applications, used in preparing reports.

These documents can be inspected and copied from three clear days before a meeting. There is no charge to inspect a document but councils will charge for making photocopies.

Step 6

The planning department may prepare a report for the planning committee, which is made up of elected councillors. Or the council may give a senior officer in the planning department the responsibility for deciding your application on its behalf. If a report has been made, then this will be presented to a meeting of the council committee, with recommendations on the decision to be made, based on the implications and objections received.

💡 You are entitled to see and have a copy of any report submitted to a local government committee. You are also entitled to see certain background papers used in the preparation of reports. The background papers will generally include the comments of consultees, objectors and supporters which are relevant to the determination of your application. Such material should normally be made available at least three working days before the committee meeting.

Step 7

The councillors or council officers who decide your application must consider whether there are any good planning reasons for refusing planning permission or for granting permission subject to conditions. The council cannot reject a proposal simply because many people oppose it. It will look at whether your proposal is consistent with the development plan for the area.

The committee will consider the merits of a proposal; ensure the proposed work meets all the conditions of any local plan or requirements for a district and that the process has been followed properly. The kinds of planning issue it can also consider include potential traffic problems; the effect on amenity and the impact the proposal may have on the appearance of the surrounding area. Moral issues, the personal circumstances of the applicant or the effect the development might have on nearby property prices are not relevant to planning and will not normally be taken into account by the council. The committee will arrive at its decision and the result will be communicated back to the applicants via the planning department.

4.14.1 How long will the council take?

You can expect to receive a decision from the planning department within eight weeks and, once granted, planning permission is valid for five years. If the work is not begun within that time, you will have to apply for planning permission again.

If the council cannot make a decision within eight weeks then it must obtain your written consent to extend the period. If it has not done so, you can appeal to the Secretary of State for the Environment, Transport and the Regions, or, in Wales, to the National Assembly for Wales (see below). But appeals can take several months to decide and it may be quicker to reach agreement with the council.

Do not be afraid to discuss the proposal with a representative of the planning department **before** you submit your application. They will do their best to help you meet the requirements.

4.14.2 What can I do if my application is refused?

If the council refuses permission or imposes conditions, it must give reasons. If you are unhappy or unclear about the reasons for refusal or the conditions imposed, talk to staff at the planning department. Ask them if changing your plans might make a difference. If your application has been refused, you may be able to submit another application with modified plans free of charge within 12 months of the decision on your first application.

The planning department will always grant planning permission unless there are very sound reasons for refusal, in which case the department must explain the decision to you so that you can amend your plans accordingly and resubmit them for further consideration.

 A second application is normally exempt from a fee.

The following are some of the main objection areas that your application may meet.

The property is a listed building

Listed buildings are protected for their special architectural or historical value. A Listed Building Consent: Planning Applications may be needed for alterations but grants could be available towards repair and restoration!

If it's a listed building, it probably has some historic importance and will have been listed by the Department of the Environment. This could apply to houses, factories, warehouses and even walls or gateways. Most alterations which affect the external appearance or design will require listed building consent in addition to other planning consents.

The property is in a Conservation Area

This is an area defined by the local authority, which is subject to special restrictions in order to maintain the character and appearance of that area. Again, other planning consents may be needed for areas designated as green belt, area of outstanding natural beauty, national park or site of specific scientific interest.

The application does not comply with the local development plan

Local authorities often publish a development plan, which sets out policies and aims for future development in certain areas. These are to maintain specific environmental standards and can include very detailed requirements such as minimum or maximum dimensions of plot sizes, number of dwellings per acre, height and style of dwellings etc. It is important to check if a plan exists for your area, as proposals can meet with some fierce objections from residents protecting their environment.

The property is subject to a covenant

This is an agreement between the original owners of the land and the persons who acquired it for development. They were implemented to safeguard residential standards and can include things like the size of out-buildings, banning use of front gardens for parking cars, or even just the colours of exterior paintwork.

Is there existing planning permission?

A previous resident or owner may have applied for planning permission, which may not have expired yet. This could save time and expense if a new application can be avoided. If you are considering a planning application, you should consider the above questions. Normally your retained expert – architect, surveyor or builder – can advise and help you to get an application passed. Most information can be collected from your local planning department, or if you need to find out about covenants, look for the appropriate land registry entry.

The proposal infringes a right of way

If your proposed development would obstruct a public path that crosses your property, you should discuss the proposals with the council at an early stage. The granting of planning permission will **not** give you the right to interfere with, obstruct or move the path. A path cannot be legally diverted or closed unless the council has made an order to divert or close it to allow the development to go ahead. The order must be advertised and anyone may object. You must not obstruct the path until any objections have been considered and the order has been confirmed. You should bear in mind that confirmation is not automatic; for example, an alternative line for the path may be proposed.

4.14.3 What matters cannot be taken into account?

- Competition
- Disturbance from construction work
- Loss of property value
- Loss of view
- Matters controlled under other legislation such as Building Regulations (e.g. structural stability, drainage, fire precautions etc.)
- Moral issues
- Need for development
- Private issues between neighbours (e.g. land and boundary disputes, damage to property, private rights of way, deeds, covenants etc.)
- Sunday trading
- The identity or personal characteristics of the applicant.

4.14.4 What are the most common stumbling blocks?

In no particular order of priority, these are:

- Adequacy of parking
- Archaeology
- Design, appearance and materials
- Effect on listed building or Conservation Area
- Government advice
- Ground contamination
- Hazardous materials
- Landscaping
- Light pollution
- Local planning policies
- Nature conservation
- Noise and disturbance from the use (but not from construction work)
- Overlooking and loss of privacy
- Previous planning decisions
- Previous appeal decisions

- Road access
- Size, layout and density of buildings
- The effect on the street or area (but not loss of private view)
- Traffic generation and overall highway safety.

4.15 Can I appeal if my application is refused?

If you think the council's decision is unreasonable, you can appeal to the Secretary of State or the National Assembly for Wales. Appeals must be made within six months of the date of the council's notice of decision. You can also appeal if the council does not issue a decision within eight weeks.

A free booklet *Planning Appeals – A Guide* is available from the Planning Inspectorate, Tollgate House, Houlton Street, Bristol BS2 9DJ or Crown Buildings, Cathays Park, Cardiff CF10 3NQ.

Appeals are intended as a last resort and they take several months to decide. It is often quicker to discuss with the council whether changes to your proposal would make it more acceptable. The planning authority will supply you with the necessary appeal forms.

Be careful not to proceed without approval, as you might find yourself obliged to restore the property to its original condition.

4.16 Before you start work

There are many kinds of alterations and additions to houses and other buildings which do not require planning permission. Whether or not you need to apply, you should think about the following before you start work.

4.16.1 What about neighbours?

Have the neighbours any rights to complain?
Many of us live in close proximity to others and your neighbours should be the first individuals you talk to.
What if your alteration infringes their access to light, or a view? Such disputes are notorious for causing bad feeling but with a little consideration, at an early stage, you can avoid a good deal of unpleasantness later.
Plans for the local area can normally be viewed at the local town hall, but most planning applications will involve consultation with neighbours and statutory consultees such as the Highways Authority and the drainage authorities. The extent of consultation will, quite naturally, reflect on the nature and scale

of the proposed development – together with its location. Applications to make an alteration to your property can also be refused because you live in an area of outstanding natural beauty, a national park, or a Conservation Area or your property is listed. Any alterations to public utilities such as drains or sewers, or changes to public access such as footpaths will require consultation with the local council. They will have to approve your plans. Even a sign on or above your property may need to be of a certain size or shape.

Some properties may also be the home of a range of protected species such as bats or owls. These animals are protected by the Wildlife and Countryside Act 1981, and the Nature Conservancy Council must give approval to any work that may potentially disturb them. Likewise many members of the public are extremely defensive of trees that grow where they live. Tree preservation orders may control the extent to which you can fell or even prune a tree, even if it is on your property. Trees in Conservation Areas are particularity protected, and you will need to supply at least six weeks' notice before working upon them.

 New street names and house numbers need approval from the council.

Let your neighbours know about the work you intend to carry out to your property. They are likely to be as concerned about work which might affect them as you would be about changes which might affect your enjoyment of your own property. For example, your building work could take away some of their light or spoil a view from their windows. If the work you carry out seriously overshadows a neighbour's window and that window has been there for 20 years or more, you may be affecting his or her 'right to light' and you could be open to legal action. It is best to consult a lawyer if you think you need advice about this.

You may be able to meet some of your neighbour's worries by modifying your proposals. Even if you decide not to change what you want to do, it is usually better to have told your neighbours what you are proposing before you apply for planning permission and before any building work starts.

 If you do need to make a planning application for the work you want to carry out, the council will ask your neighbours for their views. If you or any of the people you are employing to do the work need to go on to a neighbour's property, you will, of course, need to obtain their consent before doing so.

4.16.2 What about design?

Everybody's taste varies and different styles will suit different types of property. Nevertheless, a well-designed building or extension is likely to be much more attractive to you and to your neighbours. It is also likely to add more value to your house when you sell it. It is therefore worth thinking carefully about how your property will look after the work is finished.

Extensions often look better if they use the same materials and are in a similar style to the buildings that are there already – but good design is impossible to define and there may be many ways of producing a good result. In some areas, the council's planning department issues design guides or other advisory leaflets that may help you.

4.16.3 What about crime prevention?

You may feel that your home is secure against burglary and you may already have taken some precautions such as installing security locks to windows. However, alterations and additions to your house may make you more vulnerable to crime than you realize. For example, an extension with a flat roof, or a new porch, could give access to upstairs windows which previously did not require a lock. Similarly, a new window next to a drainpipe could give access. Ensure that all windows are secure. Also, your alarm may need to be extended to cover any extra rooms or a new garage. The crime prevention officer at your local police station can provide helpful advice on ways of reducing the risk.

4.16.4 What about lighting?

If you are planning to install external lighting for security or other purposes, you should ensure that the intensity and direction of light does not disturb others. Many people suffer extreme disturbance due to excessive or poorly-designed lighting. Ensure that beams are **not** pointed directly at windows of other houses. Security lights fitted with passive infra-red detectors (PIRs) and/or timing devices should be adjusted so that they minimize nuisance to neighbours and are set so that they are not triggered by traffic or pedestrians passing outside your property.

4.16.5 What about covenants?

Covenants or other restrictions in the title to your property or conditions in the lease may require you to get someone else's agreement before carrying out some kinds of work to your property. This may be the case even if you do not need to apply for planning permission. You can check this yourself or consult a lawyer.

You will probably need to use the professional services of an architect or surveyor when planning a loft conversion. Their service should include considerations of planning control rules.

4.16.6 What about listed buildings?

Buildings are listed because they are considered to be of special architectural or historic interest and as a result require special protection. Listing protects the whole building, both inside and out and possibly also adjacent buildings if they were erected before 1 July 1948.

The prime purpose of having a building listed is to protect the building and its surroundings from changes that will materially alter the special historic or architectural importance of the building or its setting.

The list of buildings is prepared by the Department of Culture, Media and Sport and properties are scheduled into one of three grades, Grade I, Grade II* and Grade II, with Grade I being the highest grade. Over 90% of all listed properties fall within Grade II. (In Scotland the grades are A, B and C.)

All buildings erected prior to 1700 and substantially intact are listed, as are most buildings constructed between 1700 and 1840, although some selection does take place. The selection process is more discriminating for buildings erected since 1840 because so many more properties remain today. Buildings less than 30 years old are generally only listed if they are of particular architectural or historic value and are potentially under threat. Your district council holds a copy of the statutory list for public inspection and this provides details on each of the listed properties.

See *Planning Policy Guidance Note 15 (PPG.15) – Planning and the Historic Environment*, which provides a practical understanding of the Planning (Listed Buildings and Conservation Areas) Act 1990 which can be viewed at your planning office or in main libraries, or purchased from The Stationery Office, 29 Duke Street, Norwich, NR3 1GN (Tel: 0870 600 5522; Fax: 0870 600 5533, www.tso.co.uk).

Owners responsibilities?

If you are the owner of a listed building or come into possession of one, you are tasked with ensuing that the property is maintained in a reasonable state of repair. The council may take legal action against you if they have cause to believe that you are deliberately neglecting the property, or have carried out works without consent. Enforcement action may be instigated.

There is no statutory duty to effect improvements, but you must not cause the building to fall into any worse state than it was in when you became its owner. This may necessitate some works, even if they are just to keep the building wind and watertight. However, you may need listed building consent in order to carry these works out!

A photographic record of the property when it came into your possession may be a useful asset, although you may also have inherited incomplete or unimplemented works from your predecessor, which you will become liable for.

If you are selling a listed building you may wish to indemnify yourself against future claims: speak to your solicitor.

4.16.7 What about Conservation Areas?

Tighter regulations apply to developments in Conservation Areas and to developments affecting listed buildings. Separate Conservation Area consent and/or listed building consent may be needed in addition to planning consent and Building Regulation consent.

Conservation Areas are *'areas of special architectural or historic interest the character and appearance of which it is desirable to preserve or enhance'*. (Civic Amenities Act 1967)

As the title indicates these designations cover more than just a building or property curtilage and most local authorities have designated Conservation Areas within their boundary. Although councils are not required to keep any statutory lists, you can usually identify Conservation Areas from a local plan's 'proposals maps' and appendices. Some councils may keep separate records or even produce leaflets for individual areas.

The purpose of designating a Conservation Area is to provide the council with an additional measure of control over an area that they consider being of special historic or architectural value. This does not mean that development proposals cannot take place, or that works to your property will be automatically refused. It means however that the council will have regard to the effect of your proposals on the designation in addition to their normal assessment. The council may also apply this additional tier of assessment to proposals that are outside the designated Conservation Area boundary, but which may potentially affect the character and appearance of the area.

As a result, local planning authorities may ask for more information to accompany your normal planning application concerning proposals within (or adjoining a) Conservation Area. This may include:

- a site plan to 1:1250 or 1:2500 scale showing the property in relation to the Conservation Area;
- a description of the works and the effect (if any) you think they may have on the character and appearance of the Conservation Area;
- a set of scale drawings showing the present and proposed situation, including building elevations, internal floor plans and other details as necessary.

If you live or work in a Conservation Area, grants may be available towards repairing and restoring your home or business premises.

For major works you may need to involve an architect with experience of works affecting Conservation Areas.

4.16.8 What is Conservation Area consent?

Development within Conservation Areas is dealt with under the normal planning application process, except where the proposal involves demolition.

In this case you will need to apply for Conservation Area consent on the appropriate form obtainable from the planning department.

Here again the council will assess the proposal against its effect upon the special character and appearance of the designated area.

 More details can be obtained by reference to *Planning Policy Guidance Note 15 (PPG.15) – Planning and the Historic Environment*, which provides a practical understanding of the Planning (Listed Buildings and Conservation Areas) Act 1990. These can be viewed at your planning office or in main libraries, or purchased from The Stationery Office, 29 Duke Street, Norwich, NR3 1GN (Tel: 0870 600 5522 Fax: 0870 600 5533, www.tso.co.uk).

4.16.9 What about trees in Conservation Areas?

Nearly all trees in Conservation Areas are automatically protected.

Trees in Conservation Areas are generally treated in the same way as if they were protected by a tree preservation order, i.e. it is necessary to obtain the council's approval for works to trees in Conservation Areas before they are carried out. There are certain exceptions (where a tree is dead or in a dangerous condition) but it is always advisable to seek the opinion of your council's tree officer to ensure your proposed works are acceptable. Even if you are certain that you do not need permission, notifying the council may save the embarrassment of an official visit if a neighbour contacts them to tell them what you are doing.

If you wish to lop, top or fell a tree within a Conservation Area you must give six weeks' notice, in writing, to the local authority. This is required in order that they can check to see if the tree is already covered by a Tree Preservation Order (TPO), or consider whether it is necessary to issue a TPO to control future works on that tree.

Contact your council's landscape or tree officer for further information.

4.16.10 What are tree preservation orders?

Trees are possibly the biggest cause of upset in town and country planning and many neighbours fall out over tree related issues. They may be too tall, may block out natural light, have overhanging branches, shed leaves on other property or the roots may cause damage to property. When purchasing a property the official searches carried out by your solicitor should reveal the presence of a TPO on the property or whether your property is within a Conservation Area within which trees are automatically protected.

However not all trees are protected by the planning regulations system – but trees that have protection orders on them must not be touched unless specific approval is granted. Don't overlook the fact that a preservation order could have been put on a tree on your land before you bought it and is still be enforceable.

Planning authorities have powers to protect trees by issuing a TPO and this makes it an offence to cut down, top, lop, uproot, wilfully damage or destroy any protected tree(s) without first having obtained permission from the local authority. All types of tree can be protected in this way, whether as single trees or as part of a woodland, copse or other grouping of trees. Protection does not however extend to hedges, bushes or shrubs.

TPOs are recorded in the local land charges register which can be inspected at your council offices. The local authority regularly checks to see if trees on their list still exist and are in good condition. Civic societies and conservation groups also keep a close eye on trees. Before carrying out work affecting trees, you should check if the tree is subject to a TPO. If it is, you will need permission to carry out the work.

All trees in a Conservation Area are protected, even if they are not individually registered. If you intend to prune or alter a tree in any way you must give the local authority plenty of notice so they can make any necessary checks.

Even with a preservation order it is possible to have a tree removed, if it is too decayed or dangerous, or if it stands in the way of a development, the local authority may consider its removal, but will normally want a similar tree put in or near its place.

A TPO will not prevent planning permission being granted for development. However, the council will take the presence of TPO trees into account when reaching their decision.

If you have a tree on your property that is particularly desirable – either an uncommon species or a mature specimen, then you can request a preservation order for it. However, this will mean that in years to come you, and others, will be unable to lop it, remove branches or fell it unless you apply for permission.

What are my responsibilities?

Trees covered by TPOs remain the responsibility of the landowner, both in terms of any maintenance that may be required from time to time and for any damage they may cause. The council must formally approve any works to a TPO tree. If you cut down, uproot or wilfully damage a protected tree or carry out works such as lopping or topping which could be likely to seriously damage or destroy the tree then there are fines on summary conviction of up to £20 000, or, on indictment, the fines are unlimited. Other offences concerning protected trees could incur fines of up to £2500.

What should I do if a protected tree needs lopping or topping?

Although there are certain circumstances in which permission to carry out works to a protected tree are not required, it is generally safe to say that you should always write to your council seeking their permission before undertaking any works. You should provide details of the trees on which you intend to do work, the nature of that work – such as lopping or topping – and the reasons why you think this is necessary. The advice of a qualified tree surgeon may also be helpful, see *Yellow Pages*.

💣 You may be required to plant a replacement tree if the protected tree is to be removed.

4.16.11 What about nature conservation issues?

Many traditional buildings, particularly farm buildings, provide valuable wild-life habitats for protected species such as barn owls and bats.

Planning permission will not normally be granted for conversion and re-use of buildings if protected species would be harmed. However, in many cases, careful attention to the timing and detail of building work can safeguard or re-create the habitat value of a particular building. Guidance notes pre-pared by English Nature are available from councils or from their website www.english-nature.org.uk.

4.16.12 What about bats and their roosts?

Bats make up nearly one-quarter of the mammal species throughout the world. Some houses may hold roosts of bats or provide a refuge for other protected species. The Wildlife and Countryside Act 1981 gives special protection to **all** British bats because of their roosting requirements. English Nature (EN) or the Countryside Council for Wales (CCW) must be notified of any proposed action (e.g., remedial timber treatment, renovation, demolition and extensions) which is likely to disturb bats or their roosts. EN or CCW must then be allowed time to advise on how best to prevent inconvenience to both bats and householders.

💡 Information on bats and the law is included in the booklet *Focus on Bats* which can be obtained free of charge from your local EN office. Similar booklets can be obtained from CCW local offices.

The type of stone barns and traditional buildings found in the UK have lots of potential bat roosting sites; the most likely places being gaps in stone rubble walls, under slates or within beam joints. These sites can be used throughout the year by varying numbers of bats, but could be particularly important for winter hibernation. As a result, the following points should be followed when considering or undertaking any work on a stone barn or similar building, particularly where bats are known to be in the area.

Bats might be present in gaps in stone rubble walls, under slates or within beam joints. A survey for the presence of bats should be carried out by a member of the local bat group (contact via English Nature) before any work is done to a suitable barn during the summer bat breeding period.

The pipistrelle, the smallest of the European bats, has been found lurking in many strange places including vases, under floorboards, and between the panes of double glass.

Any pointing of walls should not be carried out between mid-November and mid-April to avoid potentially entombing any bats. When walls are to be pointed, areas of the walls high up on all sides of the building should be left un-pointed to preserve some potential roosting sites. If any bats are found whilst work is in progress, work should be stopped and English Nature contacted for advice on how to proceed.

If any timber treatment is carried out, only chemicals safe for use in bat roosts should be used. A list of suitable chemicals is available from English Nature on request. Any pre-treated timber used should have been treated using the CCA method (copper chrome arsenic) which is safe for bats.

Work should not be commenced during the winter hibernation period (mid-November to mid-April). Any bats present during the winter are likely to be torpid, i.e. unable to wake up and fly away, and are therefore particularly vulnerable.

If these guidelines are followed, then the accidental loss of bat roosts and death or injury to bats will be reduced.

Although vampire bats feed primarily on domestic animals, they have been known to feed on sleeping humans on rare occasions! Vampire bats have chemicals in their saliva that prevent the blood they are drinking from clotting. They consume five teaspoons of blood each day. The vampire bat has been known to transmit rabies to livestock and to man.

4.16.13 What about barn owls?

Barn owls also use barns and similar buildings as roosting sites in some areas. These are more obvious than bats, and therefore perhaps easier to take into account. Barn owls are also fully protected by law, and should not be disturbed during their breeding season. Special owl boxes can be incorporated into walls during building work, details of which can be obtained from English Nature.

4.17 What could happen if you don't bother to obtain planning permission?

If you build something which needs planning permission without obtaining permission first, you may be forced to put things right later, which could prove troublesome and costly. You might even have to remove an unauthorized building.

4.17.1 Enforcement

If you think that works are being carried out without planning permission, or not in accord with approved plans and/or conditions of consent, then seek the advice of the local planning officer who will then investigate, and if necessary take appropriate steps to deal with the problem. Conversely, if you are carrying out development works, it is important that you stick to the approved plans and condition. If changes become necessary please contact the development control staff before they are made.

4.18 How much does it cost?

A fee is required for the majority of planning applications and the council cannot deal with your application until the correct fee is paid. The fee is not refundable if your application is withdrawn or refused.

In most cases you will also be required to pay a fee when the work is commenced. These fees are dependent on the type of work that you intend to carry out. The fees outlined in Sections 14.18.1–10 are typical of the charges made by Local Authorities during 2002, when submitting an application.

Work to provide access and/or facilities for disabled people to existing dwellings are exempt from these fees.

4.18.1 Householder applications

Outline applications (most types)	£190.00 per 0.1 ha (or part thereof) of site area, maximum £4750 (2.5 ha)
Full applications and reserved matters	
Dwellings – erection of new	£190.00 per dwelling house, maximum £9500 (=50)
Dwellings – alteration (including outline)	£95.00 per dwelling house, maximum £190.00
Approval of reserved matters where flat rate does **not** apply	A fee based upon the amount of floorspace and/or number of dwelling houses involved
Flat rate (only when maximum fee has been paid)	£190.00

4.18.2 Industrial/retail and other buildings applications

Industrial retail buildings

- Where no additional floorspace is created £95.00
- Works not creating more than $40\,m^2$ of additional floorspace £95.00
- More than $40\,m^2$ but not more than $75\,m^2$ of additional floorspace £190.00
- Each additional $75\,m^2$ (or part thereof) £190.00, maximum £9500 (=$3750\,m^2$)

Outline applications (see above)

Plant and machinery (erection, alteration, replacement)

£190.00 per 0.1 ha (0.24 acre), or part thereof, of the site area, maximum £9500 (5 ha)

4.18.3 Prior notice applications

Approvals for agricultural/forestry buildings/ operations and demolition of buildings and telecommunications works

According to type of approval required

4.18.4 Agricultural applications

Agricultural buildings

- Buildings not exceeding $465\,m^2$ £35.00
- Buildings exceeding $465\,m^2$ but not more than $540\,m^2$ £190.00
- More than $540\,m^2$ £190.00 for first $540\,m^2$ and £190.00 for each additional $75\,m^2$ (or part thereof), maximum £9500

Erection of glasshouses/polytunnels (on land used for agriculture)

- Works not creating more than $465\,m^2$ £35.00
- Works creating more than $465\,m^2$ £1085.00

4.18.5 Legal applications

Application for a certificate of lawfulness for an **existing** use or operation

Same fee payable as if making a planning application

Application for a certificate of lawfulness for an **existing** activity in breach of planning condition(s)

£95.00

Application for a certificate of lawfulness for a **proposed** use or operation

Half the fee payable as if making a planning application

4.18.6 Advertisement applications

Adverts relating to the business on the premises	£50.00
Advance signs directing the public to a business	£50.00
Advance signs directing the public to a business (unless business can be seen from the sign's position)	£190.00
Other advertisements (e.g. hoardings)	£190.00

4.18.7 Other applications

Exploratory drilling for oil or natural gas	£190.00 per 0.1 ha (or part thereof) of site area, maximum £14 250 (=7.5 ha)
Storage of minerals etc. and waste disposal	£95.00 per 0.1 ha (or part thereof) of site area, maximum £14 250 (=15 ha)
Car parks, service roads or other accesses (existing uses only)	£95.00
Other operations on land	£95.00 per 0.1 ha (or part thereof) of site area, maximum £950 (=1 ha)
Non compliance with conditions	£95.00
Renewal of temporary permissions	£95.00
Removal or variation of conditions including renewal of unimplemented consents that have not lapsed	£95.00 (full fee if consent has lapsed)
Change of use to sub-division of dwellings	£190.00 per additional dwelling created (maximum £9500)
Other changes of use except waste or minerals	£190.00

4.18.8 Concessionary fees and exemptions

Works to improve the disabled persons' access to a public building, or to improve their access, safety, health or comfort at their dwelling house	No fee
Applications by parish councils (all types)	Half the normal fee
Applications required by an Article 4 direction or removal of permitted development rights	No fee
Playing fields (for sports clubs etc.)	£190.00

Revised or fresh applications of the same character or description within 12 months of refusal, or the expiry of the statutory 8 week period where the applicant has appealed to the Secretary of State on grounds of non-determination. Withdrawn applications of the same character or description must be made within 12 months of making the earlier one	No fee
Revised or fresh application of the same character or description within 12 months of receiving permission	No fee
Duplicate applications made by the same applicant submitted within 28 days of each other	Full fee for each application
Alternative applications for one site submitted at the same time	Highest of the fees applicable for each alternative and a sum equal to half the rest
Development crossing local authority boundaries	Only one fee paid to the authority having the larger site but calculated for the whole scheme and subject to a special ceiling

4.18.9 Hazardous substances applications

Application for new consent	£200.00
New consent where maximum quantity specified exceeds twice the controlled quantity	£400.00
All other types of application	£250.00
Continuation of hazardous consent under Section 17(1) of the 1992 Regulations	£200.00

4.18.10 Examples of mixed development

Outline application for 1000 m² of office floorspace and five flats on a site of 0.5 ha: Since the application is in outline and the sites are one and the same, the fee is the same under either category of development	0.5 ha at £190.00 per 0.1 ha (or part thereof) = 5 × £190.00 = £950.00
Application for approval of reserved matters, if not subject to a flat-rate fee, for a development of 4000 m² of mixed shops and offices and 60 dwellings:	The total fee is therefore £19 000

Since the floorspace exceeds $3750\,m^2$ the maximum fee of £9500 applies, and since the number of dwelling houses exceeds 50, the maximum fee of £9500 also applies. Fees for this kind of mixed development are additions

Application for full permission for a building incorporating $200\,m^2$ of shops, four flats with a total area of $370\,m^2$ and $30\,m^2$ of common service floorspace:

The total floorspace of the development occupies $200+370+30\,m^2=600\,m^2$. The proportion occupied by shops is 200 out of $600=$ one third. One third of the common service area ($\frac{1}{3}\times30\,m^2=10\,m^2$) is added to the non-residential floorspace

Fee calculation is
$200+10=210\,m^2$ of shops at £190.00 per $75\,m^2$ (or part thereof)
4 Dwellings at £190.00

Total £1330.00

Application for use of land as a caravan site of 2 ha incorporating roads, hardstandings, a shop and service building of $150\,m^2$ of floorspace

(Another example in this category would be development of a golf course)

Change of use$=$£190.00
2 ha of 'other operations' (because the site exceeds 1 ha), the maximum fee of £950.00 applies $150\,m^2$ of non-residential floorspace at £190.00 per $75\,m^2$ (or part thereof) $=$£380.00
Of these the highest fee, £950.00, applies

Extracted from *Circular 31192 The Town & Country Planning (Fees for Applications & Deemed Applications) (Amendment) Regulations 1987*

5

Requirements for planning permission and Building Regulations approval

Before undertaking any building project, you must first obtain the approval of local-government authorities. There are two main controls that districts rely on to ensure that adherence to the local plan is ensured, namely planning permission and building regulation approval.

Whilst both of these controls are associated with gaining planning permission, actually receiving planning permission does not automatically confer Building Regulation approval and vice versa. You **may** require **both** before you can proceed. Indeed, there may be variations in the planning requirements, and to some extent the Building Regulations, from one area of the country to another.

Provided, however, that the work you are completing does not affect the external appearance of the building, you are allowed to make certain changes to your home without having to apply to the local council for permission. These are called permitted development rights, but the majority of building work that you are likely to complete will still require you to have planning permission – so be warned!

The actual details of planning requirements are complex but for most domestic developments, the planning authority is only really concerned with construction work such as an extension to the house or the provision of a new garage or new outbuildings that is being carried out. Structures like walls and fences also need to be considered because their height or siting might well infringe the rights of your neighbours and other members of the community. The planning authority will also want to approve any change of use, such as converting a house into flats or running a business from premises previously occupied as a dwelling only.

Planning consent **may** be needed for minor works such as television satellite dishes, dormer windows, construction of a new access, fences, walls, and garden extensions. You are advised to consult with Development Control staff before going ahead with such minor works.

5.1 Decoration and repairs inside and outside a building

	Requirement		
	Planning permission		Building Regulation approval
No	Unless it is of a listed building or within a Conservation Area Consult your local authority	No	Unless it is a listed building or within a Conservation Area Consult your local authority
No	As long as the use of the house is not altered	Possibly	Consult your local authority
Yes	If the alterations are major such as removing or part removing of a load bearing wall or altering the drainage system	Yes	

Generally speaking, you do not need to apply for planning permission:

- for repairs or maintenance;
- for minor improvements, such as painting your house or replacing windows;
- for internal alterations;
- for the insertion of windows, skylights or roof lights (but, if you want to create a new bay window, this will be treated as an extension of the house);
- for the installation of solar panels which do not project significantly beyond the roof slope (rules for listed buildings and houses in Conservation Areas are different however);
- to re-roof your house (but additions to the roof are treated as extensions to the house).

Occasionally, you may need to apply for planning permission for some of these works because your council has made an Article 4 direction withdrawing permitted development rights.

Do I need approval to carry out repairs to my house, shop or office?

No – if the repairs are of a minor nature – e.g. replacing the felt to a flat roof, repointing brickwork, or replacing floorboards.

Yes – if the repair work is major in nature – e.g. removing a substantial part of a wall and rebuilding it, or underpinning a building.

Do I need to apply for planning permission for internal decoration, repair and maintenance?

No.

Do I need to apply for planning permission for external decoration, repair and maintenance?

No – external work in most cases doesn't need permission, provided it does not make the building any larger.

Do I need approval to or alter the position of a WC, bath, etc. within my house, shop or flat?

No – unless the work involves new or an extension of drainage or plumbing.

Do I need approval to alter in any way the construction of fireplaces, hearths or flues within my house, shop or flat?

Yes.

Do I need to apply for planning permission if my property is a listed building?

Yes – if your property is a listed building consent will probably be needed for **any** external work, especially if it will alter the visual appearance, or use alternative materials. You also may need planning permission to alter, repair or maintain a gate, fence, wall or other means of enclosure.

Do I need to apply for planning permission if my property is in a Conservation Area?

Yes – if the building undergoing repair or decoration is in a Conservation Area, or comes under any type of covenant restricting changes you will probably need planning permission. You may also be restricted to replacing items such as roof tiles with the approved material, colour and texture, and have to use cast iron guttering rather than plastic etc.

5.2 Structural alterations inside

	Requirement		
	Planning permission	Building Regulation approval	
No	As long as the use of the house is not altered	Possibly	Consult your local authority
Yes	If the alterations are major such as removing or part removing of a load bearing wall or altering the drainage system	Yes	
Yes	If they are to an office or shop	Yes	

Do I need approval to make internal alterations within my house?

Yes – if the alterations are to the structure such as the removal or part removal of a load bearing wall, joist, beam or chimney breast, or would affect fire precautions of a structural nature either inside or outside your house. You also need approval if, in altering a house, work is necessary to the drainage system or to maintain the means of escape in case of fire.

Do I need approval to make internal alterations within my shop or office?

Yes.

Do I need approval to insert cavity wall insulation?

Yes.

Do I need approval to apply cladding?

Yes – if you live in a Conservation Area, a national park, an area of outstanding natural beauty or the Norfolk Broads. You will need to apply for planning permission before cladding the outside of your house wi.   ne, tiles, artificial stone, plastic or timber.

If you are in any doubt about whether you need to apply for permission, you should contact your local authority planning department before commencing any work to your property. They will usually give you advice but if you want to obtain a formal ruling you can apply, on payment of a fee, for a lawful development certificate. You may also require Building Regulation approval.

5.3 Replacing windows and doors

Requirement			
Planning permission		Building Regulation approval	
Yes	If they are to an office or shop	Yes	
No	Unless: • they project beyond the foremost wall of the house facing the highway • the building is a listed building • the building is in a Conservation Area	Possibly	Consult your local authority
Yes	To replace shop windows	Yes	

Do I need approval to install replacement windows in my house, shop or office?

No – provided:

- the window opening is not enlarged. If a larger opening is required, or if the existing frames are load-bearing, then a structural alteration will take place and approval will be required.
- the installation (as a replacement) is carried out by a person who is registered under the Fenestration Self-Assessment Scheme by Fensa Ltd.
- you do not remove those opening windows which are necessary as a means of escape in case of fire.

Do I need approval to replace my shop front?

Yes.

5.4 Electrical work

Planning permission	Requirement Building Regulation approval	
No	No	But it must comply with IEE Regulations

Do I need approval to install or replace electric wiring?

No – but:

- you must comply with current IEE Regulations;
- your contract with the electricity supply company has conditions about safety which must not be broken. In particular, you should not interfere with the company's equipment which includes the cables to your consumer unit or up to and including the separate isolator switch if provided; and
- whoever undertakes electrical work should be competent to do so. If you use an electrician they should be complying with the Electricity at Work Regulations, which for domestic building work are usually enforced by the local authority.

Such systems must be installed by a competent person.

5.5 Plumbing

Planning permission	Requirement	
	Building Regulation approval	
No	No	For replacements (but you will need to consult the technical services department for any installation which alters present internal or external drainage)
	Yes	For an unvented hot water system

Do I need approval to install hot water storage within my house, shop or flat?

Yes – if the water heater is unvented (i.e. supplied directly from the mains without an open expansion tank and with no vent pipe to atmosphere) and has storage capacity greater than 15 litres.

5.6 Central heating

Planning permission	Requirement	
	Building Regulation approval	
No	No	If electric
	Yes	If gas, solid fuᵣ ɪil

Do I need approval to alter the position of a heating appliance within my house, shop or flat?

- **Gas**: yes, unless the work is supervised by an approved installer under the Gas Safety (Installation and Use) Regulations 1984.
- **Solid fuel**: yes.
- **Oil**: yes.
- **Electric**: no.

5.7 Oil-storage tank

Planning permission	Requirement	Building Regulation approval
No Provided that it is in the garden and has a capacity of not more than 3500 litres (778 gallons) and **no** point is more than 3 m (9 ft 9") high and **no** part projects beyond the foremost wall of the house facing the highway		No

Oil storage tanks, and the pipes connecting them to combustion appliances, should be constructed and protected so as to reduce the risk of the oil escaping and causing pollution.

5.8 Planting a hedge

	Requirement	
	Planning permission	Building Regulation approval
No	Unless it obscures view of traffic at a junction or access to a main road	No

You do not need planning permission for hedges or trees. However, if there is a condition attached to the planning permission for your property which restricts the planting of hedges or trees (for example, on an 'open plan' estate or where a sight line might be blocked), you will need to obtain the council's consent to relax or remove the condition before planting a hedge or tree screen. If you are unsure about this, you can check with the planning department of your council.

Hedges should not be allowed to block out natural light, and the positioning of fast growing hedges should be checked with your local authority. Recent incidents regarding hedging of the fast growing Leylandii trees have led to changes in the planning rules, where hedges previously had no restrictive laws.

5.9 Building a garden wall or fence

	Requirement	
	Planning permission	Building Regulation approval
Yes	If it is more than 1 m (3 ft 3") high and is a boundary enclosure adjoining a highway	No
Yes	If it is more than 2 m (6 ft 6") high elsewhere	No

Do I need approval to build or alter a garden wall or boundary wall?

No – subject to size.

You will need to apply for planning permission if:

- your house is a listed building or in the curtilage of a listed building; or
- the fence, wall or gate would be over 1 metre high and next to a highway used for vehicles; or over 2 metres high elsewhere.

In normal circumstances, the only restriction on walls and fences is the height allowed. This is 2 metres or no more than 1 metre if the walls or fence is near a highway or road junction, where its height might obscure a driver's view of other traffic, pedestrians or road users.

If there is a valid reason for a wall or fence higher than the prescribed dimensions, then it is possible to get planning consent. There may be security issues that would support an application for a high fence. If it has no affect on other people's valid interests and does not impair any amenity qualities in an area, there is no reason why a request should be refused.

Some walls have historic value and they, as well as arches and gateways, can be listed. Modifications, extensions and removal of these must have planning consent.

5.10 Felling or lopping trees

	Requirement	
Planning permission		Building Regulation approval
No	Unless the trees are protected by a Tree Preservation Order or you live in a Conservation Area	No

Many trees are protected by Tree Preservation Orders (TPOs), which mean that, in general, you need the council's consent to prune or fell them. Nearly all trees in Conservation Areas are automatically protected.

Ask the council for a copy of the free leaflet *Protected Trees: a guide to tree preservation procedures.*

5.11 Laying a path or a driveway

	Requirement	
Planning permission		Building Regulation approval
No	Unless it provides access to a main road	No

Do I need to apply for planning permission to install a pathway?

Generally no – but you may need approval from the highways department if the pathway crosses a pavement.

Do I need to apply for planning permission to lay a driveway?

No – unless it adjoins the main road.

Driveways

Provided a pathway or drive does not meet a public thoroughfare you will not need planning consent. There are no restrictions on the area of land around your house that you can cover with hard surfaces.

You will need to apply for planning permission only if the hard surface is not to be used for domestic purposes and is to be used instead, for example, for parking a commercial vehicle or for storing goods in connection with a business.

In the case of hardstanding you do not need permission to gain access to it within the confines of your land, but you would need permission for a hardstanding leading on to a public highway.

You must obtain the separate approval of the highways department of your council if you want to make access to a roadway or if a new driveway would cross a pavement or verge. The exception is if the roadway is unclassified and the drive or footway is related to a development that does not require planning permission. Your local authority highways department will be able to tell you if a road is classified or unclassified. If the road is classified then, depending on the volumes of traffic, it is harder to get permission. The busier the road the less likely a new driveway or footway will be allowed to meet it.

If a driveway crosses a pedestrian access, pavement or roadside verge, then the planning department will gain approval from the highways department. If this is the case, highways approval is required in addition to planning consent. The basic principle is to maintain safety and eliminate hazards.

You will also need to apply for planning permission if you want to make a new or wider access for your driveway onto a trunk or other classified road. The highways department of your council can tell you if the road falls into this category.

Pathways

Pathways do not normally need planning permission and you can lay paths however you like in the confines of your own property. The exception is for any path making access to a highway or public thoroughfare, in which case certain safety aspects arise. You may also need permission if your building is listed or is in a Conservation Area, so the style and size is suitable for the area.

If a pathway crosses a pedestrian access, pavement or roadside verge, then the planning department will gain approval from the highways department. If this is the case, highways approval is required in addition to planning consent. The basic principle is to maintain safety and eliminate hazards.

5.12 Building a hardstanding for a car, caravan or boat

Requirement	
Planning permission	Building Regulation approval
No Provided that it is within your boundary and is not used for a commercial vehicle	No

Do I need to apply for planning permission to build a hardstanding for a car?

No – provided that it is within your boundary and is not used for a commercial vehicle.

Check local council rules.

Access from a new hardstanding to a highway requires planning consent. The exception is if the roadway is unclassified and the access to the hardstanding is related to a development that does not require planning permission. Your local authority highways department will be able to tell you if a road is classified or unclassified. If the road is classified then, depending on the volumes of traffic, it is harder to get permission. The busier the road the less likely a new driveway or footway will be allowed to meet it.

If the access crosses a pedestrian thoroughfare, pavement or roadside verge, then the planning department will gain approval from the highways department. If this is the case, highways approval is required in addition to planning consent. The basic principle is to maintain safety and eliminate hazards.

For a hardstanding on your own land, you do not need permission to gain access to it within the confines of your land, but you would need permission for a hardstanding leading on to a public highway.

There are different rules depending on what you use a hardstanding for. Planning permission is generally not needed provided there are no covenants limiting the installation of hardstanding for parking of cars, caravans or boats. There are still rules for commercial parking, however (e.g. taxis or commercial delivery vans) and a 'change of use' as a trade premises would probably need to be granted for this to be allowed.

You should check if there are any local covenants limiting changes in access to your premises or for hardstanding and parking of vehicles on it. If in doubt, contact the relevant local authority planning department for specific advice.

*Do I need to apply for planning permission to build
a hardstanding for a caravan and/or boat?*

Some local authorities do not allow the parking of caravans or boats on driveways or hardstandings in front of houses. Check what the local rules are with your planning department, and if there's no restriction then you don't need to apply for permission.

There are no laws to prevent you, or your family from making use of a parked caravan while it's on your land or drive, but you cannot actually live in it as this would be classed as an additional dwelling. In addition, you cannot use a parked caravan for business use as this would constitute a change of use of the property.

If you want to put a caravan on your land to lease out as holiday accommodation or for friends or family to stay in while they visit you, then this would require planning permission. Rules on siting of static caravans or mobile homes are quite stringent.

5.13 Installing a swimming pool

	Requirement		
	Planning permission		Building Regulation approval
Possibly	Consult your local planning officer	Yes	For an indoor pool

5.14 Erecting aerials, satellite dishes and flagpoles

	Requirement	
	Planning permission	Building Regulation approval
No	Unless it is a stand alone antenna or mast greater than 3 m in height	No
Possibly	If erecting a satellite dish, especially in a Conservation Area or if it is a listed building (consult your local planning officer)	No

*Do I need to apply for planning permission to erect satellite
dishes, television and radio aerials and flagpoles?*

No – unless it is a stand-alone antenna or flagpole greater than 3 m in height.

Flagpoles etc, erected in your garden are treated under the same rules as outbuildings, and cannot exceed 3 metres in height.

Normally there is no need for planning permission for attaching an aerial or satellite dish to your house or its chimneys. However, if it rises significantly higher than the roof's highest point then it may contravene local regulations or covenants.

◑ You should get specific advice if you plan to install a large satellite dish or aerial, such as a short wave mast, as the rules differ between different authorities.

In certain circumstances, you will need to apply for planning permission to install a satellite dish on your house (see DTE's free booklet *A Householder's Planning Guide for the Installation of Satellite Television Dishes*, which can be obtained from your local council).

Conservation Areas have specific local rules on aerials and satellite dishes, so you need to approach your local planning department to find out the rules for your area. Certainly, if your house is a listed building, you may need listed building consent to install a satellite dish on your house.

◑ Remember, if you are a leaseholder, you may need to obtain permission from the landlord.

5.15 Advertising

	Requirement		
	Planning permission		Building Regulation approval
No	If the advertisement is less than 0.3 m² and not illuminated	Possibly	Consult your local planning officer

Do I need to apply for planning permission to erect an advertising sign?

Advertisement signs on buildings and on land often need planning consent. Some smaller signs and non-illuminated signs may not need consent, but it is always advisable to check with development control staff.

You are allowed to display certain small signs at the front of residential premises such as election posters, notices of meetings, jumble sales, car for sale etc. but business types of display and permanent signs may need to have planning permission granted. They may come under the category of 'advertising control' for which planning consent is required.

You may need to apply for advertisement consent to display an advertisement bigger than 0.3 m² on the front of, or outside, your property. This includes your house name or number or even a sign saying 'Beware of the dog'. Temporary notices up to 0.6 m² relating to local events, such as fêtes and concerts, may be displayed for a short period. There are different rules for estate agents' boards, but, in general, these should not be bigger than 0.5 m² on each side.

It is illegal to post notices on empty shops' windows, doors, and buildings, and also on trees. This is commonly known as 'fly posting' and can carry heavy fines under the Town and Country Planning Act.

Illuminated signs and all advertising signs outside commercial premises need to be approved. Most local authorities can give advice, by way of booklets or leaflets on what kinds of sign are allowed, not allowed or need approval.

💡 You can get advice from the planning department of your local council; ask for a copy of the free booklet *Outdoor advertisements and signs*.

5.16 Building a porch

	Requirement	
	Planning permission	Building Regulation approval
No	Unless: • the floor area exceeds 3 m² (3.6 y²) • any part is more than 3 m (9 ft 9") high • any part is less than 2 m (6 ft 6") from a boundary adjoining a highway or public footpath	Yes If area exceeds 30 m² (35.9 y²)

Do I need planning permission for a porch?

Yes – depending on its size and position.

You will need to apply for planning permission if the porch:

- if your house is listed or is in a Conservation Area, national park, area of outstanding natural beauty;
- would have a ground area (measured externally) of more than 3 m²;
- would be higher than 3 m above ground level;
- would be less than 2 m away from the boundary of a dwelling house with a highway (which includes all public roads, footpaths, bridleways and byways).

 All measurements are taken externally.

However, a porch or conservatory built at ground level and under 30 m² in floor area is exempt provided that the glazing complies with the safety glazing requirements of the Building Regulations (Part N). Your local authority building control department or an approved inspector can supply further information on safety glazing. It is advisable to ensure that a conservatory is not constructed so that it restricts ladder access to windows serving a room in the roof or a loft conversion, particularly if that window is needed as an emergency means of escape in the case of fire.

The regulations are quite complicated and depend on previous works on the site, if any, so you should always check with development control staff.

5.17 Outbuildings

	Requirement		
	Planning permission		Building Regulation approval
Possibly	Provided the building is less than 10 m³ (13.08 y³) in volume, not within 5 m (16 ft 3") of the house or an existing extension.	Yes	If area exceeds 30 m² (35.9 y²)
	Erecting outbuildings can be a potential minefield and it is best to consult the local planning officer before commencing work		If it is within 1 m (3 ft 3") of a boundary, it must be built from incombustible materials

Many kinds of buildings and structures can be built in your garden or on the land around your house without the need to apply for planning permission. These can include sheds, garages, greenhouses, accommodation for pets and domestic animals (e.g. chicken houses), summer houses, swimming pools, ponds, sauna cabins, enclosures (including tennis courts) and many other kinds of structure.

Outbuildings intended to go in the garden of a house do not normally require any planning permission, so long as they are associated with the residential amenities of the house and a few requirements are adhered to such as position and size.

You can build an outbuilding up to 10 m³ (13.08 y³) in volume without planning permission if it is within 5 m (16 ft 3") of the house or an extension. Further away than this, it can be up to half the area of the garden, but the height must not exceed 4 m (13 ft).

If your new building exceeds 10 m³ (and/or comes within 5 m of the house) it would be treated as an extension and would count against your overall volume entitlement.

There are a few conditions to follow in order to avoid the need for planning consent:

- The structure should not result in more than half the original garden space being covered by the building.
- No part of the structure should extend beyond the original house limits on any side facing a public highway or footpath or service road.
- The height should not exceed 3 m (or 4 m if it has a ridged roof).

💡 If your house is listed or is in a Conservation Area, national park, or area of outstanding natural beauty, then you will more than likely need to obtain planning consent. If in doubt, contact the relevant local authority planning department for specific advice.

Permission is required, however, for:

- any building/structure nearer to a highway than the nearest part of the original house, unless more than 20 m away from a highway;
- structures not required for domestic use;
- structures over 3 m high (or 4 m if it has a ridged roof);
- propane gas (LPG) tank;
- storage tank holding more than 3500 litres;
- a building or structure which would result in more than half of the grounds of your house being covered by buildings/structures.

You will also need to apply for planning permission if any of the following cases apply:

- You want to put up a building or structure which would be nearer to any highway than the nearest part of the original house, unless there would be at least 20 m between the new building and any highway. The term 'highway' includes public roads, footpaths, bridleways and byways.
- More than half the area of land around the original house would be covered by additions or other buildings.
- The building or structure is not to be used for domestic purposes and is to be used instead, for example, for parking a commercial vehicle, running a business or for storing goods in connection with a business.
- You want to put up a building or structure which is more than 3 m high, or more than 4 m high if it has a ridged roof (measured from the highest ground next to it).
- If your house is a listed building and you want to put up a building or structure with a volume of more than $10\,\text{m}^3$.

External water storage tanks

Many years ago the demand for external tanks for capturing rainwater made their installation quite commonplace. But it is rare today to need extra storage tanks, unless you are in a rural position.

If you are considering installing an external water tank you should seek guidance from your local authority, especially if the tank is to be mounted on a roof.

Fuel storage tanks

Storage of oil, or any other liquids, especially petrol, diesel and chemicals is strictly controlled and would not be allowed on residential premises. If you are considering installing an external oil storage tank for central heating use, then no planning permission is required, provided its capacity is no more than 3500 litres, it is no more than 3 m from the ground and it does not project beyond any part of a building facing a public thoroughfare.

You will need to apply for planning permission in the following circumstances:

- You want to install a storage tank for domestic heating oil with a capacity of more than 3500 litres or a height of more than 3 m above ground level.
- You want to install a storage tank, which would be nearer to any highway than the nearest part of the 'original house', unless there would be at least 20 m between the new storage tank and any highway. The term 'highway' includes public roads, footpaths, bridleways and byways.
- You want to install a tank to store Liquefied Petroleum Gas (LPG) or any liquid fuel other than oil.

Erecting outbuildings can be a potential minefield and it is best to consult with the local planning officer before commencing work.

5.18 Garages

	Requirement	
	Planning permission	Building Regulation approval
Possibly	You can build a garage up to 10 m³ (13.08 y³) in volume without planning permission, if it is within 5 m (16 ft 3") of the house or an existing extension. Further away than this, it can be up to half the area of the garden, but the height must not exceed 4 m (13 ft)	Yes

Do I need approval to build a garage extension to my house, shop or office?

Yes – but a carport extension built at ground level, open on at least two sides and under 30 m² in floor area, is exempt.

Do I need approval for a detached garage?

Yes – but a single storey garage at ground level, under 30 m² in floor area and with no sleeping accommodation, is exempt provided it is either built mainly using non-combustible material or, when built, it has a clear space of at least 1 m from the boundary of the property.

Garages planned to go in the garden of a house do not normally require any planning permission, so long as they are associated with the residential amenities of the house and a few requirements are adhered to such as position and size.

There are a few conditions to follow in order to avoid the need for planning consent, such as:

- the structure should not result in more than half the original garden space being covered by the building;
- no part of the structure should extend beyond the original house limits, on any side facing a public highway or footpath or service road;
- the height should not exceed 3 m (or 4 m if it has a ridged roof).

Integral garages (that is, those directly attached on the side or under existing rooms in your house) will nearly always require planning consent.

5.19 Building a conservatory

	Requirement		
	Planning permission	Building Regulation approval	
Possibly	You can extend your house by building a conservatory, provided that the total of both previous and new extensions does not exceed the permitted volume	Yes	If area exceeds 30 m^2 (35.9 y^2)

Do I need permission to erect a conservatory?

Possibly – see below.

Conservatories and sun lounges attached to a house are classed as extensions. If you want a conservatory or sun lounge separated from the house, this needs planning consent under similar rules for outbuildings.

If the answer to all the following questions is **no** then it is quite likely that planning permission will not be required:

- Is the conservatory going to be used as a separate dwelling, i.e. self-contained accommodation?
- Is your property listed, in a Conservation Area, national park or area of outstanding natural beauty?
- Will the conservatory cover more than half of the original garden space?
- Will any part of the conservatory within 2 m of the plot boundary be more than 4 m above ground level?
- Will any part of the conservatory be higher than the original roof of the main building?

- Will any part of the conservatory be nearer to a service road, public road or footpath than any part of the original building?
- Will the volume of the original house be increased so that any part of the conservatory within 2 metres of the plot boundary is more than 4 metres above ground level?
- Will the conservatory be behind the building line?
- Will your conservatory be 1 m away from the boundary (although most buildings tend to be nearer than this)?
- Will the conservatory be more than 50 ft away from the nearest road?

🔔 A conservatory has to be separated from the rest of the house to be exempt (i.e. patio doors).

Another thing to keep in mind is your neighbours' reaction – always keep them informed of what's happening, and be prepared to alter the plans you had for locating the building if they object – it's better in the long run, believe me.

💡 Will you need planning permission therefore? Generally no, as the building is classed as a 'portable building', however it is *your* responsibility to check with your local planning office.

However, a porch or conservatory built at ground level and under $30\,m^2$ in floor area is exempt provided that the glazing complies with the safety glazing requirements of the Building Regulations (Part N). Your local authority building control department or an approved inspector can supply further information on safety glazing. It is advisable to ensure that a conservatory is not constructed so that it restricts ladder access to windows serving a room in the roof or a loft conversion, particularly if that window is needed as an emergency means of escape in the case of fire.

💡 The regulations are quite complicated and depend on previous works on the site, if any, so you should always check with development control staff.

5.20 Loft conversions, roof extensions and dormer windows

	Requirement	
	Planning permission	Building Regulation approval
No	Provided the volume of the house is unchanged and the highest part of the roof is not raised	Yes
Yes	For front elevation dormer windows or rear ones over a certain size	Yes

Do I need approval for a loft conversion?

Yes – see below.

Do I need to apply for planning permission to re-roof my house?

No – unless you live in a Conservation Area, a national park, an area of outstanding natural beauty or the Norfolk Broads.

Do I need to apply for permission to insert roof lights or skylights?

No.

Do I need to apply for planning permission to extend or add to my house?

Yes – in the following circumstances:

- If you want to build an addition or extension to any roof slope which faces a highway.
- If the roof extension would add more than $40\,\mathrm{m}^3$ to the volume of a terraced house or more than $50\,\mathrm{m}^3$ to any other kind of house.

  💣 These volume limits count as part of the allowance for extending the property (see extensions above).

- If the work would increase the height of the roof.
- If the intention is to create a separate dwelling, such as self-contained living accommodation or a granny flat.

If the answer to all the following questions is **no**, therefore, then it is quite likely that planning permission will not be required:

- Is the loft conversion going to be used as a separate dwelling, i.e. a self-contained flat?
- Is your property listed, in a Conservation Area, national park or area of outstanding natural beauty?
- Will any part of the loft conversion within 2 m of the plot boundary be more than 4 m above ground level?
- Will any part of the loft conversion be higher than the original roof of the main building?
- Will any part of the loft conversion be nearer to a service road, public road or footpath than any part of the original building?
- Will the roof be extended where it faces a public highway?
- Will the volume of the original house be increased beyond the following limits?

- if the house is in a terrace, a Conservation Area, a national park, or an area of outstanding natural beauty $40\,m^3$ or 10% whichever is the greater, up to a maximum of $115\,m^3$.
- for any kind of house $50\,m^3$ or 15% whichever is the greater, up to a maximum of $115\,m^3$.

 The volume is calculated from external measurement.

Do I need to apply for planning permission to alter a roof?

You will need to apply for planning permission if you live in a Conservation Area, a national park, an area of outstanding natural beauty or the Norfolk Broads and you want to build an extension to the roof of your house or any kind of addition that would materially alter the shape of the roof.

Roofs are expected to match those of the surrounding area, so consider this if you live in a protected area. Some areas require that the colour and style of the roof covering matches the original, and the pitch and construction should be the same. If you plan to save the expense of matching the roof, by opting for a flat roof, be sure that your local authority will accept this. Often high flat roofs are not desirable, due to the appearance of the house elevation. Provided the alterations to your roof do not make a noticeable change or don't increase its height you would normally not need to obtain planning permission.

 You do not normally need to apply neither for planning permission to re-roof your house nor for the insertion of roof lights or skylights.

In the case of re-roofing, if the tiles are the same type then no approval is needed. If the new tiling or roofing material is substantially heavier or lighter than the existing material, or if the roof is thatched or is to be thatched where previously it was not, then an approval under Building Regulations is probably required.

5.21 Building an extension

	Requirement		
	Planning permission		Building Regulation approval
Possibly	You can extend your house by building an extension, provided that the total of both previous and new extensions does not exceed the permitted volume	Yes	If area exceeds $30\,m^2$ $(35.9\,y^2)$

Do I need approval to build an extension to my house?

Yes – if it would 'materially alter the appearance of the building'.

Major alteration and extension nearly always need approval. However some small extensions such as porches, garages and conservatories may be 'Permitted Development' and not need planning consent.

 Building extensions can be a potential minefield and it is best to consult the local planning officer before contemplating any work.

If the answer to all the following questions is **no**, then it is quite likely that planning permission will not be required to build an extension to your house:

- Is the extension going to be used as a separate dwelling, i.e. a self-contained flat?
- Is the property listed?
- Will the extension cover more than half of the original garden space?
- Will any part of the extension within 2 m of the plot boundary be more than 4 m above ground level?
- Will any part of the extension be higher than the original roof of the main building?
- Will any part of the extension be nearer to a service road, public road or footpath than any part of the original building?
- Will the volume of the original house be increased beyond the following limits?
  - If the house is in a terrace, a Conservation Area, a national park, or an area of outstanding natural beauty $50\,m^3$ (or 10% whichever is the greater) up to maximum of $115\,m^3$.
  - For any other kind of house, $70\,m^3$ (or 15% whichever is the greater) up to a maximum of $115\,m^3$.

 The volume is calculated from external measurements.

You **will** need to apply for planning permission to extend or add to your house. You may also require planning permission if your house has previously been added to or extended. You may also require planning permission if the original planning permission for your house imposed restrictions on future development (i.e. permitted development rights may have been removed by an 'Article 4 direction'. This is often the case with more recently constructed houses.)

You **will** also require planning permission if you want to make additions or extensions to a flat or maisonette.

 Check with your local authority planning department if you are not sure.

You will, therefore, need to apply for planning permission:

- if an extension to your house comes within 5 m of another building belonging to your house (i.e. a garage or shed). The volume of that building counts against the allowance given above.
- for all additional buildings which are more than $10\,m^3$ in volume, if you live in a Conservation Area, a national park, an area of outstanding natural

beauty or the Norfolk Broads. Wherever they are in relation to the house, these buildings will be treated as extensions of the house and reduce the allowance for further extensions.

- for a terraced house, end-of-terrace house, or any house in a Conservation Area, national park, an area of outstanding natural beauty or the Broads – where the volume of the original house would be increased by more than 10% or $50\,m^3$ (whichever is the greater).
- for any other type of house (i.e. detached or semi-detached) the volume of the original house would be increased by more than 15% or $70\,m^3$ (whichever is the greater). In any case the volume of the original house would be increased by more than $115\,m^3$.
- alterations to the roof, including dormer windows (but permission is not normally required for skylights).
- extensions nearer to a highway than the nearest part of the original house (unless the house, as extended, would be at least 20 m away from the highway).
- to extend or add to your house so as to create a separate dwelling, such as self-contained living accommodation or a granny flat.
- if an extension to your house comes within 5 m of another building belonging to your house.
- to build an addition, which would be nearer to any highway than the nearest part of the original house, unless there would be at least 20 m between your house (as extended) and the highway. The term 'highway' includes all public roads, footpaths, bridleways and byways.
- if more than half the area of land around the original house would be covered by additions or other buildings – although you may not have built an extension to the house, a previous owner may have done so.
- if the extension or addition exceeds the certain limits on height and volume.
- the extension is higher than the highest part of the roof of the original house or any part of the extension is more than 4 m high and is within 2 m of the boundary of your property.

Any building which has been added to your property and which is more than $10\,m^3$ in volume and which is within 5 m of your house is treated as an extension of the house and so reduces the allowance for further extensions without planning permission.

Where the word 'original' is used above, in planning regulations terms this means the house as it was first built or as it was on 1 July 1948. Any extensions added since that date are counted towards the allowances.

Limitations
Planning permission is required if:

- **Height** – Any part is higher than the highest part of the house roof.
- **Projections** – Any part projects beyond the foremost wall of the house facing a highway.

- **Boundary** – Any part within 2 m (6 ft 6") of a boundary is more than 4 m (13 ft) high.
- **Area** – It will cover more than half the original area of the garden.
- **Dwelling** – It is to be an independent dwelling.

💡 You should measure the height of buildings from the ground level immediately next to it. If the ground is uneven, you should measure from the highest part of the surface, unless you are calculating volume.

- **Volume** – Planning permission is required if the extension results in an increase in volume of the original house by whichever is the greater of the following amounts:
  - for terraced houses 50 m³ (65.5 y³) or 10% up to a maximum of 115 m³ (150.4 y³);
  - other houses 70 m³ (91.5 y³) or 15% up to a maximum of 115 m³ (150.4 y³);
  - in Scotland, general category 24 m² (28.7 y²) or 20%.

💣 The volume of other buildings which belong to your house (such as a garage or shed) will count against the volume allowances. In some cases, this can include buildings that were built at the same time as the house or that existed on 1 July 1948.

5.22 Conversions

	Requirement		
	Planning permission		Building Regulation approval
Yes	For flats – even where construction works may not be intended	Yes	Unless you are not proposing any building work to make the change
Yes	For shops and offices unless no building work is envisaged		

Do I need approval to convert my house into flats?

Yes – even where construction works may not be intended.

Do I need approval to convert my house to a shop or office?

No – if you are not proposing any building work to make the change.

Do I need approval to convert part or all of my shop or office to a flat or house?

Yes.

Where building work is proposed you will probably need approval if it affects the structure or means of escape in case of fire. But you should check with the local fire authority and the county council, to see whether a fire certificate is actually required.

⚫ You will probably also need planning permission whether or not building work is proposed.

5.22.1 Converting an old building

	Requirement	
Planning permission	Building Regulation approval	
Yes	Yes	

Do I need planning permission to convert an old building?

Yes.

Throughout the UK there are many under-used or redundant buildings, particularly farm buildings which may no longer be required, or suitable, for agricultural use. Such buildings of weathered stone and slate contribute substantially to the character and appearance of the landscape and the built environment. Their interest and charm stems from an appreciation of the functional requirements of the buildings, their layout and proportions, the type of building materials used and their display of local building methods and skills.

In most cases traditional buildings are best safeguarded if their original use can be maintained. However with changing patterns of land use and farming methods, changes of use or conversion may have to be considered.

The conversion or re-use of traditional buildings may, in the right locations, assist in providing employment opportunities, housing for local people, or holiday accommodation. Applicants and developers are encouraged to refer to the local plan for comprehensive guidance and to seek advice from a Planning Officer if further assistance is necessary.

All councils place the highest priority to good design and proposals. Those that fail to respect the character and appearance of traditional buildings, will not be permitted. Sensitive conversion proposals should ensure that existing ridge and eaves lines are preserved; new openings avoided as far as possible; traditional matching materials are used; and the impact of parking and garden areas is minimized. Buildings that are listed as being of 'special architectural or historic interest' require skilled treatment to conserve internal and external features.

In many instances, traditional buildings that are of simple, robust form with few openings may only be suitable for use as storage or workshops. Other uses, such as residential, may be inappropriate.

5.23 Change of use

	Requirement	
	Planning permission	Building Regulation approval
Possibly	Even if no building or engineering work is proposed	Yes

The use of buildings or land for a different purpose may need consent even if no building or engineering works are proposed. Again, it is always advisable to check with development control staff.

What is meant by material change of use?

A material change of use is where there is a change in the purposes for which or the circumstances in which a building is used, so that after that change:

    (a)   the building is used as a dwelling, where previously it was not;

    (b)   the building contains a flat, where previously it did not;

    (c)   the building is used as an hotel or a boarding house, where previously it was not;

    (d)   the building is used as an institution, where previously it was not;

    (e)   the building is used as a public building, where previously it was not;

    (f)   the building is not a building described in Classes I to VI in Schedule 2, where previously it was; or

    (g)   the building, which contains at least one dwelling, contains a greater or lesser number of dwellings than it did previously.

 'Public building' means a building consisting of or containing:

- a theatre, public library, hall or other place of public resort;
- a school or other educational establishment;
- a place of public worship.

What are the requirements relating to material change of use?

Where there is a material change of use of the whole of a building, such work, if any, shall be carried out as is necessary to ensure that the building complies with the applicable requirements of the following paragraphs of Schedule 1:

(a) in all cases:

- means of warning and escape (B1)
- internal fire spread – linings (B2)
- internal fire spread – structure (B3)
- external fire spread – roofs (B4)
- access and facilities for the fire service (B5)
- ventilation (F1 and F2)
- sanitary conveniences and washing facilities (G1)
- bathrooms (G2)
- foul water drainage (H1)
- solid waste storage (H6)
- combustion appliances (J1, J2 and J3)
- conservation of fuel and power – dwellings (L1)
- conservation of fuel and power – buildings other than dwellings (L2).

In the case of a building exceeding 15 m in height:

- external fire spread – walls (B4 1).

(b) in other cases:

Material change of use	Requirement	Approved document
The building is used as a dwelling, where previously it was not	Resistance to weather and ground moisture	C4
	Resistance to the passage of sound	
The building contains a flat, where previously it did not	Resistance to the passage of sound	E1, E2 & E3
The building is used as an hotel or a boarding house, where previously it was not	Structure	A1, A2 & A3
The building is used as an institution, where previously it was not	Structure	A1, A2 & A3
The building is used as a public building, where previously it was not	Structure	A1, A2 & A3
The building is not a building described in Classes I to VI in Schedule 2, where previously it was	Structure	A1, A2 & A3
The building, which contains at least one dwelling, contains a greater or lesser number of dwellings than it did previously	Structure	A1, A2 & A3
The building, which contains at least one dwelling, contains a greater or lesser number of dwellings than it did previously	Resistance to the passage of sound	E1, E2 & E3

5.23.1 Buildings suitable for conversion

Most local plans stipulate that conversion proposals 'should relate to buildings of traditional design and construction which enhance the natural beauty of the landscape' as opposed to 'non-traditional buildings, buildings of inappropriate design, or buildings constructed of materials which are of a temporary nature'.

Isolated buildings

Planning permission will not normally be granted for the conversion or re-use of isolated buildings. Exceptionally, permission may be given for such buildings to be used for small-scale storage or workshop uses or for camping purposes.

An isolated building is normally:

- *a building, or part of a building, standing alone in the open countryside; or*
- *a building, or part of a building, comprised within a group which otherwise occupies a remote location having regard to the disposition of other buildings within the locality, to the character of the surroundings, and to the nature and availability of access and essential services.*

Assessing whether or not a particular building should be regarded as isolated may not always be straightforward and, in such instances early discussion with a planning officer at the National park authority is advised.

Structural condition

Buildings proposed for conversion should be large enough to accommodate the proposed use without the necessity for major alterations, extension or re-construction. In cases of doubt regarding the structural condition of any particular building, the authority will require the submission of a full structural survey to accompany a planning application. The authority can advise on this requirement and, if necessary, on persons who are suitably qualified to undertake such work and who practise locally.

Planning permission will not normally be granted for re-construction if substantial collapse occurs during work on the conversion of a building.

 A list of local consulting engineers can be found in *Yellow Pages*.

Workshop conversions

Redundant farm buildings and buildings of historic interest are often well suited to workshop use and such conversions require minimal alterations. Potential problems of traffic generation and unneighbourliness can usually be addressed by the imposition of appropriate conditions.

The local authority will generally favourably consider proposals that make good use of traditional buildings by promoting local employment opportunities. In some instances grants may be available from other agencies to assist the conversion of buildings to workshop use.

Residential conversions

When reviewing proposals for conversion of traditional buildings the local authority will pay particular attention to the overall objectives of the housing policies of the local plan. If land that can be used for a new housing development is limited, residential conversions can make a valuable contribution to the local housing stock and support the social and economic well being of rural communities.

The local plan will require that residential conversions should, in most instances, contribute to the housing needs of the locality. Permission for such conversions are, in some districts, only granted subject to a condition restricting occupancy to local persons.

☀ Local persons are normally defined as persons working, about to work, or having last worked in the locality or who have resided for a period of three years within the locality.

Renovation

Districts dedicate some areas as Environmentally Sensitive Areas (ESAs) and grants may be available towards the cost of renovating historical and important local buildings that have fallen into disrepair or towards the cost of renovation works to retain agricultural buildings in farming use, so as to retain their importance as landscape features. Further advice on the workings of the scheme may be obtained from the ESA project officers or the authority's building conservation officer.

Applicants are strongly advised to employ qualified architects or designers in preparing conversion proposals. Informal discussions with a planning officer at an early stage in considering design solutions are also encouraged.

5.24 Building a new house

	Requirement
Planning permission	Building Regulation approval
Yes	Yes

Do I need planning permission to erect a new house?

Yes.
All new houses or premises of any kind require planning permission.

Private individuals will normally only encounter this if they intend to buy a plot of land to build on, or buy land with existing buildings that they want to demolish to make way for a new property to be built.

In all cases like this, unless you are an architect or a builder, you must seek professional advice. If you are using a solicitor to act on your behalf in purchasing a plot on which to build, he will include the planning questions within all the other legal work, as well as investigating the presence of covenants, existing planning consent together with other constraints or conditions.

The architect, surveyor or contractor you hire will then need to take into account the planning requirements as part of their planning and design procedures. They will normally handle planning applications for any type of new development.

⬤ If you are hiring a professional, or more than one – say a building contractor to do the work and a surveyor or architect to plan and design – be sure to find out exactly who does what and that approval is obtained before going to too much expense, should a refusal arise.

5.25 Infilling

	Requirement		
	Planning permission		Building Regulation approval
Possibly	Consult your local planning officer	Yes	If a new development

Can I use an unused, but adjoining, piece of land to build a house (e.g. build a new house on land that used to be a large garden)?

Often there may be no official grounds for denying consent, but residents and individuals can impose quite some delay. It is worth testing the likelihood of a successful application by talking to the neighbours and judging opinions.

Planning consent is often quite difficult to obtain in these cases as this sort of development normally causes a lot of opposition as it is in a settled residential area and people do not like change.

New developments will undoubtedly also need to follow building regulations. This, and all site visits from inspectors, is normally arranged by your building contractor.

💡 There are plenty of substantial building projects that don't require any planning permission. However, it is undoubtedly a good idea to consult a range of people before you consider any work.

5.26 Demolition

	Requirement		
	Planning permission		Building Regulation approval
Yes	If it is a listed building or in a Conservation Area	No	For a complete detached house
	If the whole house is to be demolished	Yes	For a partial demolition to ensure that the remaining part of the house (or adjoining buildings/ extensions) are structurally sound

You must have good reasons for knocking a building down, such as making way for rebuilding or improvement (which in most cases would be incorporated in the same planning application). Penalties are severe for demolishing something illegally.

You do not need to make a planning application to demolish a listed building or to demolish a building in a Conservation Area. However, you may need listed building or Conservation Area consent.

Elsewhere, you will **not** need to apply for planning permission:

- to demolish a building such as a garage or shed of less than $50\,m^3$; or
- if the demolition is urgently necessary for health and safety reasons; or
- if the demolition is required under other legislation; or
- where the demolition is on land that has been given planning permission for redevelopment; or
- to demolish a gate, fence, wall or other means of enclosure.

In all other cases, such as demolishing a house or block of flats, the council may wish to agree the details of how you intend to carry out the demolition and how you propose to restore the site afterwards. You will need to apply for a formal decision on whether the council wishes to approve these details. This is called a 'prior approval application' and your council will be able to explain what it involves.

You are not allowed to begin any demolition work (even on a dangerous building) unless you have given the local authority notice of your intention and this has either been acknowledged by the local authority or the relevant notification period has expired. In this notice you will have to:

- specify the building to be demolished;
- state the reason(s) for wanting to demolish it;
- show how you intend to demolish it.

Copies of this notice will have to be sent to:

- the local authority;
- the occupier of any building adjacent to the building;
- British Gas;
- the area electricity board in whose area the building is situated.

💡 This regulation does not apply to the demolition of an internal part of an occupied building, or a greenhouse, conservatory, shed or prefabricated garage (that forms part of that building) or an agricultural building defined in Section 26 of the General Rate Act 1967.

5.26.1 What about dangerous buildings? (Building Act 1984 Sections 77 and 78)

If a building, or part of a building or structure, is in such a dangerous condition (or is used to carry loads that would make it dangerous) then the local authority may apply to a magistrates' court to make an order requiring the owner:

- to carry out work to avert the danger;
- to demolish the building or structure, or any dangerous part of it, and remove any rubbish resulting from the demolition.

💣 The local authority can also make an order restricting its use until such time as a magistrates' court is satisfied that all necessary works have been completed.

These works are controllable by the local authority under Sections 77 and 78 of the Building Act 1984. In inner London the legislation is under the London Building (Amendment) Act 1939.

This involves responding to all reported instances of dangerous walls, structures and buildings within each local authority's area on a 24 hour 365 days a year basis.

💡 Refer to the relevant local authority building control office during office hours or their local authority emergency switchboard, out of hours.

If the building or structure poses a potential danger to the safety of people, the local authority will take the appropriate action to remove the danger. The local authority has powers to require the owners of buildings or structures to remedy the defects or directs their own contractors to carry out works to make the building or structure safe. In addition, the local authority may provide advice on the structural condition of buildings during fire fighting to the fire brigade.

💡 If you are concerned that a building or other structure may be in a dangerous condition, then you should report it to the local council.

Emergency measures

In emergencies the local authority can make the owner take immediate action to remove the danger, or they can complete the necessary action themselves. In these cases, the local authority is entitled to recover from the owner such expenses reasonably incurred by them. For example:

- fencing off the building or structure;
- arranging for the building/structure to be watched.

5.26.2 Can I be made to demolish a dangerous building?
(Building Act 1984 Sections 81, 82 and 83)

If the local authority considers that a building is so dangerous that it should be demolished, they are entitled to issue a notice to the owner requiring the owner/occupier:

- to shore up any building adjacent to the building to which the notice relates;
- to weatherproof any surfaces of an adjacent building that are exposed by the demolition;
- to repair and make good any damage to an adjacent building caused by the demolition or by the negligent act or omission of any person engaged in it;
- to remove material or rubbish resulting from the demolition and clear the site;
- to disconnect, seal and remove any sewer or drain in or under the building;
- to make good the surface of the ground that has been disturbed in connection with this removal of drains etc.;
- in accordance with the Water Act 1945 (interference with valves and other apparatus) and the Gas Act 1972 (public safety), arranging with the relevant statutory undertakers (e.g. water board, British Gas or electricity supplier) for the disconnection of gas, electricity and water supplies to the building;
- to leave the site in a satisfactory condition following completion of all demolition work.

Before complying with this notice, the owner must give the local authority 48 hours' notice of commencement.

In certain circumstances, the owner of an adjacent building may be liable to assist in the cost of shoring up their part of the building and waterproofing the surfaces. It could be worthwhile checking this point with the local authority!

Replacing a demolished building

If you decide to demolish a building, even one that has suffered fire or storm damage, it does not automatically follow that you will get planning permission to build a replacement.

Under Section 80 of the Building Act 1984 anyone carrying out demolition work is required to notify the local authority. The local authority then has six weeks to respond with appropriate notices and consultation under Sections 81 and 82 of the Act (this does not apply to inner London).

6

Meeting the requirements of the Building Regulations

Background

The Building Regulations 2000 as amended by the Building Amendment Regulations 2001 (SI 2001/3335) replaced the Building Regulations 1991 (SI 1985 No. 1065). Since then, a series of Approved Documents have been endorsed by the Secretary of State that are intended to provide guidance to some of the more common building situations. They also provide a practical guide to meeting the requirements of Schedule 1 and Regulation 7 of the Building Regulations.

Approved Documents

The 2002 list of Approved Documents is given in Table 6.1 below.

Table 6.1 Approved documents 2002

Section	Title	Edition	Latest amendment
A	Structure	1992	2000
B	Fire safety	2000	2000
C	Site preparation and resistance to moisture	1992	2000
D	Toxic substances	1992	2000
E	Resistance to the passage of sound	1992	2000
F	Ventilation	1995	2000
G	Hygiene	1992	2000
H	Drainage and waste disposal	2002	
J	Combustion and waste disposal	2002	
K	Protection from falling, collision and impact	1998	2000
L1	Conservation of fuel and power in dwellings	2002	
L2	Conservation of fuel and power in buildings other than dwellings	2002	
M	Access and facilities for disabled people	1999	2000
N	Glazing – safety in relation to impact, opening and cleaning	1998	2000
	Approved Document to support Regulation 7 – Materials and workmanship	1999	2000

Note: All of these documents are published by the Stationery Office. For availability and further details, see www.thestationeryoffice.com

Compliance

There is no obligation to adopt any particular solution that is contained in any of these guidance documents especially if you prefer to meet the relevant requirement in some other way. However, should a contravention of a requirement be alleged, if you have followed the guidance in the relevant Approved Documents, that will be evidence tending to show that you have complied with the Regulations. If you have **not** followed the guidance, then that will be seen as evidence tending to show that you have not complied with the requirements and it will then be up to you, the builder, architect and/or client to demonstrate that you have satisfied the requirements of the Building Regulations.

This compliance may be shown in a number of ways such as using:

- a product bearing CE marking (in accordance with the Construction Products Directive – 89/1 06/EEC as amended by the CE Marking Directive (93/68/EEC) as implemented by the Construction Products Directive 1994 (SI 1994/3051);
- an appropriate technical specification (as defined in the Construction Products Directive – 89/1 06/EEC);
- a recognized British Standard;
- a British Board of Agrément Certificate;
- an alternative, equivalent national technical specification from any member state of the European economic area;
- a product covered by a national or European certificate issued by a European Technical Approval issuing body.

Limitation on requirements

In accordance with Regulation 8 of the Building Regulations, the requirements in Parts A to D, F to K and N (except for paragraphs H2 and J6) of Schedule 1 to the Building Regulations, are only intended to guarantee that reasonable standards of health and safety for persons in or about the building are ensured.

You may show that you have complied with Regulation 7 in a number of ways, for example, by the appropriate use of a product bearing a CE marking in accordance with the Construction Products Directive (89/106/EEC) as amended by the CE Marking Directive (93/68/EEC), or by following an appropriate technical specification (as defined in that Directive), a British Standard, a British Board of Agrément Certificate, or an alternative national technical specification of any member state of the European Community which, in use, is equivalent. You will find further guidance in the Approved Document supporting Regulation 7 on materials and workmanship.

Materials and workmanship

As stated in the Building Regulations, *'Any building work which is subject to requirements imposed by Schedule 1 of the Building Regulations should, in accordance with Regulation 7, be carried out with proper materials and in a workmanlike manner'.*

What materials can I use?

Other than the two exceptions below, provided the materials and components that you have chosen to use are from an approved source and are of approved quality (e.g. CE marked in the case of equipment) then the choice is fairly unlimited.

Short lived materials

Even if a plan for building work complies with the Building Regulations, if this work has been completed using short lived materials (i.e. materials that are, in the absence of special care, liable to rapid deterioration) the local authority can:

- reject the plans;
- pass the plans subject to a limited use clause (on expiration of which they will have to be removed);
- restrict the use of the building.

(Building Act 1984 Section 19)

Unsuitable materials

If, once building work has begun, it is discovered that it has been made using materials or components that have been identified by the Secretary of State (or his nominated deputy) as being unsuitable materials, the local authority have the power to:

- reject the plans;
- fix a period in which the offending work must be removed;
- restrict the use of the building.

(Building Act 1984 Section 20)

If the person completing the building work fails to remove the unsuitable material or component(s), then that person is liable to be prosecuted and, on summary conviction, faces a heavy fine.

Technical specifications

Building Regulations are made for specific purposes such as:

- health and safety;
- conservation of fuel and power;
- prevention of contamination of water;
- welfare and convenience of disabled people.

Standards and technical approvals, as well as providing guidance, also address other aspects of performance such as serviceability and/or other aspects related to health and safety not covered by the Regulations.

When an Approved Document makes reference to a named standard, the relevant version of the standard is the one listed at the end of that particular Approved Document. However, if this version of the standard has been revised or updated by the issuing standards body, the new version may be used as a source of guidance provided it continues to address the relevant requirements of the Regulations.

The Secretary of State has agreed with the British Board of Agrément on the aspects of performance that it needs to assess in preparing its certificates in order that the board may demonstrate the compliance of a product or system that has an Agrément Certificate with the requirements of the Regulations. An Agrément Certificate issued by the board under these arrangements will give assurance that the product or system to which the certificate relates (if properly used in accordance with the terms of the certificate) will meet the relevant requirements.

Standards and technical approvals

Standards and technical approvals provide guidance related to the Building Regulations and address other aspects of performance such as serviceability or aspects which although they relate to health and safety are not covered by the Regulations.

European pre-standards (ENV)

The British Standards Institution (BSI) will be issuing Pre-standard (ENV) Structural Eurocodes as they become available from the European Standards Organisation, Comité Europeen de Normalisation Electrotechnique (CEN).

DD ENV 1992-1-1: 1992 Eurocode 2: Part 1 and DD ENV 1993-1-1: 1992 Eurocode 3: Part 1-1 *General Rules* and *Rules for Buildings in concrete and steel* have been thoroughly examined over a period of several years and are considered to provide appropriate guidance when used in conjunction with their national application documents for the design of concrete and steel buildings respectively.

When other ENV Eurocodes have been subjected to a similar level of examination they may also offer an alternative approach to Building Regulation compliance and, when they are eventually converted into fully approved EN standards, they will be included as referenced standards in the guidance documents.

House – construction

There are two main types of buildings in common use today: those made of brick and those made of timber. There are many different styles of brick-built houses and, equally there are various methods of construction.

Brickwork, as well as giving a building character, provides the main load bearing element of a brick-built house. Timber-framed houses, on the other hand, are usually built on a concrete foundation with a 'strip' or 'raft' construction to spread the weight and differ from their brick-built counterparts in that the main structural elements are timber frames.

6.1 Foundations

To support the weight of the structure, most brick-built buildings are   )orted on a solid base called foundations. Timber framed houses are usuall)   lt on a concrete foundation with a 'strip' or 'raft' construction to spread the weight.

6.1.1 Requirements

The building shall be constructed so that:

- *the combined dead, imposed and wind loads are sustained and transmitted by it to the ground, safely and without causing any building deflection/ deformation or ground movement that will affect the stability of any part of the building;*
- *ground movement caused by swelling, shrinkage or freezing of the subsoil; land-slip or subsidence will not affect the stability of any part of the building.*

(Approved Document B)

Buildings with five or more storeys (each basement level being counted as one storey) shall be constructed so that:

- *in the event of an accident, the building will not collapse to an extent inconsistent to the cause.*

(Approved Document B)

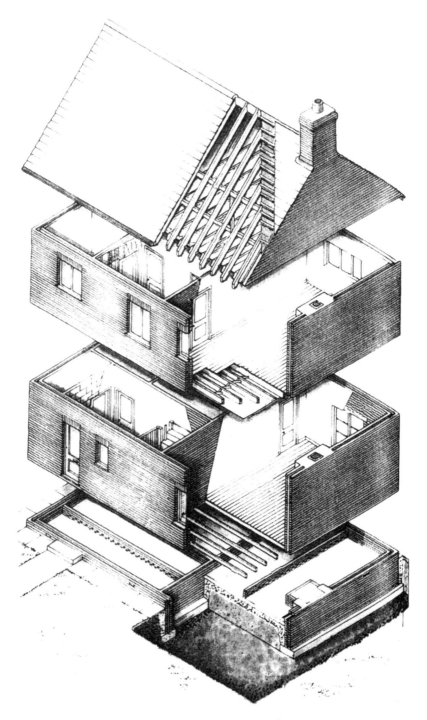

Figure 6.1 Brick built house – typical components

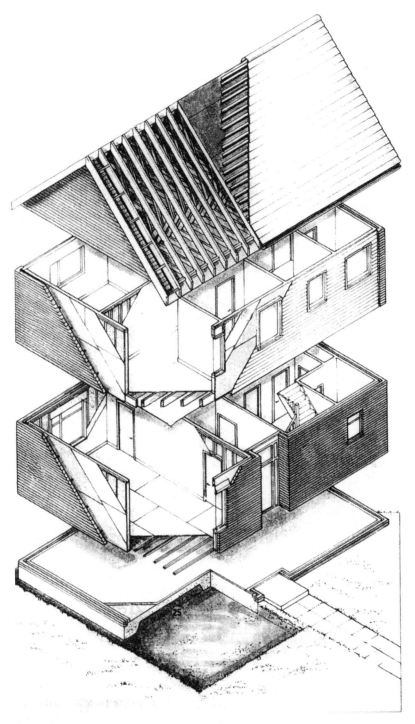

Figure 6.2 Timber framed house – typical components

Precautions shall be taken to reduce risks to the health and safety of persons in buildings by safeguarding them and the buildings against the adverse effects of:

- *vegetable matter*
- *contaminants on or in the ground to be covered by the building*
- *ground water.*

(Approved Document C)

There may be known and/or recorded conditions of ground instability, such as geological faults, landslides, disused mines, or unstable strata of similar nature which affect (or may potentially affect) a building site or its environs. These should be taken into account before proceeding with the design of a building or its foundations.

What about contaminated ground?

Potential building sites that are likely to contain contaminants can be identified at an early stage from planning records or from local knowledge (e.g. previous uses). In addition to solid and liquid contaminants, problems can also arise from natural contamination such as methane and the radioactive gas radon (and its decay product). There are many possibilities and the following list identifies sites most likely to contain contaminants:

- asbestos works
- chemical works
- gas works, coal carbonization plants and ancillary
- by-product works
- industries making or using wood preservatives
- landfill and other waste disposal sites
- metal mines, smelters, foundries, steel works and metal finishing works
- munitions production and testing sites
- nuclear installations
- oil storage and distribution sites
- paper and printing works
- railway land, especially the larger sidings and depots
- scrap yards
- sewage works, sewage farms and sludge disposal sites
- tanneries.

If any signs of possible contaminants are present, the environmental health officer should be told at once. If he confirms the presence of any of these contaminants, then he will require that the actions as listed in Table 6.2 are completed before any planning permission for building work can be sought.

In the most hazardous conditions, only the total removal of contaminants from the ground to be covered by the building can provide a complete remedy. In other cases remedial measures can reduce the risks to acceptable levels.

Table 6.2 Possible contaminants and actions

Signs of possible contaminants	Possible contaminant	Relevant action
Vegetation (absence, poor or unnatural growth)	Metals Metal compounds	None
	Organic compounds Gases	Removal
Surface materials (unusual colours and contours may indicate wastes and residues)	Metals Metal compounds	None
	Oily and tarry wastes	Removal, filling or sealing
	Asbestos (loose)	Filling or sealing
	Other mineral fibres	None
	Organic compounds including phenols	Removal or filling
	Combustible material including coal and coke dust	Removal or filling
	Refuse and waste	Total removal
Fumes and odours (may indicate organic chemicals at very low concentrations)	Flammable explosive and asphyxiating gases including methane and carbon dioxide	Removal
	Corrosive liquids	Removal, filling or sealing
	Faecal animal and vegetable matter (biologically active)	Removal or filling
Drums and containers (whether full or empty)	Various	Removal with all contaminated ground

These measures should only be undertaken with the benefit of expert advice. Where the removal would involve handling large quantities of contaminated materials, then you are advised to seek expert advice.

Even when these actions have been successfully completed, the ground to be covered by the building will still need to have at least 100 mm of concrete laid over it.

6.1.2 Gaseous contaminants

Radon

Radon is a naturally occurring, radioactive, colourless and odourless gas, which is formed in small quantities by radioactive decay wherever uranium and radium are found. It can move through the subsoil and then into buildings and exposure to high levels over long periods would increase the risk of developing lung cancer. Some parts of the country (in particular the West Country) have higher natural levels than elsewhere and precautions against radon may be necessary.

Guidance on the construction of dwellings in areas susceptible to radon have been published by the Building Research Establishment as a report (*Radon: guidance on protective measures for new dwellings*).

6.1.3 Landfill gas and methane

Landfill gas is generated by the action of anaerobic microorganisms on bio-degradable material and generally consists of methane and carbon dioxide with small quantities of other gases. It can migrate under pressure through the subsoil and through cracks and fissures into buildings.

Methane is an asphyxiant, will burn, and can explode in air. Carbon dioxide is non-flammable and toxic. Many of the other components of landfill gas are flammable and some are toxic.

6.1.4 Meeting the requirement

Foundations

Table 6.3 provides guidance on determining the type of soil on which it is intended to lay a foundation.

Table 6.3 Types of subsoil

Type	Applicable field test
Rock (being stronger/ denser than sandstone, limestone or firm chalk)	Requires at least a pneumatic or other mechanically operated pick for excavation.
Compact gravel and/or sand	Requires a pick for excavation. Wooden peg 50 mm square in cross section hard to drive beyond 150 mm.
Stiff clay or sandy clay	Cannot be moulded with the fingers and requires a pick or pneumatic or other mechanically operated spade for its removal.
Firm clay or sandy clay	Can be moulded by substantial pressure with the fingers and can be excavated with a spade.
Loose sand, silty sand or clayey sand	Can be excavated with a spade. Wooden peg 50 mm square in cross section can be easily driven.
Soft silt, clay, sandy clay or silty clay	Fairly easily moulded in the fingers and readily excavated.
Very soft silt, clay, sandy clay or silty clay	Natural sample in winter conditions exudes between the fingers when squeezed in fist.

Subsoil

There should not be a wide variation in type of subsoil within the loaded area, or a weaker type of subsoil below the soil on which the foundation rests that could affect the stability of the structure.

A1/2 (1E1)

The foundations should be situated centrally under the wall.

A1/2 (1E2a)

Strip foundations should have the minimum widths given in Table 6.4.

A1/2 (1E2b and 1E3)

For foundations in chemically aggressive soil conditions, guidance in BS 5328: Part 1 should be followed.

A1/2 (1E2c)

For foundations in non-aggressive soils, concrete should be composed of cement (to BS 12: 1989) and fine and coarse aggregate (conforming to BS 882: 1983). The mix should either be in the proportion of 50 kg of cement to not more than 0.1 m^3 of fine aggregate and 0.2 m^3 of coarse aggregate or be Grade ST1 concrete conforming to BS 5328 Part 2.

A1/2 (1E2c)

Minimum thickness (T) of concrete foundation should be 150 mm or P, whichever is the greater where P is derived using Table 6.4.

A1/2 (1E2d)

Foundations stepped on elevation should overlap by twice the height of the step, by the thickness of the foundation, or 300 mm, whichever is greater (see Figure 6.3).

A1/2 (1E2e)

The foundation of piers, buttresses and chimneys should project as indicated in Figure 6.4 and the projection X should never be less than P.

A1/2 (1E2g)

Table 6.4 Minimum width of strip foundations

Type	Minimum width of strip foundation according to the type of load bearing walling encountered (in kN/linear meter)					
	20	30	40	50	60	70
Rock (being stronger/denser than sandstone, limestone or firm chalk)	In each case, equal to the width of the wall					
Compact gravel and/or sand	250	300	400	500	600	650
Stiff clay or sandy clay	250	300	400	500	600	650
Firm clay or sandy clay	300	350	450	600	750	850
Loose sand, silty sand or clayey sand	400	600				
Soft silt, clay, sandy clay or silty clay	450	650				
Very soft silt, clay, sandy clay or silty clay	600	850				

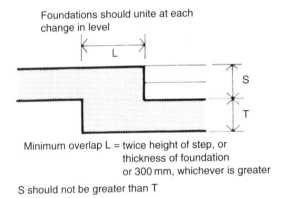

Foundations should unite at each change in level

L

S

T

Minimum overlap L = twice height of step, or
thickness of foundation
or 300 mm, whichever is greater

S should not be greater than T

Figure 6.3 Elevation of stepped foundation

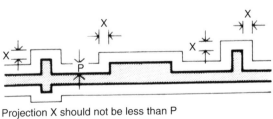

Projection X should not be less than P

Figure 6.4 Piers and chimneys

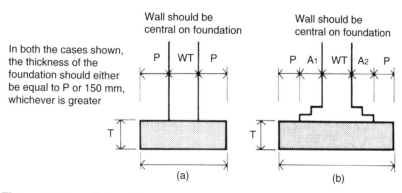

Figure 6.5 Foundation dimensions. (a) Strip foundation. (b) Strip foundation with footing

Vertical loading

Vertical loading caused by concrete floor slabs, precast concrete floors, and timber floors on the walls, should be distributed.	A1/2 (1C25a)
The combined dead and imposed load should not exceed 70 kN/m at base of wall (see Figure 6.38).	A1/2 (1C25c)
To reduce the sensitivity of the building to disproportionate collapse, horizontal and vertical ties should be provided.	A1/2 (A3 5.1)

💡 If it is not feasible to provide effective horizontal and vertical tying of any of the load bearing members, then each support member should be considered to be notionally removed, one at a time from each storey in turn, to check that even when it is removed, the risk of the structure within that storey (and the storeys immediately adjacent) is limited to 15% of the area of the storey or 70 m^2, whichever is the less (see Figure 6.6).

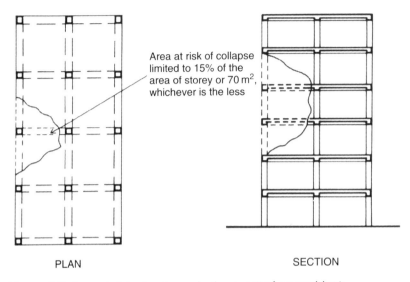

PLAN   SECTION

Figure 6.6 Area at risk of collapse in the event of an accident

Maximum floor area

No floor enclosed by structural walls on all sides shall exceed 70 m² (see Figure 6.7).	A1/2 (1C15)
No floor with a structural wall on one side shall exceed 30 m² (see Figure 6.7).	A1/2 (1C15)

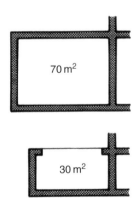

Figure 6.7 Maximum floor area that is enclosed by structural walls

Maximum height of buildings

The maximum height of a building shall not exceed the heights given in Table 6.5 with regard to the relevant wind speed.   A1/2 (1C17)

Table 6.5 Maximum height of building on normal or slightly sloping sites

Location	Normal or slightly sloping sites							Steeply sloping sites, hills, cliffs and escarpments						
Wind speed	36	38	40	42	44	46	48	36	38	40	42	44	46	48
Unprotected site, open countryside with no obstructions	15	15	15	15	15	11	9	8	6	4	3	0	0	0
Open countryside with scattered windbreaks	15	15	15	15	15	15	13	11	9	7.5	6	5	4	3
Country with many windbreaks, small towns, outskirts of large cities	15	15	15	15	15	15	15	15	15	14	12	10	8	6.5
Protected sites and country centres	15	15	15	15	15	15	15	15	15	15	15	15	15	14

Imposed loads on roofs, floors and ceilings

The imposed loads on roofs, floors and ceilings shall not exceed those shown in Table 6.6.   A1/2 (1C16)

Table 6.6 Imposed loads

Element	Distributed loads	Concentrated load
Roofs	1.00 kN/m^2 for spans not exceeding 12 m 1.50 kN/m^2 for spans not exceeding 6 m	
Floors	2.00 kN/m^2	
Ceilings	0.25 kN/m^2	0.9 kN/m^2

Subsoil

All turf and other vegetable matter should be removed from the ground to be covered by the building at least to a depth sufficient to prevent later growth.	C1 (1.2)
💡 This provision does not apply to a building which is to be used wholly for storing goods.	
All building services (such as below-ground drainage) should be sufficiently robust or flexible to accommodate the presence of any roots and all joints should be made so that roots will not penetrate them.	C1 (1.3)
In areas where the water table can rise to within 0.25 m of the lowest floor of the building, or where surface water could enter or adversely affect the building, then the ground to be covered by the building should be drained by gravity or other effective means.	C1 (1.4–1.7) C4 (3.1–3.8) C4 (4.1–4.15) C4 (5.1–5.9)
If an active subsoil; drain is cut during excavation, then it shall be either re-laid or re-routed.	C1 (1.6)

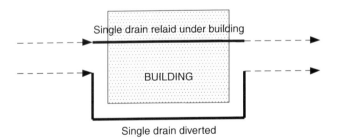

Single drain relaid under building

BUILDING

Single drain diverted

Figure 6.8 Subsoil drain cut during excavation

Landfill and gas

If the ground to be covered by a building is on, or within 250 m of, landfill where gas production is possible, further investigation should be made to determine what, if any, protective measures are necessary.	C2 (2.1–2.11)

If the level of methane in the ground is unlikely to exceed 1% by volume then the ground floor of a house or similar small building should be on suspended concrete and ventilated.	C2 (2.1–2.11)
If there is a carbon dioxide concentration of greater than 1.5% by volume in the ground, then you need to consider measures to prevent gas ingress.	C2 (2.1–2.11)
If there is a carbon dioxide concentration of greater than 5% by volume in the ground, then specific design measures are required.	C2 (2.1–2.11)

The use of permanent continuous mechanical ventilation to ensure that methane or carbon dioxide does not accumulate at any time in or under a house is not usually feasible since there is no management system to look after it. Passive protection is generally the only viable alternative and is effective only where gas concentrations in the ground are low and likely to remain so.

6.2 Buildings – size

6.2.1 Classification of purpose groups

Many of the provisions in Approved Documents are related to the use of the building. The use classifications are termed purpose groups and represent different levels of hazard. They can apply to a whole building, or (where a building is compartmented) to a compartment in the building, and the relevant purpose group should be taken from the main use of the building or compartment. Table 6.7 sets out the purpose group classification.

Table 6.7 Classification of purpose groups

Title	Group	Purpose for which the building or compartment of a building is intended to be used
Residential* (dwellings)	1(a)	Flat or maisonette.
	1(b)	Dwelling house which contains a habitable storey with a floor level which is more than 4.5 m above ground level.
	1(c)	Dwelling house which does not contain a habitable storey with a floor level which is more than 4.5 m above ground level.
Residential (institutional)	2(a)	Hospital, home, school or other similar establishment used as living accommodation for, or for the treatment, care or maintenance of persons suffering from disabilities due to illness or old age or other physical or mental incapacity, or under the age of five years, or place of lawful detention, where such persons sleep on the premises.

Table 6.7 Classification of purpose groups (*continued*)

Title	Group	Purpose for which the building or compartment of a building is intended to be used
Other	2(b)	Hotel, boarding house, residential college, hall of residence, hostel, and any other residential purpose not described above.
Office	3	Offices or premises used for the purpose of administration, clerical work (including writing, book keeping, sorting papers, filing, typing, duplicating, machine calculating, drawing and the editorial preparation of matter for publication, police and fire service work), handling money (including banking and building society work), and communications (including postal, telegraph and radio communications) or radio, television, film, audio or video recording, or performance (not open to the public) and their control.
Shop and commercial	4	Shops or premises used for a retail trade or business (including the sale to members of the public of food or drink for immediate consumption and retail by auction, self-selection and over-the-counter wholesale trading, the business of lending books or periodicals for gain and the business of a barber or hairdresser) and premises to which the public is invited to deliver or collect goods in connection with their hire, repair or other treatment, or (except in the case of repair of motor vehicles) where they themselves may carry out such repairs or other treatments.
Assembly and recreation	5	Place of assembly, entertainment or recreation; including bingo halls, broadcasting, recording and film studios open to the public, casinos, dance halls; entertainment, conference, exhibition and leisure centres; funfairs and amusement arcades; museums and art galleries; non-residential clubs, theatres, cinemas and concert halls; educational establishments, dancing schools, gymnasia, swimming pool buildings, riding schools, skating rinks, sports pavilions, sports stadia; law courts; churches and other buildings of worship, crematoria; libraries open to the public, non-residential day centres, clinics, health centres and surgeries; passenger stations and termini for air, rail, road or sea travel; public toilets; zoos and menageries.
Industrial	6	Factories and other premises used for manufacturing, altering, repairing, cleaning, washing, breaking-up, adapting or processing any article; generating power or slaughtering livestock.
Storage and other non-industrial**	7(a)	Place for the storage or deposit of goods or materials (other than described under 7(b)) and any non-residential building not within any of the purpose groups 1 to 6.
	7(b)	Car parks designed to admit and accommodate only cars, motorcycles and passenger or light goods vehicles weighing no more than 2500 kg gross.

Notes:
*Includes any surgeries, consulting rooms, offices or other accommodation, not exceeding 50 m² in total, forming part of a dwelling and used by an occupant of the dwelling in a professional or business capacity.
**A detached garage not more than 40 m² in area is included in purpose group 1(c); as is a detached open carport of not more than 40 m², or a detached building which consists of a garage and open carport where neither the garage nor open carport exceeds 40 m² in area.

Conditions – building

Limitations on size and proportion and parts of the building.	A1/2 (1C14)
Maximum allowable floor areas.	A1/2 (1C15)
Maximum imposed load and wind loads.	A1/2

Size of residential buildings

The maximum height of the building measured from the lowest finished ground level to the highest point of any wall or roof should be less than 15 m (see Figure 6.9).	A1/2 (1C14ai)
The height of the building should not exceed twice the least width of the building (see Figure 6.9).	A1/2 (1c14aii)
The height of the wing H^2 should not be greater than twice the least width of the wing W^2.	A1/2 (1c14aiii)

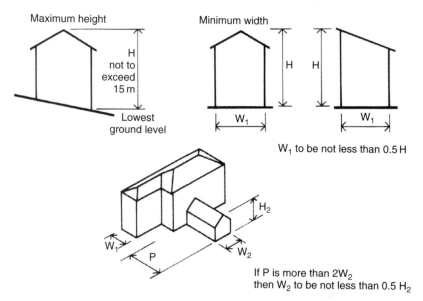

Figure 6.9 Residential buildings not more than three storeys

Size of single storey non residential buildings

The height should not exceed 3 m and the width should not exceed 9 m (see Figure 6.10).	A1/2 (1C14b)

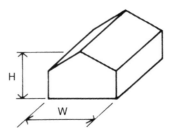

Figure 6.10 Single storey non residential building

Size of annexes

The height should not exceed 3 m (see Figure 6.11).	A1/2 (1c14c)

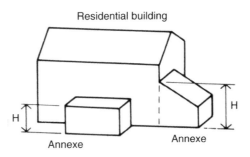

Figure 6.11 Size of annexes

6.2.2 Requirements – ventilation

Ventilation (mechanical and/or air-conditioning systems designed for domestic buildings) shall be capable of restricting the accumulation of moisture and pollutants originating within a building.

(Approved Document F1)

Ventilation

Buildings (or spaces within buildings) other than those:

- into which people do not normally go; or
- which are used solely for storage; or
- which are garages used solely in connection with a single dwelling,

shall be provided with ventilation to:

- extract water vapour from non-habitable areas where it is produced in significant quantities (e.g. kitchens, utility rooms and bathrooms).   F1 (1.1–1.5)

- extract pollutants (which are a hazard to health) from areas where they are produced in significant quantities (e.g. rooms containing processes which generate harmful contaminants and rest rooms where smoking is permitted).   F1 (2.3–2.5)

- rapidly dilute (when necessary) pollutants and water vapour produced in habitable rooms, occupiable rooms and sanitary accommodation.   F1 (1.1–1.4) F1 (1.6–1.8)

- provide a minimum supply of fresh air for occupants.   F1 (2.6–2.8)

Table 6.8 Ventilation of rooms containing openable windows (i.e. located on an external wall)

Room	Rapid ventilation (e.g. opening windows)	Background ventilation	Extract ventilation fan rates or passive stack (PSV)
Habitable room	1/20th of floor area	8000 mm²	
Kitchen	Opening window (no minimum size)	4000 mm²	30 litres/second adjacent to a hob or 60 litres/second elsewhere or PSV
Utility room	Opening window (no minimum size)	4000 mm²	30 litres/second or PSV
Bathroom (with or without WC)	Opening window (no minimum size)	4000 mm²	15 litres/second or PSV
Sanitary accommodation (separate from bathroom)	1/20th of floor area or mechanical extract at 6 litres/second	4000 mm²	

6.3 Drainage

6.3.1 The requirement (Building Act 1984 Sections 21 and 22)

All plans for building work need to show that drainage of refuse water (e.g. from sinks) and rainwater (from roofs) have been adequately catered for. Failure to do so will mean that these plans will be rejected by the local authority.

(Building Act 1984 Section 26)

All plans for buildings must include at least one (or more) water or earth closets **unless** the local authority are satisfied in the case of a particular building that one is not required (for example in a large garage separated from the house).

If you propose using an earth closet, the local authority cannot reject the plans unless they consider that there is insufficient water supply to that earth closet.

What are the rules about drainage? (Building Act 1984 Section 59)

The Building Act requires that all drains are connected either with a sewer (unless the sewer is more than 120 ft away or the person carrying out the building work is not entitled to have access to the intervening land) or is able to discharge into a cesspool, settlement tank or other tank for the reception or disposal of foul matter from buildings.

The local authorities view this requirement very seriously and will need to be satisfied that:

- satisfactory provision has been made for drainage;
- all cesspools, private sewers, septic tanks, drains, soil pipes, rain water pipes, spouts, sinks or other appliances are adequate for the building in question;
- all private sewers that connect directly or indirectly to the public sewer are not capable of admitting subsoil water;
- the condition of a cesspool is not detrimental to health, or does not present a nuisance;
- cesspools, private sewers and drains previously used, but now no longer in service, do not prejudice health or become a nuisance.

This requirement can become quite a problem if it is not recognized in the early planning stages and so it is always best to seek the advice of the local authority. In certain circumstances, the local authority might even help to pay for the cost of connecting you up to the nearest sewer!

The local authority has the authority to make the owner renew, repair or cleanse existing cesspools, sewers and drains etc.

Can two buildings share the same drainage?

Usually the local authority will require every building to be drained separately into an existing sewer but in some circumstances they may deicide that it would be more cost effective if the buildings were drained in combination. On occasions, they might even recommend that a private sewer is constructed.

What about ventilation of soil pipes? (Building Act 1984 Section 60)

A major requirement of the Building Regulations is that all soil pipes from water closets shall be properly ventilated and that no use shall be made of:

- an existing or proposed pipe designed to carry rain water from a roof to convey soil and drainage from a sanitary convenience;
- an existing pipe designed to carry surface water from a premises to act as a ventilating shaft to a drain or a sewer conveying foul water.

What happens if I need to disconnect an existing drain? (Building Act 1984 Section 662)

If, in the course of your building work, you need to:

- reconstruct, renew or repair an existing drain that is joined up with a sewer or another drain;
- alter the position of an existing drain that is joined up with a sewer or another drain;
- seal off an existing drain that is joined up with a sewer or another drain,

then, provided that you give 48 hours' notice to the local authority, the person undertaking the reconstruction may break open any street for this purpose.

You do not need to comply with this requirement if you are demolishing an existing building.

Can I repair an existing water closet or drain? (Building Act 1984 Section 63)

Repairs can be carried out to water closets, drains and soil pipes, but if that repair or construction work is prejudicial to health and/or a public nuisance, then the person who completed the installation or repair is liable, on conviction, to a heavy fine.

In the Greater London area, a 'water closet' can **also** be taken to mean a urinal.

Can I repair an existing drain? (Building Act 1984 Section 61)

Only in extreme emergencies are you allowed to repair, reconstruct or alter the course of an underground drain that joins up with a sewer, cesspool or other drainage method (e.g. septic tank).

💣 If you have to carry out repairs etc. in an emergency, then make sure that you do **not** cover over the drain or sewer without notifying the local authority of your intentions!

Drains – Fire protection

Drains should also provide a degree of fire protection as shown by the following requirement:

- *all openings in fire-separating elements shall be suitably protected in order to maintain the integrity of the continuity of the fire separation,*
- *any hidden voids in the construction shall be sealed and subdivided to inhibit the unseen spread of fire and products of combustion, in order to reduce the risk of structural failure, and the spread of fire.*

(Approved Document B3)

Foul water drainage

The foul water drainage system shall:

- *convey the flow-off foul water to a foul water outfall (i.e. sewer, cesspool, septic tank or settlement (i.e. holding) tank),*
- *minimise the risk of blockage or leakage,*
- *prevent foul air from the drainage system from entering the building under working conditions,*
- *be ventilated,*
- *be accessible for clearing blockages,*
- *not increase the vulnerability of the building to flooding.*

(Approved Document H1)

Wastewater treatment systems and cesspools

Wastewater treatment systems shall:

- *have sufficient capacity to enable breakdown and settlement of solid matter in the wastewater from the buildings;*
- *be sited and constructed so as to prevent overloading of the receiving water.*

Cesspools shall have sufficient capacity to store the foul water from the building until they are emptied.

Wastewater treatment systems and cesspools shall be sited and constructed so as not to:

- *be prejudicial to health or a nuisance;*
- *adversely affect water sources or resources;*
- *pollute controlled waters;*
- *be in an area where there is a risk of flooding.*

Septic tanks and wastewater treatment systems and cesspools are constructed and sited so as to:

- *have adequate ventilation;*
- *prevent leakage of the contents and ingress of subsoil water;*
- *have regard to water table levels at any time of the year and rising groundwater levels.*

Drainage fields are sited and constructed so as to:

- *avoid overloading of the soakage capacity, and*
- *provide adequately for the availability of an aerated layer in the soil at all times.*

(Approved Document H2)

Rainwater drainage

Rainwater drainage systems shall:

- *minimise the risk of blockage or leakage;*
- *be accessible for clearing blockages;*
- *ensure that rainwater soaking into the ground is distributed sufficiently so that it does not damage foundations of the proposed building or any adjacent structure;*
- *ensure that rainwater from roofs and paved areas is carried away from the surface either by a drainage system or by other means;*
- *ensure that the rainwater drainage system carries the flow of rainwater from the roof to an outfall (e.g. a soakaway, a watercourse, a surface water or a combined sewer).*

(Approved Document H3)

Building over existing sewers

Building or extension or work involving underpinning shall:

- *be constructed or carried out in a manner which will not overload or otherwise cause damage to the drain, sewer or disposal main either during or after the construction;*
- *not obstruct reasonable access to any manhole or inspection chamber on the drain, sewer or disposal main;*
- *in the event of the drain, sewer or disposal main requiring replacement, not unduly obstruct work to replace the drain, sewer or disposal main, on its present alignment;*
- *reduce the risk of damage to the building as a result of failure of the drain, sewer or disposal main.*

(Approved Document H4)

Separate systems for drainage

Separate systems of drains and sewers shall be provided for foul water and rainwater where:

(a)   *the rainwater is not contaminated; and*

(b)   *the drainage is to be connected either directly or indirectly to the public sewer system and either –*

    (i)   *the public sewer system in the area comprises separate systems for foul water and surface water; or*

    (ii)   *a system of sewers which provides for the separate conveyance of surface water is under construction either by the sewerage undertaker or by some other person (where the sewer is the subject of an agreement to make a declaration of vesting pursuant to section 104 of the Water Industry Act 1991).*

(Approved Document H5)

Solid waste storage shall be:

- *designed and sited so as not to be prejudicial to health,*
- *of sufficient capacity having regard to the quantity of solid waste to be removed and the frequency of removal,*
- *sited so as to be accessible for use by people in the building and of ready access from a street for emptying and removal.*

(Approved Document H6)

(Building Act 1984 Section 84)

You are required by the Building Act 1984 to ensure that all courts, yards and passageways giving access to a house, industrial or commercial building (not maintained at the public expense) are capable of allowing satisfactory drainage of its surface or subsoil to a proper outfall.

The local authority can require the owner of any of the buildings to complete such works as may be necessary to remedy the defect.

6.3.2 Meeting the requirement

Enclosures for drainage and/or water supply pipes

The enclosure should: B3 (11.8)

- be bounded by a compartment wall or floor, an outside wall, an intermediate floor, or a casing
- have internal surfaces (except framing members) of Class 0

- not have an access panel which opens into a circulation space or bedroom
- be used only for drainage, or water supply, or vent pipes for a drainage system

The casing should: B3 (11.8)

- be imperforate except for an opening for a pipe or an access pane
- not be of sheet metal
- have (including any access panel) not less than 30 minutes' fire resistance

The opening for a pipe, either in the structure or the   B3 (11.8)
casing, should be as small as possible and fire-stopped
around the pipe.

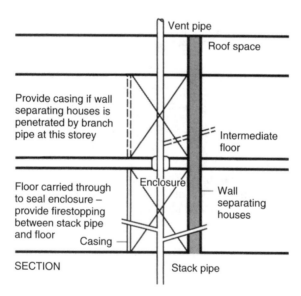

Figure 6.12 Enclosure for drainage or water supply pipes (house with any number of storeys)

Protection of openings for pipes

Pipes which pass through a compartment wall or compartment floor (unless the pipe is in a protected shaft), or through a cavity barrier, should conform to one of the following alternatives:

Proprietary seals (any pipe diameter) that maintain the fire resistance of the wall, floor or cavity barrier.	B3 (11.5–11.6)
Pipes with a restricted diameter where fire-stopping is used around the pipe, keeping the opening as small as possible.	B3 (11.5 and 11.7)
Sleeving – a pipe of lead, aluminium, aluminium alloy, fibre-cement or UPVC, with a maximum nominal internal diameter of 160 mm, may be used with a sleeving of non-combustible pipe as shown in Figure 6.13.	B3 (11.5 and 11.8)

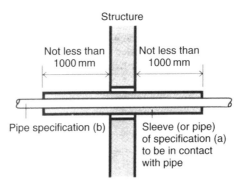

Figure 6.13 Pipes penetrating a structure

📖 Make the opening in the structure as small as possible and provide fire-stopping between pipe and structure

Foul water drainage

The capacity of the system should be large enough to carry the expected flow at any point (*BS 5572, BS 8301*).	H1 (0.1)
All pipes, fittings and joints should be capable of withstanding an air test of positive pressure of at least 38 mm water gauge for at least 3 minutes.	H1 (1.38)
Every trap should maintain a water seal of at least 25 mm.	H1 (1.38)

Traps

All points of discharge into the system should be fitted with a trap (e.g. a water seal) to prevent foul air from the system entering the building.	H1 (1.3–1.4)
All traps should be fitted directly over an appliance and should be removable or be fitted with a cleaning eye.	H1 (1.6)

Table 6.9 Minimum trap sizes and seal depths

Appliance	Diameter of trap (mm)	Depth of seal (mm of water or equivalent)
Washbasin Bidet	32	75
Bath Shower	40	50
Food waste disposal unit Urinal bowl	40	75
Sink Washing machine Dishwashing machine		
WC pan (outlet <80 mm)	75	50
WC pan (outlet >80 mm)	100	50

Branch discharge pipes

Branch pipes should either discharge into another branch pipe or a discharge stack (unless the appliances discharge into a gully on the ground floor or at basement level).	H1 (1.5)
If the appliances are on the ground floor the pipe(s) may discharge to a stub stack, discharge stack, directly to a drain, or (if the pipe carries only waste water) to a gully.	H1 (1.5–1.17) H1 (1.11) H1 (1.30)
A branch pipe from a ground floor closet should only discharge directly to a drain if the depth from the floor to the drain is 1.3 m or less (see Figure 6.14).	H1 (1.9)

A branch pipe serving any ground floor appliance may discharge direct to a drain or into its own stack.

H1 (A5)

A branch pipe should not discharge into a stack in a way which could cause cross flow into any other branch pipe (see Figure 6.15).

H1 (1.10)

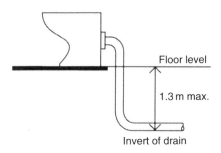

Floor level

1.3 m max.

Invert of drain

Figure 6.14 Direct connection of ground floor WC to drain

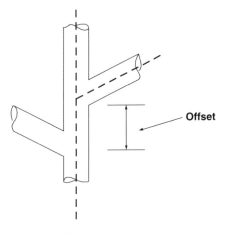

Offset

Figure 6.15 Branch connections

Branch discharge pipes

A branch discharge pipe should not discharge into a stack lower than 450 mm above the invert of the tail of the bend at the foot of the stack in single dwellings up to 3 storeys (see Figure 6.16).

H1 (1.8)
H1 (A3, A4)
H1 (1.21)

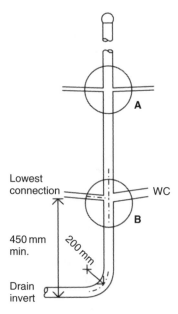

Figure 6.16 Branch discharge stack

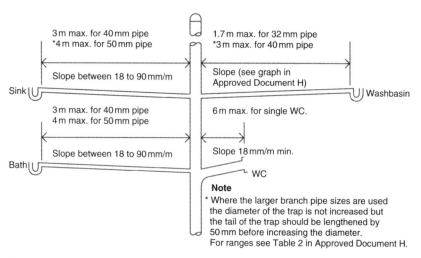

Figure 6.17 Branched connections

Branch discharge pipes

Branch pipes may discharge into a stub stack.	H1 (1.12)
	H1 (1.30)
A branch pipe discharging to a gully should terminate between the grating or sealing plate and the top of the water seal.	H1 (1.13)

Bends in branch pipes should be avoided if possible.	H1 (1.16)
Junctions on branch pipes should be made with a sweep of 25 mm radius or at 45°.	H1 (1.17)
Rodding points should be provided to give access to any lengths of discharge pipes which cannot be reached by removing traps or appliances with integral traps.	H1 (1.25) H1 (1.6)
A branch pipe discharging to a gully should terminate between the grating or sealing plate and the top of the water seal.	H1 (1.13)
Condensate drainage from boilers may be connected to sanitary pipework provided:	H1 (1.14)

(a) The connection should preferably be made to an internal stack with a 75 mm condensate trap.
(b) If the connection is made to a branch pipe, the connection should be made downstream of any sink waste connection.
(c) All sanitary pipework receiving condensate should be made from materials resistant to a pH value of 6.5 and lower and be installed in accordance with BS 6798.

Pipes serving a single appliance should have at least the same diameter as the appliance trap (see Table 6.9).

💡 A separate ventilating stack is only likely to be preferred where the numbers of sanitary appliances and their distance to a discharge stack are large.

Branch ventilation stacks

Should be connected to the discharge pipe within 750 mm of the trap and should connect to the ventilating stack or the stack vent, above the highest 'spillover' level of the appliances served.	H1 (1.22)
The ventilating pipe should have a continuous incline from the discharge pipe to the point of connection to the ventilating stack or stack vent.	H1 (1.22)

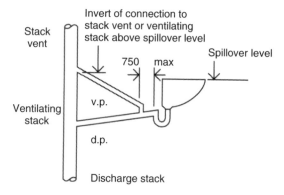

Figure 6.18 Branch ventilation pipes

Branch ventilating pipes which run direct to outside air should finish at least 900 mm above any opening into the building nearer than 3 m (see Figure 6.19).	H1 (1.23)
A dry stack may provide ventilation for branch ventilation pipes as an alternative to carrying them to outside air or to a ventilated discharge stack (ventilated system).	H1 (A7 and 1.21)
Ventilation stacks serving buildings with not more than 10 storeys and containing only dwellings should be at least 32 mm diameter (for all other buildings see paragraph 1.29).	H1 (A8) H1 (1.21 and 1.29)
A separate ventilating stack is only likely to be preferred where the numbers of ventilating pipes and their distance to a discharge stack are large.	H1 (1.19) H1 (Table 2)

Discharge stacks

All stacks should discharge to a drain.	H1 (1.26)
The bend at the foot of the stack should have as large a radius (i.e. at least 200 mm) as possible.	H1 (1.26)
Discharge stacks should be ventilated.	H1 (1.29)
Offsets in the 'wet' portion of a discharge stack should be avoided.	H1 (1.27)

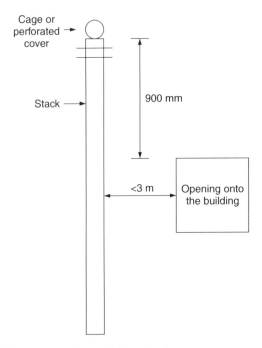

Figure 6.19 Termination of ventilation stacks

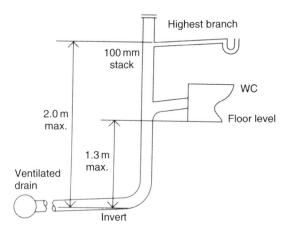

Figure 6.20 Stub stack

Stacks serving urinals should be not less than 50 mm.	H1 (1.28)
Stacks serving closets with outlets less than 80 mm should be not less than 75 mm.	H1 (1.28)

Stacks serving closets with outlets greater than 80 mm should be not less than 100 mm.	H1 (1.28)
The internal diameter of the stack should be not less than that of the largest trap or branch discharge pipe.	H1 (1.28)
Ventilating pipes open to outside air should finish at least 900 mm above any opening into the building within 3 m and should be fitted with a perforated cover or cage (see Figure 6.19) which should be metal if rodent control is a problem.	H1 (1.31)
Ventilating pipes open to outside air should finish at least 900 mm above any opening into the building within 3 m.	H1 (1.31)
Ventilating pipes should be finished with a wire cage (metallic in areas with a rodent problem) or other perforated cover, fixed to the end of the ventilating pipe.	H1 (1.31)
Stack ventilation pipes should be not less than 75 mm.	H1 (1.32)
Ventilated discharge stacks may be terminated inside a building when fitted with air admittance valves complying with prEN 12380.	H1 (1.33)
Discharge stacks may terminate inside a building when fitted with air admittance valves.	H1 (1.29)
Rodding points should be provided to give access to any lengths of pipe that cannot be reached from any other part of the system.	H1 (1.30–1.31)
Pipes should be firmly supported without restricting thermal movement.	H1 (1.31)
The pipes, fittings and joints should be airtight.	H1 (1.32)
A stub stack may be used if it connects into a ventilated discharge stack or into a ventilated drain not subject to surcharging.	H1 (1.30)
Air admittance valves should be located in areas that have adequate ventilation.	H1 (1.33)
Air admittance valves should not be used outside buildings or in dust laden atmospheres.	H1 (1.33)
Rodding points should be provided in discharge stacks.	H1 (1.34)

Pipes should be firmly supported without restricting thermal movement.	H1 (1.35)
Sanitary pipework connected to WCs should not allow light to be visible through the pipe wall, as this is believed to encourage damage by rodents.	H1 (1.36)

💡 Drainage serving kitchens in commercial hot food premises should be fitted with a grease separator complying with prEN 1825-1.

Foul drainage

Some public sewers may carry foul water and rainwater in the same pipe. If the drainage system is also to carry rainwater to such a sewer these combined systems should not be capable of discharging into a cesspool or septic tank.	H1 (2.1)
Foul drainage should be connected to either:	
• a public foul or combined sewer (wherever this is reasonably practicable)	H1 (2.3)
• an existing private sewer that connects with a public sewer, or	H1 (2.6)
• a wastewater treatment system or cesspool should be provided.	H1 (2.7)
Combined and rainwater sewers are designed to surcharge (i.e. the water level in the manhole rises above the top of the pipe) in heavy rainfall.	H1 (2.8)
Basements containing sanitary appliances, where the risk of flooding due to sewer surcharge of the sewer is possible should either use an anti-flooding valve (if the risk is low) or be pumped.	H1 (2.9) H1 (2.36–2.39) H1 (2.10)
💡 For other low lying sites (i.e. not basements) where the risk is considered low, a gully (at least 75 mm below the floor level) can be dug outside the building.	

Anti-flooding valves should preferably be the double valve type that complies with prEN 13564.	H1 (2.11)
The layout of the drainage system should be kept simple.	H1 (2.13)
Pipes should (wherever possible) be laid in straight lines. Changes of direction and gradient should be minimized.	
Access points should be provided only if blockages could not be cleared without them.	H1 (2.13)
Connections should be made using prefabricated components.	H1 (2.15)
Connection of drains to other drains or private or public sewers and of private sewers to public sewers should be made obliquely, or in the direction of flow.	H1 (2.14)
The system should be ventilated by a flow of air.	H1 (2.18) H1 (1.27–1.29)
Ventilating pipes should not finish near openings in buildings.	H1 (2.18) H1 (1.31)
Pipes should be laid to even gradients and any change of gradient should be combined with an access point.	H1 (2.19) H1 (2.49)
Pipes should also be laid in straight lines where practicable.	H1 (2.20) H1 (2.49)

Rodent control

If the site has been previously developed, the local authority should be consulted to determine whether any special measures are necessary for control of rodents. Special measures which may be taken include the following:

Sealed drainage – should have access covers to the pipework in the inspection chamber instead of an open channel.	H1 (2.22a)
Intercepting traps – should be of the locking type that can be easily removed from the chamber surface and securely replaced.	H1 (2.22b)

Rodent barriers – including enlarged sections on discharge stacks to prevent rats climbing, flexible downward facing fins in the discharge stack, or one-way valves in underground drainage.	H1 (2.22c)
Metal cages on ventilator stack terminals – to discourage rats from leaving the drainage system.	H1 (2.22d) H1 (1.31)
Covers and gratings to gullies – used to discourage rats from leaving the system.	H1 (2.22e)
During construction, drains and sewers that are left open should be covered when work is not in progress to prevent entry by rats.	H1 (2.56)
Disused drains or sewers less than 1.5 m deep that are in open ground should as far as is practicable be removed. Other pipes should be sealed at both ends and at any point of connection, and grout filled to ensure that rats cannot gain access.	H1 (B18)

Protection from settlement

• A drain may run under a building if at least 100 mm of granular or other flexible filling is provided round the pipe.	H1 (2.23)
• Where pipes are built into a structure (e.g. inspection chamber, manhole, footing, ground beam or wall) suitable measures (such as using rocker joints or a lintel) should be taken to prevent damage or misalignment (see Figures 6.21 and 6.22).	H1 (2.24)
The depth of cover will usually depend on the levels of the connections to the system, the gradients at which the pipes should be laid and the ground levels.	H1 (2.27) H1 (2.41–2.45)
All drain trenches should not be excavated lower than the foundations of any building nearby (see Figure 6.23).	H1 (2.25)

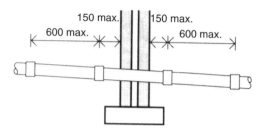

Figure 6.21 Pipe imbedded in the wall. Short length of pipe bedded in a wall with joints 150 mm of either wallface. Additional rocker pipes (max length 600 mm) with flexible joints are then added

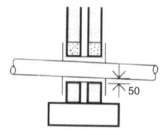

Figure 6.22 Pipe shielded by a lintel. Both sides are masked with rigid sheet material (to prevent entry of fill or vermin) and the void is filled with a compressible sealant to prevent entry of gas

Pipe gradients and sizes

Drains should have enough capacity to carry the anticipated maximum flow (see Table 6.10).	H1 (2.29)
Sewers (i.e. a drain serving more than one property) should have a minimum diameter of 100 mm when serving 10 dwellings or diameter of 150 mm if more than 10.	H1 (2.30)

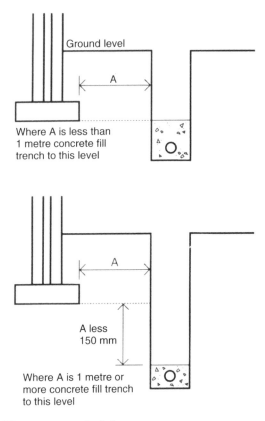

Figure 6.23 Pipe runs near buildings

Drains carrying foul water should have an internal diameter of at least 75 mm.	H1 (2.33)
Drains carrying effluent from a WC or trade effluent should have an internal diameter of at least 100 mm.	H1 (2.33)

Table 6.10 Flow rates from dwellings

Number of dwellings	Flow rate (litres/sec)
1	2.5
5	3.5
10	4.1
15	4.6
20	5.1
26	5.4
30	5.8

Pumping installations

Where gravity drainage is impracticable, or protection against flooding due to surcharge in downstream sewers is required, a pumping installation will be needed.	H1 (2.36)
Where foul water drainage from a building is to be pumped, the effluent receiving chamber should be sized to contain 24-hour inflow to allow for disruption in service.	H1 (2.39)
The minimum daily discharge of foul drainage should be taken as 150 litres per head per day for domestic use.	H1 (2.39)

Table 6.11 Materials for below-ground gravity drainage

Material	British Standard
Rigid pipes	
Vitrified clay	BS 65, BSEN 295
Concrete	BS 5911
Grey iron	BS 437
Ductile iron	BSEN 598
Flexible pipes	
UPVC	BSEN 1401
PP	BSEN 1852
Structured walled plastic pipes	BSEN 13476

Materials for pipes and jointing

To minimize the effects of any differential settlement pipes should have flexible joints.	H1 (2.40)
All joints should remain watertight under working and test conditions.	H1 (2.40)
Nothing in the pipes, joints or fittings should project into the pipe line or cause an obstruction.	H1 (2.40)
Different metals should be separated by non-metallic materials to prevent electrolytic corrosion.	H1 (2.40)

Bedding and backfill

The choice of bedding and backfilling depends on   H1 (2.41)
the depth at which the pipes are to be laid and the
size and strength of the pipes.

Special precautions should be taken to take account
of the effects of settlement where pipes run under or
near buildings.

The depth of the pipe cover will usually depend on
the levels of the connections to the system and the
gradients at which the pipes should be laid and the
ground levels.

Pipes need to be protected from damage particularly
pipes which could be damaged by the weight of
backfilling.

Rigid pipes should be laid in a trench as shown in   H1 (2.42)
Figure 6.24.

Flexible pipes shall be supported to limit   H1 (2.44)
deformation under load.

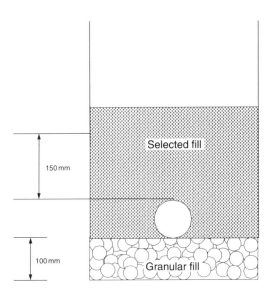

Figure 6.24 Bedding for rigid pipes

Flexible pipes with very little cover shall be protected from damage by a reinforced cover slab with a flexible filler and at least 75 mm of granular material between the top of the pipe and the underside of the flexible filler below the slabs (see Figure 6.26).

H1 (2.42–2.44)

Trenches may be backfilled with concrete to protect nearby foundations. In these cases a movement joint (as shown in Figure 6.27) formed with a compressible board should be provided at each socket or sleeve joint.

H1 (2.45)

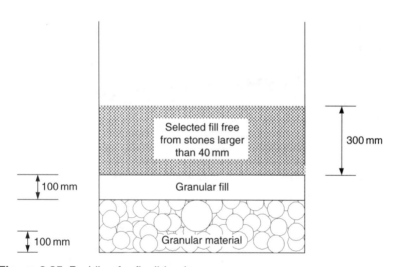

Figure 6.25 Bedding for flexible pipes

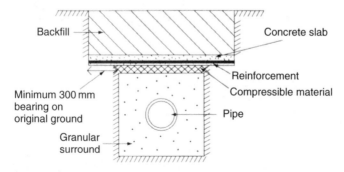

Figure 6.26 Protection of pipes laid in shallow depths

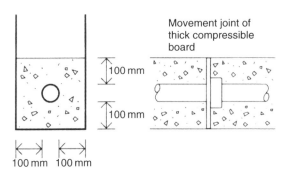

Figure 6.27 Joints for concrete encased pipes

Access points

Access should be provided to long runs.	H1 (2.50)
Sufficient and suitable access points should be provided for clearing blockages from drain runs that cannot be reached by any other means.	H1 (2.46)
Access points should be provided:	H1 (2.49)
• on or near the head of each drain run • at a bend • at a change of gradient or pipe size • at a junction.	
Access points should be either:	H1 (2.48)
• rodding eyes – capped extensions of the pipes; • access fittings – small chambers on (or an extension of) the pipes but not with an open channel; • inspection chambers – chambers with working space at ground level; • manholes – deep chambers with working space at drain level.	
Access points should be constructed so as to resist the ingress of ground water or rainwater.	H1 (2.52)
Inspection chambers and manholes should have removable non-ventilating covers of durable material (such as cast iron, cast or pressed steel, precast concrete or UPVC).	H1 (2.54)

Access points to sewers (serving more than one property) should be located in places where they are accessible and apparent for use in an emergency (e.g. highways, public open space, unfenced front gardens, and shared or unfenced driveways).	H1 (2.51)
Inspection chambers and manholes in buildings should have mechanically fixed airtight covers unless the drain itself has watertight access covers.	H1 (2.54)
Manholes deeper than 1 m should have metal step irons or fixed ladders.	H1 (2.54)

General

Drains and sewers should be protected from damage by construction traffic and heavy machinery.	H1 (2.57)
Heavy materials should not be stored over drains or sewers.	H1 (2.57)
After laying, including any necessary concrete or other haunching or surrounding and backfiring gravity drains and private sewers should be tested for watertightness.	H1 (2.59)
All pipework carrying greywater for reuse should be clearly marked with the word 'GREYWATER'.	H1 (A11)
Material alterations to existing drains and sewers are subject to (and covered by) the Building Regulations.	H1 (b7)
Repairs, reconstruction and alterations to existing drains and sewers should be carried out to the same standards as new drains and sewers.	H1 (B15)

Wastewater treatment systems and cesspools

A notice giving information as to the nature and frequency of maintenance required for the cesspool or wastewater treatment system to continue to function satisfactorily is displayed within each of the buildings.

The use of non-mains foul drainage, such as wastewater treatment systems, septic tanks or cesspools, should only be considered where connection to mains drainage is **not** practicable.

💡 Any discharge from a wastewater treatment system is likely to require a consent from the Environment Agency. For the detailed design and installation of small sewage treatment works, specialist knowledge is advisable. Guidance is also given in BS 6297: 1983 *Code of practice for design and installation of small sewage treatment works and cesspools.*

Septic tanks

Septic tanks with some form of secondary treatment (such as from a drainage field/mound or constructed wetland such as a reed bed) will normally be the most economic means of treating wastewater from small developments (e.g. 1 to 3 dwellings). They provide suitable conditions for the settlement, storage and partial decomposition of solids which need to be removed at regular intervals.

Septic tanks should be sited at least 7 m from any habitable parts of buildings, and preferably down a slope.	H2 (1.16)
Septic tanks should only be used in conjunction with a form of secondary treatment (e.g. a drainage field, drainage mound or constructed wetland).	H2 (1.15)
Septic tanks should be sited within 30 m of a vehicle access to enable the tank to be emptied and cleaned without hazard to the building occupants and without the contents being taken through a dwelling or place of work.	H2 (1.17 and 1.64)
Septic tanks and settlement tanks should have a capacity below the level of the inlet of at least 2700 litres (2.7 m³) for up to 4 users. This size should be increased by 180 litres for each additional user.	H2 (1.18)
Septic tanks may be constructed in brickwork or concrete (roofed with heavy concrete slabs) or factory-manufactured septic tanks (made out of glass reinforced plastics, polyethylene or steel) can be used.	H2 (1.19–20 and 1.65–66)
The brickwork should consist of engineering bricks at least 220 mm thick. The mortar should be a mix of 1:3 cement sand ratio and in-situ concrete should be at least 150 mm thick of C/25/P mix (see BS 5328).	H2 (1.20 and 1.66)
Septic tanks should be ventilated.	H2 (1.21)

Septic tanks should incorporate at least two chambers or compartments operating in series.	H2 (1.22)
Septic tanks should be provided with access for emptying and cleaning.	H2 (1.24)
A notice should be fixed within the building describing the necessary maintenance.	H2 (1.25)
Septic tanks should be inspected monthly to check they are working correctly.	H2 (A.11)
The septic tank should be emptied at least once a year.	H2 (A.13)

Cesspools

A cesspool is a watertight tank, installed underground, for the storage of sewage. No treatment is involved.

Cesspools should be sited at least 7 m from any habitable parts of buildings and preferably downslope.	H2 (1.58)
Provided with access for emptying and cleaning.	H2 (1.60)
Cesspools should be inspected fortnightly for overflow.	H2 (A.20)
Cesspools should be emptied on a monthly basis by a licensed contractor.	H2 (1.60) H2 (A.21)
A filling rate of 150 litres per person is assumed and if the cesspool does not fill within the estimated period, the tank should be inspected for leakage.	H2 (A.22)
Cesspools should be ventilated.	H2 (1.63)
The inlet of a cesspool should be provided with access for inspection.	H2 (1.67)
Cesspools and settlement tanks if they are to be desludged using a tanker, should be sited within 30 m of a vehicle access.	H2 (1.64)
Cesspools and settlement tanks should prevent leakage of the contents and ingress of subsoil water.	H2 (1.63)
Cesspools should have a capacity below the level of the inlet of at least 18 000 litres (18 m^3) for 2 users increased by 6800 litres (6.8 m^3) for each additional user.	H2 (1.61)

Cesspools, septic tanks and settlement tanks may be constructed in brickwork, concrete, or glass reinforced concrete.	H2 (1.65–66)
📖 Factory-made cesspools and septic tanks are available in glass reinforced plastic, polyethylene or steel.	
The brickwork should consist of engineering bricks at least 220 mm thick. The mortar should be a mix of 1:3 cement sand ratio and in-situ concrete should be at least 150 mm thick of C/25/P mix (see BS 5328).	H2 (1.66)
Cesspools should be covered (with heavy concrete slabs) and ventilated.	
Cesspools should have no openings except for the inlet, access for emptying and ventilation.	H2 (1.62)
Cesspools should be inspected fortnightly for overflow and emptied as required.	H2 (A.20)

Packaged treatment works

This term is applied to a range of systems designed to treat a given hydraulic and organic load using prefabricated components which can be installed with minimal site work. They are capable of treating effluent more efficiently than septic tank systems and this normally allows the product to be directly discharged to a watercourse.

The discharge from the wastewater treatment plant should be sited at least 10 m away from watercourses and any other buildings.	H2 (1.54)
Regular maintenance and inspection should be carried out in accordance with the manufacturer's instructions.	H2 (A.17)

Drainage fields and mounds

Drainage fields (or mounds) serving a wastewater treatment plant or septic tank should be located:	H2 (1.27)

- at least 10 m from any watercourse or permeable drain
- at least 50 m from the point of abstraction of any groundwater supply
- at least 15 m from any building

- sufficiently far from any other drainage fields, drainage mounds or soakaways so that the overall soakage capacity of the ground is not exceeded.

No water supply pipes or underground services other than those required by the disposal system itself should be located within the disposal area.	H2 (1.29)
No access roads, driveways or paved areas should be located within the disposal area.	H2 (1.30)
The ground water table should not rise to within 1 m of the invert level of the proposed effluent distribution pipes.	H2 (1.33)
An inspection chamber should be installed between the septic tank and the drainage field.	H2 (1.43)
Constructed wetlands should not be located in the shade of trees or buildings.	H2 (1.47)
The drainage field/mound should be checked on a monthly basis to ensure that it is not waterlogged and that the effluent is not backing up towards the septic tank.	H2 (A.15)

Under Section 50 (overflowing and leaking cesspools) of the Public Health Act 1936 action could be taken against a builder who had caused the problem, and **not** just against the owner.

Under Section 59 (drainage of building) of the Building Act 1984, local authorities can require either the owner or the occupier to remove (or otherwise make innocuous) any disused cesspool, septic tank or settlement tank.

Greywater and rainwater tanks

Greywater and rainwater tanks should: - prevent leakage of the contents and ingress of subsoil water, and should be ventilated; - have an anti-backflow device; - be provided with access for emptying and cleaning.	H2 (1.70)

Rainwater drainage

The capacity of the drainage system should be large enough to carry the expected flow at any point in the system.	H3 (0.3)
Rainwater or surface water should not be discharged to a cesspool or septic tank.	H3 (0.6)

Gutters and rainwater pipes

Although this part of the Building Regulations only actually applies to draining the rainfall from areas of 6 m^2 or more (unless they receive a flow from a rainwater pipe or from paved and other hard surfaces), each case should be considered separately and a decision made. This particularly applies to small roofs and balconies. Table 6.12 shows the largest effective area that should be drained into the gutter sizes most often used.

For eaves gutters the design rainfall intensity should be 0.021 litres/second/m^2. In some cases, eaves drop systems may be used (H3 (1.13)).

Table 6.12 Gutter and outlet sizes

Max effective roof area (m^2)	Gutter size (mm dia)	Outlet size (mm dia)	Flow capacity (litres/sec)
6.0	–	–	–
18.0	75	50	0.38
37.0	100	63	0.78
53.0	115	63	1.11
65.0	125	75	1.37
103.0	150	89	2.16

Gutters should be laid with any fall towards the nearest outlet.	
Gutters should be laid so that any overflow in excess of the design capacity (e.g. above normal rainfall) will be discharged clear of the building.	H3 (1.7)
Rainwater pipes should discharge into a drain or gully (but may discharge to another gutter or onto another surface if it is drained).	H3 (1.8)
Any rainwater pipe which discharges into a combined system should do so through a trap.	H3 (1.8)

The size of a rainwater pipe should be at least the size of the outlet from the gutter.   H3 (1.10)

A down pipe which serves more than one gutter should have an area at least as large as the combined areas of the outlets.   H3 (1.10)

On flat roofs, valley gutters and parapet gutters additional outlets may be necessary.   H3 (1.7)

Where a rainwater pipe discharges onto a lower roof or paved area, a pipe shoe should be fitted to divert water away from the building.   H3 (1.9)

Gutters and rainwater pipes should be firmly supported without restricting thermal movement.

The materials used should be of adequate strength and durability, and   H3 (1.16)

- all gutter joints should remain watertight under working conditions
- pipework in siphonic roof drainage systems should be able to resist to negative pressures in accordance with the design
- gutters and rainwater pipes should be firmly supported
- different metals should be separated by non-metallic material to prevent electrolytic corrosion.

Drainage of paved areas

Surface gradients should direct water draining from a paved area away from buildings.   H3 (2.2)

Gradients on impervious surfaces should be designed to permit the water to drain quickly from the surface. A gradient of at least 1 in 60 is recommended.   H3 (2.3)

Paths, driveways and other narrow areas of paving should be free draining to a pervious area such as grassland, provided that: H3 (2.6)

- the water is not discharged adjacent to buildings where it could damage foundations; and
- the soakage capacity of the ground is not overloaded.

Where water is to be drained onto the adjacent ground the edge of the paving should be finished above or flush with the surrounding ground to allow the water to run-off.	H3 (2.7)
• Where the surrounding ground is not sufficiently permeable to accept the flow, filter drains may be provided.	H3 (2.8 and 3.33)
• Pervious paving should not be used where excessive amounts of sediment are likely to enter the pavement and block the pores.	H3 (2.11)
• Pervious paving should not be used in oil storage areas, or where runoff may be contaminated with pollutants.	H3 (2.12)
• Gullies should be provided at low points where water would otherwise pond.	H3 (2.15)
• Gully gratings should be set approximately 5 mm below the level of the surrounding paved area in order to allow for settlement.	H3 (2.16)
• Provision should be made to prevent silt and grit entering the system, either by provision of gully pots of suitable size, or catchpits.	H3 (2.17)

Surface water drainage

Discharge to a watercourse may require a consent from the Environment Agency, who may limit the rate of discharge. Where other forms of outlet are not practicable, discharge should be made to a sewer (H3 (3.2–3.3)). For design purposes a rainfall interval of 0.014 litres/second/m^2 can be assumed as normal.

Some drainage authorities have sewers that carry both foul water and rainwater (i.e. combined systems) in the same pipe. Where they do, they can allow rainwater to discharge into the system if the sewer has enough capacity to take the added flow. Some private sewers (drains serving more than one property) also carry both foul water and rainwater. If a sewer (or private sewer) operated as a combined system does not have enough capacity, the rainwater should be run in a separate system with its own outfall.

Surface water drainage should discharge to a soakaway or other infiltration system where practicable.	H3 (3.2)
Surface water drainage connected to combined sewers should have traps on all inlets.	H3 (3.7)

Drains should be at least 75 mm diameter.	H3 (3.14)
Where any materials that could cause pollution are stored or used, separate drainage systems should be provided.	H3 (3.21)
On car parks, petrol filling stations or other areas where there is likely to be leakage or spillage of oil, drainage systems should be provided with oil interceptors.	H3 (3.22) H3 (A)
Separators should be leak tight and comply with the requirements of the Environmental Agency and prEN858.	H3 (A.9–10)
Infiltration devices (including soakaways, swales, infiltration basins, and filter drains) should not be built:	H3 (3.23–26)

* within 5 m of a building or road or in areas of unstable land;
* in ground where the water table reaches the bottom of the device at any time of the year;
* sufficiently far from any drainage fields, drainage mounds or other soakaways;
* where the presence of any contamination in the runoff could result in pollution of groundwater source or resource.

Soakaways should be designed to a return period of once in ten years.	H3 (3.27)
Soakaways for areas less than 100 m^2 shall consist of square or circular pits, filled with rubble or lined with dry jointed masonry or perforated ring units. Soakaways serving larger areas shall be lined pits or trench type soakaways.	H3 (3.26)
The storage volume should be calculated so that, over the duration of the storm, it is sufficient to contain the difference between the inflow volume and the outflow volume.	H3 (3.29)
Soakaways serving larger areas should be designed in accordance with BS EN 752-4.	H3 (3.30)

💡 Under Section 85 (offences of polluting controlled waters) of the Water Resources Act 1991 it is an offence to discharge any noxious or polluting material into a watercourse, coastal water, or underground water. Most surface water sewers discharge to watercourses.

💡 Under Section 111 (restrictions on use of public sewers) of the Water Industry Act 1991 it is an offence to discharge petrol into any drain or sewer connected to a public sewer.

Building over existing sewers

Where it is proposed to construct a building over or near a drain or sewer shown on any map of sewers, the developer should consult the owner of the drain or sewer, if the owner is not the developer himself.	H4 (0.3)
A building constructed over or within 3 m of any	H4 (1.2)
• rising main • drain or sewer constructed from brick or masonry • drain or sewer in poor condition shall not be constructed in such a position unless special measures are taken.	
Buildings or extensions should not be constructed over a manhole or inspection chamber or other access fitting on any sewer (serving more than one property).	H4 (1.3)
A satisfactory diversionary route should be available so that the drain or sewer could be reconstructed without affecting the building.	H4 (1.4)
The length of drain or sewer under a building should not exceed 6 m except with the permission of the owners of the drain or sewer.	H4 (1.5)
Buildings or extensions should not be constructed over or within 3 m of any drain or sewer more than 3 m deep, or greater than 225 mm in diameter except with the permission of the owners of the drain or sewer.	H4 (1.60)
Where a drain or sewer runs under a building at least 100 mm of granular or other suitable flexible filling should be provided round the pipe.	H4 (1.9)
Where a drain or sewer running below a building is less than 2 m deep, the foundation should be extended locally so that the drain or sewer passes through the wall.	H4 (1.10)

Where the drain or sewer is more than 2 m deep to invert and passes beneath the foundations, the foundations should be designed as a lintel spanning over the line of the drain or sewer. The span of the lintel should extend at least 1.5 m either side of the pipe and should be designed so that no load is transmitted onto the drain or sewer.	H4 (1.12)
A drain trench should not be excavated lower than the foundations of any building nearby.	H4 (1.13)

Separate systems for drainage

Separate systems of drains and sewers shall be provided for foul water and rainwater where:

(a)   the rainwater is not contaminated; and
(b)   the drainage is to be connected either directly or indirectly to the public sewer system, which has separate systems for foul water and surface water.

Solid waste storage

💡 Although the requirements of the Building Regulations do not cover the recycling of household and other waste, H6 sets out general requirements for solid waste storage.

For domestic developments space should be provided for storage of containers for separated waste (i.e. waste that can be recycled is stored separately from waste that cannot) and having a combined capacity of $0.25\,m^2$ per dwelling.	H6 (1.1)
In low-rise domestic developments (houses, bungalows and flats up to the 4th floor) any dwelling should have, or have access to, a location with at least two movable, individual or communal waste containers.	H6 (1.2)
In multistorey domestic developments, dwellings above the 4th storey should share a container fed by a chute unless siting or operation of a chute is impracticable. In such a case a satisfactory management arrangement for conveying refuse to the storage area should be assured.	H6 (1.6)

In multistorey domestic developments, dwellings up to the 4th floor may each have their own waste container or may share a waste container.	H6 (1.5)
For waste containers up to 250 litres, steps should be avoided between the container store and collection point wherever possible.	H6 (1.10)
Containers and chutes should be sited so that householders are not required to carry refuse further than 30 m.	H6 (1.8)
Containers should be within 25 m of the vehicle access.	H6 (1.8)
Containers should be sited so that they can be collected without being taken through a building, unless it is a garage, carport or other open covered space.	H6 (1.10)

💡 This provision applies only to new buildings.

The collection point should be reasonably accessible to the size of waste collection vehicles typically used by the waste collection authority.	H6 (1.11)
External storage areas for waste containers should be away from windows and ventilators and preferably be in shade or under shelter.	H6 (1.12)
Storage areas should not interfere with pedestrian or vehicle access to buildings.	H6 (1.12)
Where enclosures, compounds or storage rooms are provided they should allow room for filling and emptying and provide a clear space of 150 mm between and around the containers.	H6 (1.13)

• Enclosures, compounds or storage rooms for communal containers should be a minimum of 2 m high.	H6 (1.13)
• Enclosures for individual containers should be sufficiently high to allow the lid to be opened for filling.	H6 (1.13)
• The enclosure should be permanently ventilated at the top and bottom and should have a paved impervious floor.	H6 (1.13)
• Communal storage areas should have provision for washing down and draining the floor into a system suitable for receiving a polluted effluent.	H6 (1.14)
• Gullies should incorporate a trap that maintains a seal even during prolonged periods of disuse.	H6 (1.14)
• Any room (or compound) for the open storage of waste should be secure to prevent access by vermin.	H6 (1.15)

- Where storage rooms are provided, separate rooms should be provided for the storage of waste that cannot be recycled, and waste that can be recycled.   H6 (1.16)
- Where the location for storage is in a publicly accessible area or in an open area around a building (e.g. a front garden) an enclosure or shelter should be considered.   H6 (1.17)
- In high-rise domestic developments, where chutes are provided they should be at least 450 mm in diameter and should have a smooth non-absorbent surface and close-fitting access doors at each storey that has a dwelling and be ventilated at the top and bottom.   H6 (1.18)

6.4 Water supplies

6.4.1 The requirement (Building Act 1984 Sections 25 and 69)

The Building Act stipulates that plans for proposed buildings will ensure that all occupants of the house will be provided with a supply of '*wholesome water, sufficient for their domestic purposes*'. This can be achieved by either:

- connecting the house to water supplies from the local water authority (normally referred to as the '*statutory water undertaker*');
- by otherwise taking water into the house by means of a pipe (e.g. from a local recognized supply);
- by providing a supply of water within a reasonable distance from the house (e.g. such as from a well).

If an occupied house is not within a reasonable distance of a supply of 'wholesome water' or if the local authority is not satisfied that the water supply is capable of supplying 'wholesome water', then they can give notice that the owner of the building must provide water within a specified time. They also have the authority to prohibit the building from being occupied.

Fire precautions (construction)

Any hidden voids in the construction shall be sealed and subdivided to inhibit the unseen spread of fire and products of combustion, in order to reduce the risk of structural failure, and the spread of fire.

(Approved Document B3)

What happens if there is more than one property?

Where the local authority are satisfied that two or more houses can most conveniently be met by means of a joint supply, they may give notice accordingly.

Can I ask the local authority to provide me with a supply of water?

If you are unable to provide a suitable supply of water, the local authority can themselves provide, or secure the provision of, a supply of water to the house or houses in question and then recover any expenses reasonably incurred from the owner of the house, or (where two or more houses are concerned), the owners of those houses.

💡 The maximum amount that a local authority can charge for providing a suitable supply of water is £3000 in respect of any one house.

Where a supply of water is provided to a house by statutory water undertakers, water rates will be included in the normal rateable value of the house.

Where two or more houses are supplied with water by a common pipe belonging to the owners or occupiers of those houses, the local authority may, when necessary, repair or renew the pipe and recover any expenses reasonably incurred by them from the owners or occupiers of the houses.

6.4.2 Meeting the requirement

Protection of openings for pipes

Pipes that pass through a compartment wall or compartment floor (unless the pipe is in a protected shaft), or through a cavity barrier, should be one of the following alternatives:

Proprietary seals (any pipe diameter) that maintain the fire resistance of the wall, floor or cavity barrier.	B3 (11.5–11.6)
Pipes with a restricted diameter where fire-stopping is used around the pipe, keeping the opening as small as possible.	B3 (11.5 and 11.7)
Sleeving – a pipe of lead, aluminium, aluminium alloy, fibre-cement or UPVC, with a maximum nominal internal diameter of 160 mm, may be used with a sleeving of non-combustible pipe as shown in Figure 6.13.	B3 (11.5 and 11.8)

6.5 Cellars

6.5.1 The requirement (Building Act 1984 Section 74)

Unless you have the consent of the local authority, you are not allowed to construct a cellar or room *in (or as part of) a house, an existing cellar, a shop, inn, hotel or office if the floor level of the cellar or room is lower than the ordinary level of the subsoil water on, under or adjacent to the site of the house, shop, inn, hotel or office.*

 This does not apply to:

- the construction of a cellar or room carried out in accordance with plans deposited on an application under the Licensing Act 1964;
- the construction of a cellar or room in connection with a shop, inn, hotel or office that forms part of a railway station.

💣 If the owner of the house, shop, inn, hotel or office allows a cellar or room forming part of it to be used in a manner that he knows to be in contravention with the Building Regulations, he is liable, on summary conviction, to a fine.

Fire precautions

Facilities for venting for heat and smoke from basement areas shall be made available.

(Approved Document B5)

6.5.2 Meeting the requirements

Venting of heat and smoke from basements

Smoke outlets (also referred to as smoke vents) should be available so as to provide a route for heat and smoke to escape to the open air from the basement level.	B4 (19.2)
Where practicable, each basement space should have one or more smoke outlets (see Figure 6.28).	B4 (19.3)

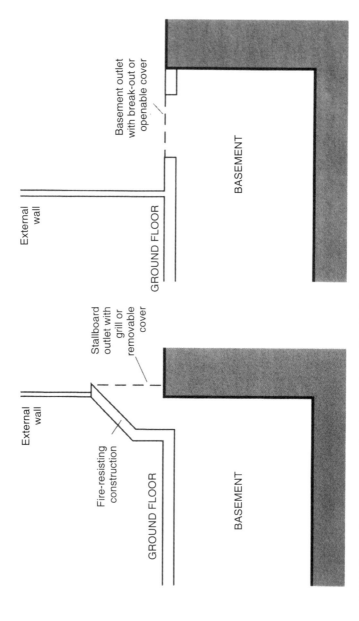

Figure 6.28 Fire resistant construction for smoke outlet shafts

Smoke outlets

Smoke outlets, connected directly to the open air, should be provided from every basement storey, except for a basement in a single family dwellinghouse.	B4 (19.4)
Smoke outlets should be sited at high level, either in the ceiling or in the wall of the space they serve.	B4 (19.7)
Outlet ducts or shafts, including any bulkheads over them (see Figure 6.28), should be enclosed in non-combustible construction having not less fire resistance than the element through which they pass.	B4

6.6 Floors and ceilings

The ground floor of a building is either solid concrete or a suspended timber type. With a concrete floor, a Damp Proof Membrane (DPM) is laid between walls. With timber floors, sleeper walls of honeycomb brickwork are built on oversite concrete between the base brickwork; a timber sleeper plate rests on each wall and timber joists are supported on them. Their ends may be similarly supported, let into the brickwork or suspended on metal hangers. Floorboards are laid at right angles to joists. First-floor joists are supported by the masonry or hangers.

Similar to a brick built house, the floors in a timber-framed house are either solid concrete or suspended timber. In some cases, a concrete floor may be screeded or surfaced with timber or chipboard flooring. Suspended timber floor joists are supported on wall plates and surfaced with chipboard.

6.6.1 Requirements

Precautions shall be taken to reduce risks to the health and safety of persons in buildings by safeguarding them and the buildings against the adverse effects of:

- *vegetable matter*
- *contaminants on or in the ground to be covered by the building*
- *ground water.*

(Approved Document C)

The building shall be designed and constructed so that there are appropriate provisions for the early warning of fire, and appropriate means of escape in case of fire from the building to a place of safety outside the building capable of being safely and effectively used at all material times.

(Approved Document B1)

💡 For a typical one- or two-storey dwelling, the requirement is limited to the provision of smoke alarms and to the provision of openable windows for emergency exit (see B1.i).

As a fire precaution, all materials used for internal linings of a building should have a low rate of surface flame spread and (in some cases) a low rate of heat release.

(Approved Document B2)

Fire precautions

- *all loadbearing elements of structure of the building shall be capable of withstanding the effects of fire for an appropriate period without loss of stability;*
- *ideally the building should be subdivided by elements of fire-resisting construction into compartments;*
- *all openings in fire-separating elements shall be suitably protected in order to maintain the integrity of the continuity of the fire separation;*
- *any hidden voids in the construction shall be sealed and subdivided to inhibit the unseen spread of fire and products of combustion, in order to reduce the risk of structural failure, and the spread of fire.*

(Approved Document B3)

Airborne and impact sound

Dwellings shall be designed and built so that the noise from normal domestic activity in an adjoining dwelling (or other building) is kept down to a level that:

- *does not affect the health of the occupants of the dwelling;*
- *will allow them to sleep, rest and engage in their normal domestic activities without disruption.*

(Approved Document E)

6.6.2 Meeting the requirements

A solid or suspended floor shall be built next to the ground to prevent undue moisture from reaching the upper surface of the floor (see Figure 6.29)	C3
• Unless used purely for storage or for housing plant and/or machinery, all floors next to the ground should be capable of resisting ground moisture from reaching the upper surface of the floor.	C3 (3.2)

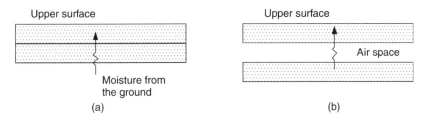

Figure 6.29 Floor, resistance to moisture. (a) Solid floor. (b) Suspended floor

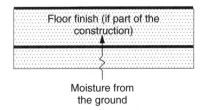

Figure 6.30 Ground floor supported principle

Floors next to the ground

In order to resist ground moisture from reaching the upper surface of the floor, a concrete ground supported floor must be built as follows:

The floor next to the ground should be a hardcore bed of clean broken brick or similar inert material.	C4 (3.4)
The concrete should be at least 100 mm thick, composed of 50 kg of cement to not more than 0.11 m³ of fine aggregate and 0.16 m³ of coarse aggregate or BS 5328 mix ST2.	C4 (3.4)
If there is embedded steel, the concrete should be composed of 50 kg of cement to not more than 0.08 m³ of fine aggregate and 0.13 m³ of coarse aggregate or BS 5328 mix ST4.	C4 (3.4)
A damp-proof membrane of either polyethylene (if laid below the concrete) or three coats of cold applied bitumen solution (if laid above the concrete) shall be put down.	C4 (3.5–3.6)

Timber floors laid directly on concrete   C4 (3.7)
may be bedded in a material that
may also serve as a damp-proof
membrane.

Timber fillets laid in the concrete   C4 (3.7)
as a fixing for a floor finish should be
treated with an effective preservative
unless they are above the damp-proof
membrane.

The ground (of a ground supported floor)
shall be covered with dense concrete laid
on a hardcore bed with a damp-proof membrane
(see Figure 6.32).

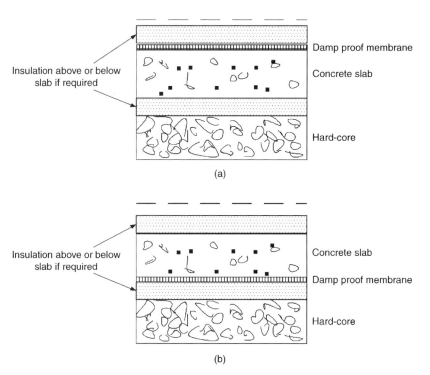

Figure 6.31 Ground supported floor construction. Damp-proof membrane (a) above slab and (b) below slab

Suspended timber ground floors

For all suspended timber ground floors:

The ground shall be covered so as to resist moisture and prevent plant growth.	C4 (3.9–3.11)
There shall be a ventilated air space between the ground covering and the timber.	C4 (3.9–3.11)
There shall be a damp-proof course between the timber and any material that can carry moisture from the ground (see Figure 6.32).	C4 (3.9–3.11)

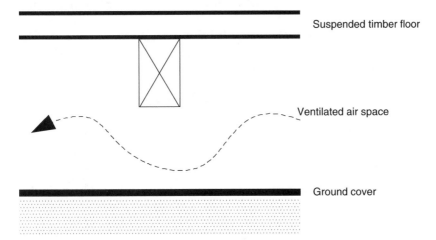

Suspended timber floor

Ventilated air space

Ground cover

Figure 6.32 Suspended timber floor principle

Suspended concrete ground floors

All suspended in-situ or precast concrete floors shall include a ventilated air space in those situations where there is a risk that accumulated gas could lead to an explosion.	C4 (3.12–3.14)
All floors next to the ground (walls and roof) shall not be damaged by moisture from the ground, rain or snow and shall not carry that moisture to any part of the building that it would damage.	C3

💡 This damage can be avoided either by preventing moisture from getting to materials that would be damaged or by using materials that will not be damaged.

Means of escape

Floors more than 7.5 m above ground level (where, in the case of a fire, the risk of the stairway becoming impassable before occupants of the upper parts of the house have escaped is appreciable) will be provided with an alternative route.	B1 (2.1)
In the event of fire, suitable means shall be provided for emergency egress from each storey.	B1 (2.1)
Floors more than 7.5 m above ground level (where, in the case of a fire, the risk that the stairway will become impassable before occupants of the upper parts of the house have escaped is appreciable) will be provided with an alternative route.	B1 (2.1)
Except for kitchens, all habitable rooms in the ground storey should either open directly onto a hall leading to the entrance or other suitable exit, or be provided with a window (or door).	B1 (2.8)
Where a sleeping gallery is provided:	B1 (2.9a–c)

- the gallery should be not more than 4.5 m above ground level;
- the distance between the foot of the access stair to the gallery and the door to the room containing the gallery should not exceed 3 m;
- galleries longer than 7.5 m should be provided with a separate alternative exit.

Houses with floors above 4.5 m above ground level

The top storey should be separated from the lower storeys by fire-resisting construction and be provided with an alternative escape route leading to its own final exit.	B1 (2.13b)
If a house has two or more storeys with floors more than 4.5 m above ground level (typically a house of four or more storeys), then an alternative escape route should be provided from each storey or level situated 7.5 m or more above ground level.	B1 (2.14)

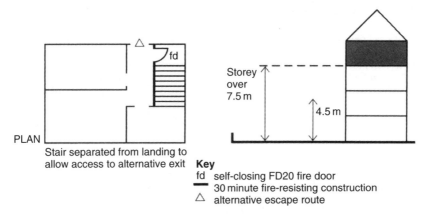

PLAN

Stair separated from landing to
allow access to alternative exit

Key
fd   self-closing FD20 fire door
──   30 minute fire-resisting construction
△   alternative escape route

Figure 6.33 Fire separation in houses with more than one floor over 4.5 m
above ground level

Floorboarding

Floorboarding shall comply with BS 1297: 1987.   A1/2 (1BBc)

💡 Floorboard strength classes, species, grades
and species combinations referred to are defined
in BS 5268: Part 2.

Floor joists

Floor joists spanning in excess of 2.5 m should be   A1/2 (1B10)
strutted by one or more rows of solid or herringbone
strutting in accordance with Table 6.13.

Solid strutting should be at least 38 mm timber   A1/2 (1B10)
thickness extending at least 0.75 times the depth
of the joists.

Herringbone strutting should be of at least   A1/2 (1B10)
38 mm × 38 mm timber size but should not be used
where the distance between joists is greater than
3 times the depth of the joists.

Table 6.13 Strutting of joists

Joist span (m)	Number of strutting rows
Less than 2.5 m	None
2.5–4.5 m	1 at mid span
More than 4.5 m	2 at one third span positions

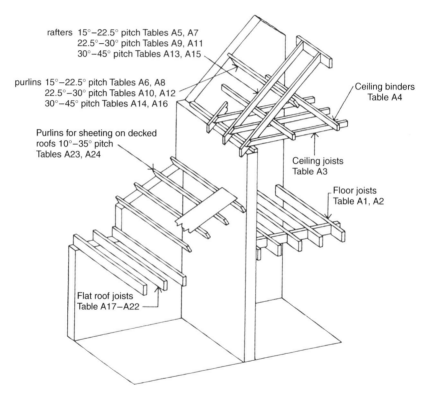

rafters 15°–22.5° pitch Tables A5, A7
         22.5°–30° pitch Tables A9, A11
         30°–45° pitch Tables A13, A15

purlins 15°–22.5° pitch Tables A6, A8
        22.5°–30° pitch Tables A10, A12
        30°–45° pitch Tables A14, A16

Purlins for sheeting on decked
roofs 10°–35° pitch
Tables A23, A24

Flat roof joists
Table A17–A22

Ceiling binders
Table A4

Ceiling joists
Table A3

Floor joists
Table A1, A2

Figure 6.34 Timber floor, ceiling and roof members

Maximum height of domestic buildings

The maximum height of a building shall not exceed the heights given in Table 6.14 with regard to the relevant wind speed.	A1/2 (1C17)

Table 6.14 Maximum height of building on normal or slightly sloping sites

Location		Normal or slightly sloping sites							Steeply sloping sites, hills, cliffs and escarpments						
	Wind speed	36	38	40	42	44	46	48	36	38	40	42	44	46	48
Unprotected site, open countryside with no obstructions		15	15	15	15	15	11	9	8	6	4	3	0	0	0
Open countryside with scattered windbreaks		15	15	15	15	15	15	13	11	9	7.5	6	5	4	3
Country with many windbreaks, small towns, outskirts of large cities		15	15	15	15	15	15	15	15	15	14	12	10	8	6.5
Protected sites and country centres		15	15	15	15	15	15	15	15	15	15	15	15	15	14

Imposed loads on roofs, floors and ceilings

The imposed loads on floors and ceilings shall not exceed those shown in Table 6.15.	A1/2 (1C16)

Table 6.15 Imposed loads

Element	Distributed loads	Concentrated load
Roofs	1.00 kN/m^2 for spans not exceeding 12 m 1.50 kN/m^2 for spans not exceeding 6 m	
Floors	2.00 kN/m^2	
Ceilings	0.25 kN/m^2	0.9 kN/m^2

Lateral support by floors

Floors should act to transfer lateral forces (see Table 6.16) from walls to buttressing walls, piers or chimneys and be secured to the supported wall as shown.	A1/2 (1C34)

Table 6.16 Lateral support by floors

Type of wall	Wall length	Floor lateral support required
External, compartment separating, solid or cavity wall	Any length Greater than 3 m	By every floor forming a junction with the supporting wall
Internal load bearing wall (not being a compartment or separating wall)	Any length	At the top of each storey

Maximum floor area

No floor enclosed by structural walls on all sides shall exceed 70 m² (see Figure 6.35).	A1/2 (1C15)
No floor with a structural wall on one side shall exceed 30 m² (see Figure 6.35).	A1/2 (1C15)

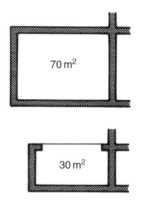

Figure 6.35 Maximum floor area that is enclosed by structural walls

Loading on walls

The maximum span for any floor supported by a wall is 6 m measured centre to centre of bearing (see Figure 6.36).	A1/2 (1C24)
Vertical loading caused by concrete floor slabs, precast concrete floors, and timber floors on the walls, should be distributed.	A1/2 (1C25a)
Differences in the level of ground or other solid construction between one side of the wall and the other should be less than four times the thickness of the wall (see Figure 6.37).	A1/2 (1C25b)
The combined dead and imposed load should not exceed 70 kN/m at the base of the wall (see Figure 6.37).	A1/2 (1C25c)

Walls should **not** be subjected to lateral load other than from wind, and that caused by the differences in the level of the ground or other solid construction between one side of the wall and the other.   A1/2 (1C25d)

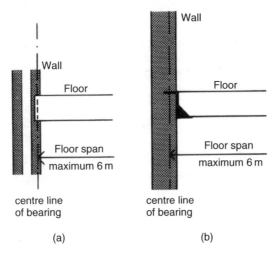

Figure 6.36 Maximum floor span. (a) Floor member bearing on wall. (b) Floor member bearing on joist hanger

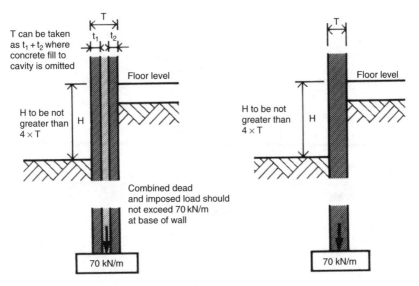

Figure 6.37 Differences in ground level

Lateral support by floors

Where contact between floors and walls on both sides of the wall is at intervals no greater than 2 m, floors should be at or about the same level on each side of the wall.	A1/2 (1C36)
Where lateral support is intermittent, the point of contact should be in line or nearly in line.	A1/2 (1C36)
Walls should be strapped to floors above ground level, at intervals not exceeding 2 m by galvanized mild steel or other durable metal straps which have a minimum cross-section of 30 mm × 5 mm as indicated in Figure 6.38.	A1/2 (1C36)

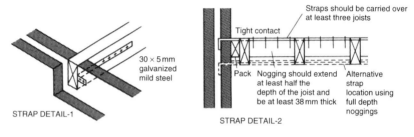

Figure 6.38 Lateral support by floors

Straps need not be provided in the longitudinal direction of joists in houses of not more than 2 storeys:	A1/2 (1C36)
1. If the joists are at not more than 1.2 m centres and have at least 90 mm bearing on the supported walls or 75 mm bearing on a timber wall-plate at each end.	
2. If the joists are carried on the supported wall by joist hangers of the restraint type (described in BS 5628 Part 1 and shown in Figure 6.39 and are incorporated at not more than 2 m centres.	
3. When a concrete floor has at least 90 mm bearing on the supported wall (see Figure 6.40).	A1/2 (1C36)
4. Where floors are at or about the same level on each side of a supported wall, and contact between the floors and wall is either continuous or at intervals not exceeding 2 m.	A1/2 (1C36)

Where contact is intermittent, the points of contact should be in line or nearly in line on plan.

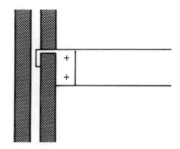

Figure 6.39 Restraint type joist hanger

X to be not less than 90 mm

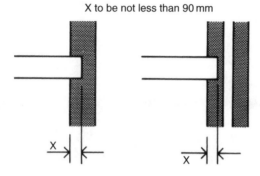

Figure 6.40 Restraint by concrete floor or roof

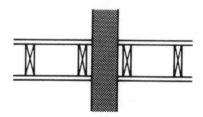

Figure 6.41 Restraint of internal walls

Interruption of lateral support

Where an opening in a floor for a stairway or the like   A1/2 (1C38)
adjoins a supported wall and interrupts the continuity
of lateral support, then:

(a) the maximum permitted length of the opening
     is to be 3 m, measured parallel to the supported
     wall, and

(b) where a connection is provided by means
     other than by anchor, this should be provided
     throughout the length of each portion of the
     wall situated on each side of the opening, and
(c) where connection is provided by mild steel
     anchors, these should be spaced closer than
     2 m on each side of the opening to provide
     the same number of anchors as if there were
     no opening, and
(d) there should be no other interruption of lateral
     support.

Floor joists

The minimum bearing length at supports   A1/2 Appendix A
for floor joists should be 35 mm.

The maximum clear span of joists (i.e. 400 mm,   A1/2 Appendix A
450 mm or 600 mm) used for calculations
depends upon the size of the joist, the spacing
between the joists, the type (and strength) of
timber used and the loading. For examples,
see Table 6.17.

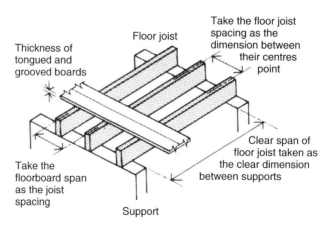

Figure 6.42 Floor joists

Table 6.17 Maximum clear span for floor joists

Timber strength (BS 5268: 1991 Part 2)	Size of joist (mm)	Joist spacing dead load								
		Less than 0.25 kN/m^3			Between 0.25 and 0.50 kN/m^3			Between 0.50 and 1.25 kN/m^3		
		400	450	600	400	450	600	400	450	600
		Clear span of floor joists (mm)								
SC3	38×97	1.83	1.69	1.30	1.72	1.56	1.21	1.42	1.30	1.04
SC4	38×97	1.94	1.83	1.59	1.84	1.74	1.51	1.64	1.55	1.36
SC3	75×220	5.27	5.13	4.79	5.11	4.97	4.64	4.74	4.6	4.07
SC4	75×220	5.42	5.27	4.93	5.25	5.11	4.78	4.88	4.74	4.35

The above table shows the range of choice available for two sizes of joist of two different strengths. For full details, see Tables A1 and A2 of Appendix A to Approved Document A.

These tables can be used when a bath is to be installed provided joists supporting the bath are duplicated.

Floorboards

Softwood tongued and grooved floorboards, if supported at a joist spacing of up to 500 mm, should be at least 16 mm finished thickness. For wider spacing (i.e. up to 600 mm) they should be 19 mm finished thickness.	A1/2 Appendix A

Ceiling joists

The minimum bearing length at supports for ceiling joists and binders should be 35 mm.	A1/2 Appendix A
No notches or holes should be cut in binders unless checked by a competent person.	A1/2 Appendix A
In calculating the ceiling joist sizes no account shall be taken of trimming (e.g. around the flues) or other loads (e.g. water tanks).	A1/2 Appendix A

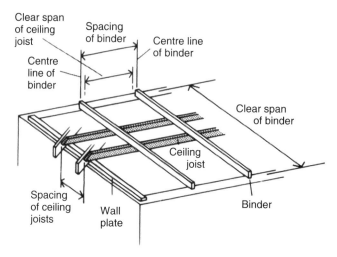

Figure 6.43 Ceiling joists and binders for ceiling joists

> The maximum clear span of joists (i.e. 400 mm,   A1/2 Appendix A
> 450 mm or 600 mm) used for calculations
> depends upon the size of the joist, the spacing
> between the joists, the type (and strength)
> of timber used and the loading.
> For example see Table 6.18.

Table 6.18 Maximum clear span for ceiling joists

Timber strength (BS 5268: 1991 Part 2)	Size of joist (mm)	Joist spacing dead load					
		Less than 0.25 kN/m³			Between 0.25 and 0.50 kN/m³		
		400	450	600	400	450	600
		Clear span of ceiling joists (mm)					
SC3	38 × 97	1.74	1.72	1.67	1.67	1.64	1.58
SC4	38 × 97	1.84	1.82	1.76	1.76	1.73	1.66
SC3	50 × 220	5.22	5.41	5.14	5.14	5.03	4.73
SC4	50 × 220	5.77	5.66	5.37	5.37	5.25	4.95

Tables 6.18 and 6.19 show the range of choice available for two sizes of joist
of two different strengths. For full details, see Tables A3 and A4 of Appendix
A to Approved Document A.

Table 6.19 Binders supporting ceiling joists

Timber strength (BS 5268: 1991 Part 2)	Size of binder (mm)	Binder supporting ceiling joists											
		Dead load less than 0.25 kN/m³						Dead load between 0.25 and 0.50 kN/m³					
		Spacing of binders (mm)											
		1200	1500	1800	2100	2400	2700	1200	1500	1800	2100	2400	2700
SC3	47 × 150	2.17	2.05	1.96	1.88	1.81		1.99	1.87				
SC4	47 × 150	2.28	2.16	2.06	1.98	1.90	1.84	2.09	1.97	1.87			
SC3	75 × 225	4.08	3.85	3.66	3.51	3.37	3.26	3.71	3.50	3.31	3.16	3.03	2.92
SC4	75 × 225	4.26	4.06	3.82	3.66	3.52	3.40	3.88	3.65	3.46	3.30	3.17	3.06

Internal fire spread (structure)

Loadbearing elements of structure

All loadbearing elements of a structure shall have a minimum standard of fire resistance.	B3 (8.1)
Structural frames, beams, floor structures and gallery structures, should have at least the fire resistance given in Appendix A of Approved Document B.	B3 (8.2)
When altering an existing two-storey, single-family dwellinghouse to provide additional storeys, the floor(s), both old and new, shall have the full 30 minute standard of fire resistance.	B3 (8.7)

Fire resistance – compartmentation

To prevent the spread of fire within a building, whenever possible, the building should be sub-divided into compartments separated from one another by walls and/or floors of fire-resisting construction.	B3 (9.1)
Parts of a building that are occupied mainly for different purposes, should be separated from one another by compartment walls and/or compartment floors.	B3 (9.11)
The wall and any floor between the garage and the house shall have a 30 minute fire resistance. Any opening in the wall to be at least 100 mm above the garage floor level with an FD30 door.	B3
In buildings containing flats or maisonettes compartment walls or compartment floors shall be constructed between:	B3 (9.15)

- every floor (unless it is within a maisonette)
- one storey and another within one dwelling
- every wall separating a flat or maisonette from any other part of the building
- every wall enclosing a refuse storage chamber.

Every compartment floor should:	B3 (9.22)

- form a complete barrier to fire between the compartments they separate; and
- have the appropriate fire resistance as indicated in Appendix A of Approved Document B, Tables A1 and A2.

Where a compartment wall or compartment floor meets another compartment wall, or an external wall, the junction should maintain the fire resistance of the compartmentation.	B3 (9.27)

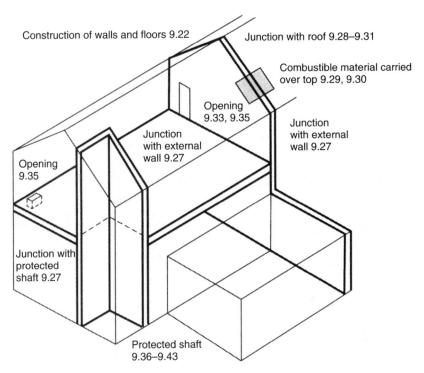

Construction of walls and floors 9.22

Junction with roof 9.28–9.31

Combustible material carried over top 9.29, 9.30

Opening 9.33, 9.35

Junction with external wall 9.27

Junction with external wall 9.27

Opening 9.35

Junction with protected shaft 9.27

Protected shaft 9.36–9.43

Figure 6.44 Compartment walls and compartment floors with reference to the relevant paragraphs in Approved Document B

Concrete

With a concrete intermediate floor:

The ground floor may be a solid slab, laid on the ground, or a suspended concrete floor.	E (p13)
Concrete floors with a mass of less than 365 kg/m² should not bear on the core.	E (p13)
If core type D (see Table 6.22) is used, the cavity should not be bridged.	E (p13)

The floor base may only be carried through if it has a mass of at least 365 kg/m².	E (p13)
If it is suspended it may only pass through the wall if it has a mass of at least 365 kg/m².	E (p13)
Seal the junction between ceiling and panel with tape or caulking.	E (p13)

Enclosures for drainage and/or water supply pipes

The enclosure should: • be bounded by a compartment wall or floor, an outside wall, an intermediate floor, or a casing • have internal surfaces (except framing members) of Class 0 • not have an access panel that opens into a circulation space or bedroom • be used only for drainage, or water supply, or vent pipes for a drainage system.	B3 (11.8)
The casing should: • be imperforate except for an opening for a pipe or an access pane • not be of sheet metal • have (including any access panel) not less than 30 minutes fire resistance.	B3 (11.8)

Suspended ceilings

Table 6.20 sets out criteria appropriate to the suspended ceilings that can be accepted as contributing to the fire resistance of a floor.

Table 6.20 Limitations on fire protected suspended ceilings

Height of building or separated part	Type of floor	Provision for fire resistance of floor	Description of suspended ceiling
<18 m	Not compartment	60 mins or less	Type A, B, C or D
	Compartment	<60 mins	Type A, B, C or D
		60 mins	Type B, C or D
18 m or more	Any	60 mins or less	Type C or D
No limit	Any	60 mins	Type D

Floors that separate: E2

- a dwelling from another dwelling
- a dwelling from another part of the same building (which is not used exclusively as part of that dwelling

shall resist the transmission of airborne sounds.

Floors above a dwelling: E3

- that separates it from another dwelling
- or separates it from another part of the same building (which is not used exclusively as part of the dwelling)

shall resist the transmission of impact sound such as speech, musical instruments and loudspeakers and impact sources such as footsteps and furniture moving.

The flow of sound energy through walls and floors   E (0.6)
should be restricted.

Walls should reduce the level of airborne sound.   E (0.7)

Floors should reduce airborne sound and also, if they   E (0.9)
are above a dwelling, impact sound.

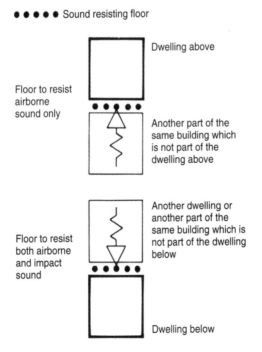

Figure 6.45 Sound resisting floors

Air paths, including those due to shrinkage, must be avoided – porous materials and gaps at joints in the structure must be sealed.	E (0.10)
The possibility of resonance in parts of the structure (such as a dry lining) should be avoided.	E (0.10)
Flanking transmission (i.e. the indirect transmission of sound from one side of a wall or floor to the other side) should be minimized.	E (0.11–0.13)

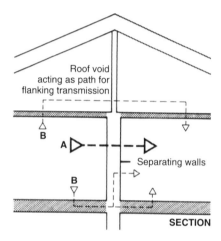

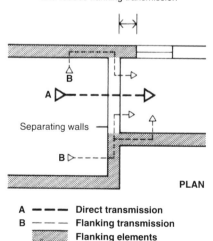

Figure 6.46 Direct and flanking transmission
For clarity not all flanking paths have been shown

Types of separating walls for new buildings

The three most common wall constructions are shown below.

Solid masonry walls (Wall type 1)

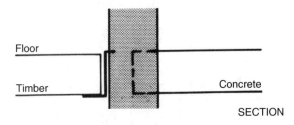

Figure 6.47 Floor junctions – solid masonry walls (Wall type 1)

Floor joists may be supported on hangers or built in.	E (p9)
All gaps should be sealed and there should be no airpaths through the wall.	E (p9)

Cavity masonry walls (Wall type 2)

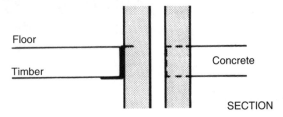

Figure 6.48 Floor junctions – cavity masonry walls (Wall type 2)

Floor joists may be supported on hangers or built in (care should be taken to ensure that there are no airpaths through the wall).	E (p11)
A suspended concrete intermediate or ground floor should be carried through to the cavity face of each leaf.	E (p11)
A concrete slab on the ground may be continuous.	E (p11)

Masonry walls between isolated panels (Wall type 3)

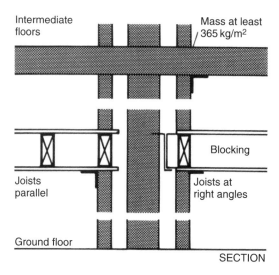

Figure 6.49 Floor junctions – masonry walls between isolated panels (Wall type 3)

Timber

With a timber intermediate floor:

Use joist hangers for any joists supported on the wall.	E (p13)
Seal the spaces between joists with full depth timber blocking.	E (p13)

Concrete

With a concrete intermediate floor:

The ground floor may be a solid slab, laid on the ground, or a suspended concrete floor.	E (p13)
Concrete floors with a mass of less than 365 kg/m^2 should not bear on the core.	E (p13)
If core type D is used, the cavity should not be bridged.	E (p13)

The floor base may only be carried through if it has a mass of at least 365 kg/m².	E (p13)
If it is suspended it may only pass through the wall if it has a mass of at least 365 kg/m².	E (p13)
Seal the junction between ceiling and panel with tape or caulking.	E (p13)

Timber framed walls (Wall type 4)

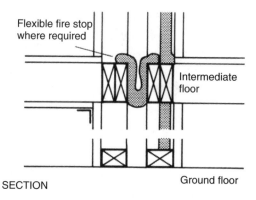

Figure 6.50 Floor junctions – timber framed walls (Wall type 4)

Block all air paths to the wall cavity either by carrying the cladding through the floor or by using a solid timber edge to the floor.	E (p15)
Where the joists are at right angles to the wall, seal spaces between joists with full depth timber blocking.	E (p15)

Types of separating floors

As shown in Table 6.21, there are three main types of separating floors.

Table 6.21 Types of separating floor

Concrete base with soft covering (Floor type 1)	Soft covering Concrete base	The resistance to airborne sound depends on the mass of the concrete base and on eliminating air paths. The soft covering reduces impact sound at source.
Concrete base with floating layer (Floor type 2)	Floating layer Resilient layer Concrete base	The resistance to airborne sound depends mainly on the mass of the concrete base and partly on the mass of the floating layer. Resistance to impact sound depends on a resilient layer isolating the floating layer from the base and from the surrounding construction.
Timber base with floating layer (Floor type 3)	floating layer resilient layer timber base pugging ceiling	The resistance to airborne sound depends partly on the structural floor plus absorbent blanket or pugging and partly on the floating layer. Resistance to impact sound depends on a resilient layer isolating the floating layer from the base and the surrounding construction.

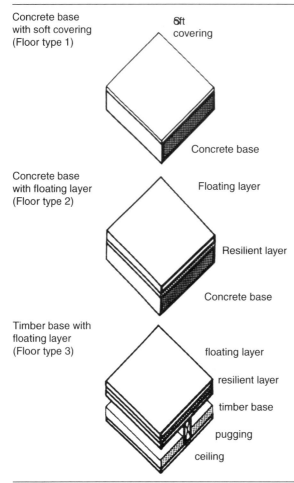

Concrete base with soft covering (Floor type 1)

The resistance to airborne sound depends on the mass of the concrete base and on eliminating air paths. The soft covering reduces impact sound at source.

Where resistance to airborne sound only is required, the soft covering (only) may be omitted.	E (p18)
All joints between parts of the floor should be filled to avoid air paths.	E (p18)
Sound paths around the floor should be controlled to reduce flanking transmission.	E (p18)

To reduce flanking transmission and to avoid air paths,   E (p18)
special attention to detail should be paid to floor perimeters
and wherever the floor is penetrated by a pipe or duct.

As shown in Table 6.22, there are four types of floor bases (A, B, C & D) that give
suitable resistance to direct transmission of airborne sound and one type of soft covering (E) that can be added to give suitable resistance to impact sound transmission.

Table 6.22 Types of floor bases

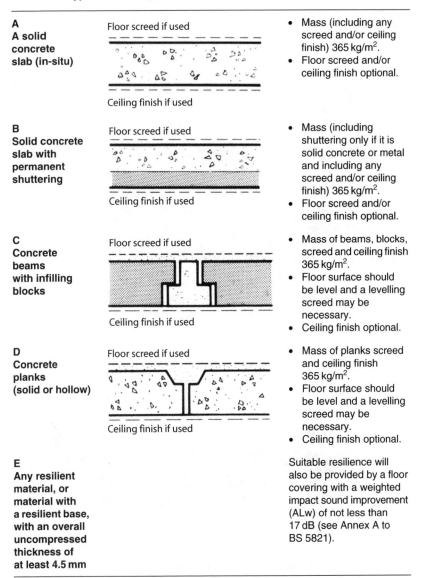

A **A solid** **concrete** **slab (in-situ)**	Floor screed if used Ceiling finish if used	• Mass (including any screed and/or ceiling finish) 365 kg/m². • Floor screed and/or ceiling finish optional.
B **Solid concrete** **slab with** **permanent** **shuttering**	Floor screed if used Ceiling finish if used	• Mass (including shuttering only if it is solid concrete or metal and including any screed and/or ceiling finish) 365 kg/m². • Floor screed and/or ceiling finish optional.
C **Concrete** **beams** **with infilling** **blocks**	Floor screed if used Ceiling finish if used	• Mass of beams, blocks, screed and ceiling finish 365 kg/m². • Floor surface should be level and a levelling screed may be necessary. • Ceiling finish optional.
D **Concrete** **planks** **(solid or hollow)**	Floor screed if used Ceiling finish if used	• Mass of planks screed and ceiling finish 365 kg/m². • Floor surface should be level and a levelling screed may be necessary. • Ceiling finish optional.
E **Any resilient** **material, or** **material with** **a resilient base,** **with an overall** **uncompressed** **thickness of** **at least 4.5 mm**		Suitable resilience will also be provided by a floor covering with a weighted impact sound improvement (ALw) of not less than 17 dB (see Annex A to BS 5821).

Junctions for concrete based (Type 1) floors

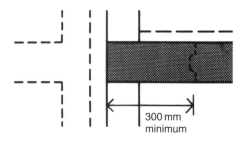

Figure 6.51 External walls or cavity separating walls (Floor type 1)

The mass of the wall leaf adjoining the floor should be at least 120 kg/m² (including any finish), unless it is an external wall having openings of at least 20% of its area in each room, in which case there is no minimum requirement.	E (p19)
The mass of the wall leaf adjoining the floor should be at least 120 kg/m² (including any finish), unless it is an external wall having openings of at least 20% of its area in each room, in which case there is no minimum requirement.	E (p19)
The floor base (excluding any screed) should pass through the leaf whether spanning parallel to or at right angles to the wall.	E (p19)
The cavity should not be bridged.	E (p19)
In type C or D floor bases, where the beams are parallel to the wall, the first joint should be 300 mm from the cavity face of the wall leaf.	E (p19)

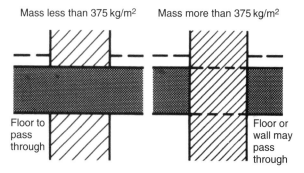

Figure 6.52 Internal walls or solid separating walls (Floor type 1)

In the Gas Safety Regulations 1972 (S1, 197211178) and the Gas Safety
stallation and use) Regulations 1984 (S1, 198411358) there are require-
nts for ventilation of ducts at each floor where they contain gas pipes.
s pipes may be contained in a separate ventilated duct or they can remain
lucted.

ncrete base with floating layer (Floor type 2)

e resistance to airborne sound depends partly on the mass of the concrete
.e and partly on the mass of the floating layer. Resistance to impact sound
)ends mainly on a resilient layer isolating the floating layer from the base
i from the surrounding construction.

Where resistance to airborne sound only is required, the full construction should still be used.	E (p20)
All joints between parts of the floor base should be filled to avoid air paths.	E (p20)
Sound paths round the floor should be controlled to reduce flanking transmission.	E (p20)
To reduce flanking transmission and to avoid air paths, special attention to detail should be paid to floor perimeters and wherever the floor is penetrated by a pipe or duct.	E (p20)
Care should be taken so as not to create a bridge between the floating layer and the base, surrounding walls, or adjacent screed.	E (p20)
With bases C and D a screed may be required to seal the floor and to accommodate surface irregularities.	E (p20)

s shown in Table 6.23, there are four types of floor bases (A, B, C & D), two
oating layers (E & F) and one resilient layer (G). Any combination of base,
silient layer and floating layer will give suitable resistance to direct trans-
ission.

Two additional resilient layers that may be used under screeds only are also
)ecified (H and I).

If the wall mass is less than 375 kg/m² (including any plaster or plasterboard) then the floor base excluding screed should pass through.

If the wall mass is more than 375 kg/m² (including an plaster or plasterboard) either the wall excluding any finishes or the floor base excluding any screed may pass through.

Where the wall does pass through, tie the floor base to the wall and grout the joint.

Floor penetrations (excluding gas pipes)

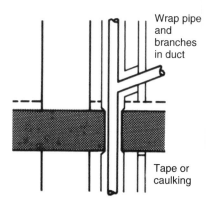

Figure 6.53 External wall junctions (Floor type 1)

Ducts or pipes which penetrate a floor separating habitabl in an enclosure (both above and below the floor). In all cas\

- enclosure material should have a mass of 15 kg/m².
- enclosure should either be lined or the duct (or pipe) wrapped in 25 mm unfaced mineral wool.

Penetrations through a separating floor by ducts and pipe\ should have fire protection in accordance with Approved Document B, Fire safety.

Fire stopping should be flexible and prevent rigid contact between the pipe and floor.

Types of floor bases

Four types are available, as shown in Table 6.23

Table 6.23 Floor bases

A **A solid** **concrete slab** **(in-situ)**	Floor screed if used 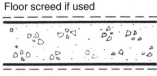 Ceiling finish if used	• Mass (including shuttering only if it is solid concrete or metal, and including any bonded screed and/or ceiling finish) 300 kg/m². • Floor screed and/or ceiling finish optional.
B **Solid concrete** **slab with** **permanent** **shuttering**	Floor screed if used 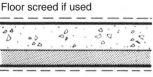 Ceiling finish if used	• Mass (including shuttering only if it is solid concrete or metal and including any screed and/or ceiling finish) 300 kg/m². • Floor screed and/or ceiling finish optional.
C **Concrete** **beams with** **infilling blocks**	Floor screed if used 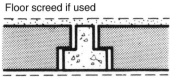 Ceiling finish if used	• Mass of beams, blocks, screed and ceiling finish 300 kg/m². • The floor base should be reasonably level (maximum 5 mm step between units). • A levelling screed may be required. • Ceiling finish optional.
D **Concrete** **planks** **(solid or** **hollow)**	Floor screed if used 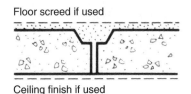 Ceiling finish if used	• Mass of planks and any bonded screed or ceiling finish 300 kg/m². • The floor base should be reasonably level (maximum 5 mm step between units). • A levelling screed may be required. • Ceiling finish is optional.

Two types of floating layer are shown in Table 6.24.

Table 6.24 Floating layers

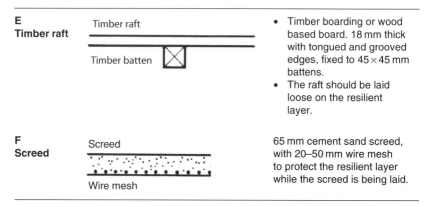

E Timber raft	Timber raft Timber batten	• Timber boarding or wood based board. 18 mm thick with tongued and grooved edges, fixed to 45 × 45 mm battens. • The raft should be laid loose on the resilient layer.
F Screed	Screed Wire mesh	65 mm cement sand screed, with 20–50 mm wire mesh to protect the resilient layer while the screed is being laid.

The resilient layer (G) and additional resilient layers for use under screeds only (H and I) are shown in Table 6.25.

Table 6.25 Resilient layer (G) and additional resilient layers (H and I)

G **Resilient layer** **25 mm mineral fibre density 36 kg/m³**	• A 13 mm thickness may be used under a timber raft if the battens used have an integral closed cell resilient foam strip. • Lay the fibre tightly butted and turned up at the edges of the floating layer. 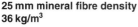 • Under a timber raft, the fibre may be paper faced on the underside. • Under a screed, fibre should be paper faced on the upper side to prevent screed entering the layer.
H **13 mm pre-compressed expanded polystyrene board (impact sound duty grade)**	• Lay boards tightly butted. • Use board on edge as a resilient strip at edges of floating screed.
I **5 mm extruded (closed cell) polyethylene foam, density 30–45 kg/m³**	• Lay material over a levelling screed (to protect the material from puncture). • Lay with joints lapped and turn up at edges of the floating screed.

Junctions for concrete based (Type 2) floor junctions

External walls or cavity separating walls

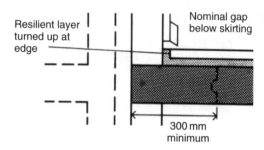

Figure 6.54 External walls or cavity separating walls (Floor type 2)

The mass of the wall leaf adjoining the floor should be at least 120 kg/m² (including any finish), unless it is an external wall having openings of at least 20% of its area in each room, in which case there is no minimum requirement.	E (p21)
The floor base (excluding any screed) should pass through the leaf whether spanning parallel to or at right angles to the wall.	E (p21)
The cavity should not be bridged.	E (p21)
If the floor base is Type C or D, where the beams are parallel to the wall the first joint should be 300 mm from the cavity face of the wall leaf.	E (p21)
Carry the resilient layer up at all edges to isolate the floating layer.	E (p21)
Leave a nominal gap between skirting and floating layer or turn resilient layer under skirting (a seal is not necessary but if used it should be flexible).	E (p21)

Internal wall or solid separating wall

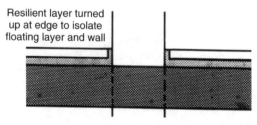

Figure 6.55 Internal walls or solid separating walls (Floor type 2)

If the wall mass is less than 375 kg/m^2 (including any plaster or plasterboard) then the floor base excluding any screed should pass through.	E (p21)
If the wall mass is more than 375 kg/m^2 (including any plaster or plasterboard) either the wall excluding any finishes or the floor base excluding any screed may pass through.	E (p21)
Where the wall does pass through, tie the floor base to the wall and grout the joint.	E (p21)

Floor penetrations (excluding gas pipes)

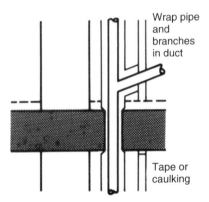

Figure 6.56 External wall junctions (Floor type 2)

Ducts or pipes that penetrate a floor separating habitable rooms should be in an enclosure (both above and below the floor). In all cases, the:

• enclosure material should have a mass of 15 kg/m^2.	E (p21)
• enclosure should either be lined or the duct (or pipe) wrapped in 25 mm unfaced mineral wool.	E (p21)
Penetrations through a separating floor by ducts and pipes should have fire protection in accordance with Approved Document B, Fire safety.	E (p21)
Fire stopping should be flexible and prevent rigid contact between the pipe and floor.	E (p21)

In the Gas Safety Regulations 1972 (S1, 197211178) and the Gas Safety (installation and use) Regulations 1984 (S1, 198411358) there are requirements for ventilation of ducts at each floor where they contain gas pipes. Gas pipes may be contained in a separate ventilated duct or they can remain unducted.

Timber base with floating layer (Floor type 3)

The resistance to airborne sound depends partly on the structural floor plus absorbent material or pugging, and partly on the floating layer.

Resistance to impact sound depends mainly on a resilient layer isolating the floating layer from the base and the surrounding construction.

Where resistance to airborne sound only is required the full construction should still be used.	E (p20)
The correct density of resilient layer should be used to ensure that it can carry the anticipated load.	E (p20)
Sound paths round the floor should be controlled to reduce flanking transmission.	E (p20)
To reduce flanking transmission and to avoid air paths, special attention to detail should be paid to floor perimeters.	E (p20)
Take care not to bridge between the floating layer and the base or surrounding walls (e.g. with services or fixings which penetrate the resilient layer).	E (p20)
Allow for movement of materials, e.g. expansion of chipboard after laying (to maintain isolation).	E (p20)

As shown in Table 6.26, there are three complete constructions (A, B & C) which give suitable resistance to direct sound transmission.

Types of pugging are described (D). Details of how junctions should be made to limit flanking transmission follow.

Types of floor bases

Three types are available, as shown in Table 6.26.

Pugging

The pugging between joists should be one of the types given in Table 6.27.

Table 6.26 Floor bases

**A
Platform floor with
absorbent material**

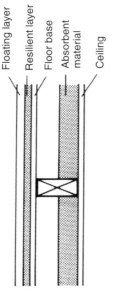

Floating layer
Resilient layer
Floor base
Absorbent material
Ceiling

This should consist of a:

- floating layer of either timber (or wood based board) 18 mm thick with tongued and grooved edges, all joints glued, spot bonded to substrate of 19 mm plasterboard; or
- two thicknesses of cement bonded particle board with joints staggered, glued and screwed together, total thickness 24 mm.
- resilient layer of 25 mm mineral fibre, density 60–100 kg/m^3.
- floor base of 12 mm timber boarding (or wood-based board) nailed to timber joists (sized to suit the structure).
- ceiling of two layers of plasterboard with joints staggered, total thickness 30 mm with an absorbent material, 100 mm unfaced mineral wool, density at least 10 kg/m^3, laid on the ceiling.

**B
Ribbed floor with
absorbent material**

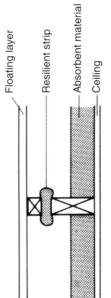

Floating layer
Resilient strip
Absorbent material
Ceiling

This should consist of a:

- floating layer of timber or wood based board 18 mm thick with tongued and grooved edges and all joints glued, spot bonded to substrate of 19 mm plasterboard, nailed to 45 mm × 45 mm timber battens placed over the joists.
- resilient strips of 25 mm mineral fibre, density 80–140 kg/m^3 laid on joists.
- floor base of 45 mm wide timber joists.
- ceiling of two layers of plasterboard with joints staggered, total thickness 30 mm, with an absorbent blanket of 100 mm unfaced rock fibre, density at least 10 kg/m^3 laid on the ceiling.

C
Ribbed floor with heavy pugging

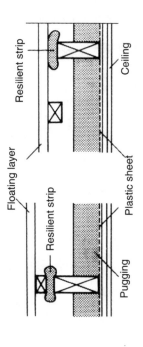

Floating layer

Resilient strip

Resilient strip

Ceiling

Pugging

Plastic sheet

This should consist of a:

- floating layer of timber or wood based board 18 mm thick with tongued and grooved edges and all joints glued, nailed or screwed to 45 mm × 45 mm timber battens placed on (or between) joists.
- resilient strips of 25 mm mineral fibre density 80–140 kg/m³ laid on joists.
- floor base of 45 mm wide timber joists.
- ceiling of either 19 mm dense plaster on expanded metal lath or 6 mm plywood fixed under the joists plus two layers of plasterboard with joints staggered, total thickness 25 mm.
- both types of ceiling to have pugging of mass 80 kg/m² laid on a polyethylene liner.

Table 6.27 Pugging

D Pugging	Traditional ash (75 mm) 2 mm–10 mm, limestone chips (60 mm) 2 mm–10 mm whin aggregate (60 mm) dry sand (50 mm) Figures in brackets show approximate thickness required to achieve 80 kg/m² (other figures denote sieve size).
	⬤ Do **no**t use sand in kitchens, bathrooms, shower rooms or water closet compartments where it may become wet and overload the ceiling.

Junctions for timber-based (Type 3) floors

Timber framed walls

Example with floor A

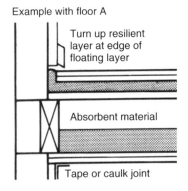

Turn up resilient
layer at edge of
floating layer

Absorbent material

Tape or caulk joint

Figure 6.57 Timber framed walls (Floor type 3)

Seal the gap between wall and floating layer with a resilient strip glued to the wall.	E (p23)
Leave a 3 mm gap between skirting and floating layer (a seal is not necessary but if used it should be flexible).	E (p23)
Block air paths between the floor base and the wall, including the space between joists.	E (p23)
When joists are at right angles to the wall seal the junction of ceiling and wall lining with tape or caulking.	E (p23)

Heavy masonry leaf

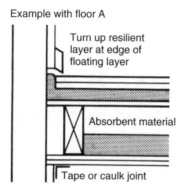

Example with floor A

Turn up resilient layer at edge of floating layer

Absorbent material

Tape or caulk joint

Figure 6.58 Heavy masonry leaf (Floor type 3)

Mass of leaf (including any finish) 375 kg/m², both above and below floor.	E (p23)
Seal the gap between wall and floating layer with a resilient strip.	E (p23)
Leave a 3 mm gap between skirting and floating layer (a seal is not necessary but if used it should be flexible).	E (p23)
Use any normal method of connecting floor base to wall.	E (p23)
Seal the junction of ceiling and wall lining with tape or caulking.	E (p23)

Light masonry leaf

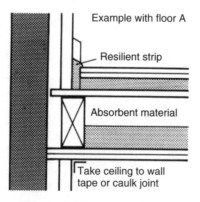

Example with floor A

Resilient strip

Absorbent material

Take ceiling to wall tape or caulk joint

Figure 6.59 Light masonry leaf (Floor type 3)

If the mass including any plaster or plasterboard is less than 375 kg/m² a free-standing panel as specified in wall type 3 should be used.	E (p23)
Seal the gap between panel and floating layer with a resilient strip.	E (p23)
Leave a 3 mm gap between skirting and floating layer (a seal is not necessary but if used it should be flexible).	E (p23)
Use any normal method of connecting floor base to wall but block air paths between floor and wall cavities.	E (p23)
Take ceiling through to masonry, seal junction with freestanding panel with tape or caulking.	E (p23)

Floor penetrations – Floor type 3

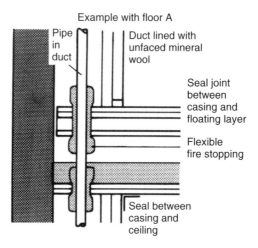

Figure 6.60 External wall junctions (Floor type 3)

Ducts or pipes which penetrate a floor separating habitable rooms should be in an enclosure (both above and below the floor). In all cases, the:

• enclosure material should have a mass of 15 kg/m².	
• enclosure should either be lined or the duct (or pipe) wrapped in 25 mm unfaced mineral wool.	E (p19)
Leave a 3 mm gap between enclosure and floating layer, seal with acrylic caulking or neoprene.	E (p24)

The enclosure may go down to the floor base if specification A is used, but in these cases, the enclosure needs to be isolated from the floating layer.	E (p24)
Penetrations through a separating floor by ducts and pipes should have fire protection in accordance with Approved Document B, Fire safety.	E (p24)
Fire stopping should be flexible and prevent rigid contact between the pipe and floor.	E (p24)

In the Gas Safety Regulations 1972 (S1, 197211178) and the Gas Safety (installation and use) Regulations 1984 (S1, 198411358) there are requirements for ventilation of ducts at each floor where they contain gas pipes. Gas pipes may be contained in a separate ventilated duct or they can remain unducted.

Remedial work in conversions

Where it cannot be shown that the existing construction (i.e. wall, floor or stair) meets the sound insulation requirements one of the following treatments shall be used in order to improve the level of sound insulation.

As the implementation of upgrading measures will impose additional loads on the existing structure, it (i.e. the existing structure) should be assessed to ensure that the additional loading can be carried safely with appropriate strengthening where necessary.	E (5.3)

Table 6.28 Floor treatments

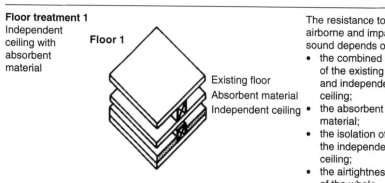

Floor treatment 1
Independent ceiling with absorbent material

Floor 1

Existing floor
Absorbent material
Independent ceiling

The resistance to airborne and impact sound depends on:
• the combined mass of the existing floor and independent ceiling;
• the absorbent material;
• the isolation of the independent ceiling;
• the airtightness of the whole construction.

Table 6.28 Floor treatments (*continued*)

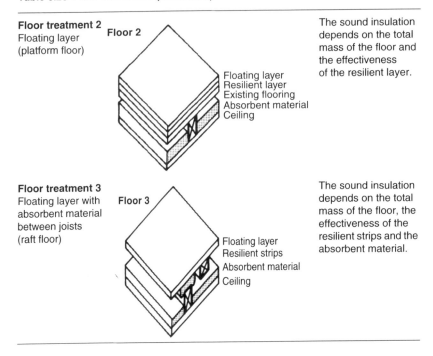

Floor treatment 2 Floating layer (platform floor) Floor 2 Floating layer Resilient layer Existing flooring Absorbent material Ceiling	The sound insulation depends on the total mass of the floor and the effectiveness of the resilient layer.
Floor treatment 3 Floating layer with absorbent material between joists (raft floor) Floor 3 Floating layer Resilient strips Absorbent material Ceiling	The sound insulation depends on the total mass of the floor, the effectiveness of the resilient strips and the absorbent material.

Floor treatment 1 (independent ceiling with absorbent material)

The construction may be used on one side of the existing wall only where the existing wall is masonry, has a thickness of at least 100 mm and is plastered on both faces. With other types of existing wall the construction should be built on both sides.

If existing ceiling is lath and plaster it may be retained if it provides acceptable fire resistance. See Approved Document B: Fire safety.	E (p30)
A gap of 25 mm should be provided between the top of the independent ceiling joists and the underside of the existing floor construction.	E (p30)
Gaps in floor boarding should be sealed either with caulking or by overlaying the floor with hardboard.	E (p30)
Where the existing ceiling is not lath and plaster it should be upgraded to a thickness of 30 mm plasterboard in 2 or 3 layers with joints staggered.	E (p30)
The independent ceiling should comprise two layers of plasterboard having a total thickness of at least 30 mm with the joints staggered, fixed to independent joists.	E (p30)

The perimeter of the independent ceiling should be sealed with tape or mastic.   E (p30)

The independent ceiling should be at least 100 mm below the existing ceiling.   E (p30)

Where a window head is near to the existing ceiling, the new independent ceiling may be raised to form a pelmet recess (see Figure 6.62).   E (p30)

The mineral wool absorbent material should be at least 100 mm thick and have a density of at least 10 kg/m³.   E (p30)

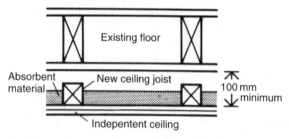

Figure 6.61 Independent ceiling with absorbent material – positioning of joists etc (Floor treatment 1)

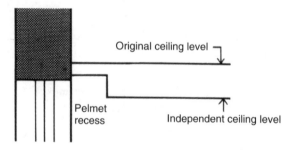

Figure 6.62 Floor treatment 1 – high window head details

Floor treatment 2 *(floating layer – platform floor)*

The sound insulation depends on the total mass of the floor and the effectiveness of the resilient layer.

If the existing ceiling is lath and plaster it may be retained if it provides acceptable fire resistance.   E (p31)

Use correct density of resilient layer and ensure it can carry the anticipated load.   E (p31)

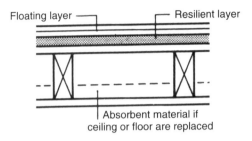

Figure 6.63 Use of existing ceiling (Floor treatment 2)

Where the existing ceiling is not lath and plaster the ceiling should be upgraded to a thickness of 30 mm of plasterboard in 2 or 3 layers with joints staggered.	E (p31)
Where possible insert a 100 mm thick absorbent layer of mineral wool between the joists.	E (p31)
Where the existing floorboards are removed they should be replaced with 12 mm thick boarding with a 100 mm thick absorbent layer of mineral wool laid between the joists (if not already done during the ceiling replacement).	E (p31)
The floating layer should be either:	E (p31)

- a timber or wood-based board 18 mm thick with tongued and grooved edges and all joints glued, spot bonded to a substrate of 19 mm plasterboard, or
- a single or double layer of material having a mass of at least 25 kg/ml and with all joints glued.

The loadbearing resilient layer should be 25 mm thick mineral wool having a density of between 60 and 100 kg/m^3.	E (p31)
At junctions with abutting construction	E (p31)

- a movement gap of 10 mm should be left around the floating layer and should be filled with resilient material.
- a 3 mm gap should be left between the skirting and the floating layer.

A seal is not necessary but if used it should be flexible.

The perimeter of any new ceiling should be sealed with tape or caulking.	E (p31)

Approved Document E also provides details of two alternative floor treatments when treatments 1, 2 and 3 are not practical.

Floor treatment 3 (ribbed floor with absorbent material or heavy pugging)

The sound insulation depends on the mass of the floor, the effectiveness of the resilient strips and the absorbent material or pugging.

If the existing ceiling is lath and plaster it may be retained if it provides acceptable fire resistance.	E (p32)
The existing joists should be 45 mm wide.	E (p32)
Additional strutting may be needed between the existing joists to ensure stability of the floor after removal of the floor boarding.	E (p32)
Where the existing ceiling is not lath and plaster it should be upgraded to a thickness of 30 mm plasterboard in 2 or 3 layers with joints staggered.	E (p32)

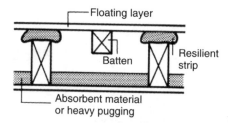

Figure 6.64 Using existing joists (Floor treatment 3)

Floating layer

Two methods are approved:

(a) A floating layer of either:

- timber or wood-based board, 18 mm thick, with tongued and grooved edges and all joints glued, spot bonded to a substrate of 19 mm plaster-board; or
- a single or double layer of material having a mass of at least 25 kg/m³, and with all joints glued.
 (for use when floor levels are critical)

(b) A timber or wood based board, 18 mm thick, with tongued and grooved edges and all joints glued, nailed or screwed to 45 mm×45 mm timber battens (for use where floor levels are critical or where the existing structure can support the additional loading imposed by the heavy pugging).

Method (a)

(A floating layer of either timber or wood-based board or a single or double layer of material having a mass of at least 25 kg/m^3)

The floating layer should be nailed or screwed to 45 mm × 45 mm timber battens.	E (p32)
Where the floating layer incorporates plasterboard, the timber battens should be placed directly over the joists.	E (p32)
The resilient strips should be 25 mm thick mineral fibre, with a density between 80 and 140 kg/m^3 laid on the joists.	E (p32)
The absorbent material laid between the joists should be 100 mm thick mineral wool of 10 kg/m^3 density.	E (p32)

Method (b)

(A timber or wood based board)

The battens should run in the direction of and between or directly over the joists.	E (p32)
The resilient strips should be of 25 mm thick mineral fibre, density between 80 and 140 kg/m^3, laid on the joists.	E (p32)

Pugging

The pugging between joists may be of the following types.

- Traditional ash (75 mm)
- 2 mm–10 mm, limestone chips (60 mm)
- 2 mm–10 mm aggregate (60 mm)
- Dry sand (50 mm).

Figures in brackets show approximate thickness required to achieve 80 kg/m^2 (other figures denote sieve size).

Do **not** use sand in kitchens, bathrooms, shower rooms or water closet compartments where it may become wet and overload the ceiling.

Junctions with abutting construction

A movement gap of 10 mm should be left around the floating layer and should be filled with resilient material.	E (p32)
A 3 mm gap should be left between the skirting and the floating layer.	E (p32)
A seal is not necessary but if used should be flexible.	E (p32)
The perimeter of any new ceiling should be sealed with tape or caulking.	E (p32)

6.7 Walls

In a brick built house, the external walls are loadbearing elements that support the roof, floors and internal walls. These walls are normally cavity walls comprising of two leaves braced with metal ties but older houses will have solid walls, at least 225 mm (9″) thick. Bricks are laid with mortar in overlapping bonding patterns to give the wall rigidity and a Damp-Proof Course (DPC) is laid just above ground level to prevent the moisture rising. Window and door openings are spanned above with rigid supporting beams called lintels. The internal walls of a brick built house are either non-loadbearing divisions (made from lightweight blocks, manufactured boards or timber studding) or loadbearing structures made of brick or block.

Modern timber-framed house walls are constructed of vertical timber studs with horizontal top and bottom plates nailed to them. The frames, which are erected on a concrete slab or a suspended timber platform supported by cavity brick walls, are faced on the outside with plywood sheathing to stiffen the structure. Breather paper is fixed over the top to act as a moisture barrier. Insulation quilt is used between studs. Rigid timber lintels at openings carry the weight of the upper floor and roof. Brick cladding is typically used to cover the exterior of the frame. It is attached to the frame with metal ties. Weatherboarding often replaces the brick cladding on upper floors.

When reading this section, you will probably notice that a few of the requirements have already been covered in Section 6.6 Floors and ceilings. This has been done in order to save the reader having to constantly turn back and re-read a previous page.

6.7.1 Requirements

Precautions shall be taken to reduce risks to the health and safety of persons in buildings by safeguarding them and the buildings against the adverse effects of:

- *vegetable matter*
- *contaminants on or in the ground to be covered by the building*
- *ground water.*

(Approved Document C)

As a fire precaution, all materials used for internal linings of a building should have a low rate of surface flame spread and (in some cases) a low rate of heat release.

(Approved Document B2)

Fire precautions

- *all loadbearing elements of structure of the building shall be capable of withstanding the effects of fire for an appropriate period without loss of stability;*
- *ideally the building should be subdivided by elements of fire-resisting construction into compartments;*
- *all openings in fire-separating elements shall be suitably protected in order to maintain the integrity of the continuity of the fire separation.*
- *any hidden voids in the construction shall be sealed and subdivided to inhibit the unseen spread of fire and products of combustion, in order to reduce the risk of structural failure, and the spread of fire*

(Approved Document B3)

- *external walls shall be constructed so that the risk of ignition from an external source, and the spread of fire over their surfaces, is restricted*
- *the amount of unprotected area in the side of the building shall be restricted so as to limit the amount of thermal radiation that can pass through the wall*
- *the roof shall be constructed so that the risk of spread of flame and/or fire penetration from an external fire source is restricted*
- *the risk of a fire spreading from the building to a building beyond the boundary, or vice versa shall be limited*

(Approved Document B4)

Cavity insulation

Fumes given off by insulating materials such as by Urea Formaldehyde (UF) foams should not be allowed to penetrate occupied parts of buildings to an extent where it could become a health risk to persons in the building by becoming an irritant concentration.

(Approved Document D)

Airborne and impact sound

Dwellings shall be designed and built so that the noise from normal domestic activity in an adjoining dwelling (or other building) is kept down to a level that:

- *does not affect the health of the occupants of the dwelling*
- *will allow them to sleep, rest and engage in their normal domestic activities without disruption.*

(Approved Document E)

6.7.2 Meeting the requirements

Conditions – wall

Maximum allowable length and height of wall	A1/2 (1C18–1C19)
Construction materials and workmanship	A1/2 (1C20–1C23)
Loading on walls	A1/2 (1C24–1C25)
End restraints	A1/2 (1C26–1C28)
Openings, recesses, overhangs and chases	A1/2 (1C29–1C32)
Lateral support by roof and floors	A1/2 (1C33–1C38)
Conditions relating to external walls of small single storey, non-residential buildings and annexes	A1/2 (1C39)

Exceptions – Any walls containing a bay window or that are constructed as a bay for (or as a gable over) a bay window above ground floor cill level.

Thickness of masonry walls

External walls, internal loadbearing walls, compartment walls and separating walls, residential buildings (of up to three storeys) and external walls and internal loadbearing walls of small single-storey non-residential buildings should comply with the relevant requirements of BS 5628: Part 3: 1985.	A1/2 (1C3)

The thickness of the wall depends on the general conditions relating to the building of which the wall forms a part (e.g. such as floor area, roof loading, and wind speed) and the design conditions relating to the wall (e.g. type of materials, loading, end restraints, openings, recesses, overhangs and lateral floor support requirements, etc.). These conditions are covered by the following sections:

External walls

Solid walls constructed of coursed brickwork or blockwork should be at least as thick as 1/16 of the storey height (see Table 6.29).	A1/2 (1C6)
Solid walls constructed in uncoursed stone, flints, clunches of bricks or other burnt or vitrified material should not be less than 1.33 times the thickness of the storey height.	A1/2 (1C7)
All cavity walls should have leaves at least 90 mm thick and cavities at least 50 mm wide (see also Table 6.30).	A1/2 (1C8)
For external walls, compartment walls and separating walls constructed with a cavity, the combined thickness of the two leaves plus 10 mm should be at least as thick as 1/16 of the storey height (see also Table 6.30).	A1/2
Irrespective of the materials used in the construction, walls providing vertical support to other walls should not be less in thickness than any part of the wall to which it gives vertical support.	A1/2 (1C9)
The single leaf of external walls of small single storey non-residential buildings and of annexes need be only 90 mm thick.	A1/2 (1C12)

Table 6.29 Minimum thickness of external walls, compartment walls and separating walls

Height of wall	Length of wall	Minimum thickness of wall
Less than 3.5 m	Less than 12 m	190 mm for whole of its height
3.5 m to 9 m	Less than 9 m	190 mm for whole of its height
	9 m to 12 m	290 mm from the base for the height of one storey and 190 mm for the rest of its height
9 m to 12 m	Less than 9 m	290 mm from the base for the height of one storey and 190 mm for the rest of its height
	9 m to 12 m	290 mm from the base for the height of two storeys and 190 mm for the rest of its height

Table 6.30 Maximum spacing of cavity wall ties

Width of cavity	Horizontal spacing	Vertical spacing	Notes
50–75 mm	900 mm	450 mm	The horizontal and vertical spacing of wall ties may be varied to suit the construction provided that the number of ties per unit is maintained. Wall ties spaced not more than 300 mm apart vertically should be provided within 225 mm from the sides of all openings with unbonded jambs.
76–100 mm	750 mm	450 mm	The horizontal and vertical spacing of wall ties may be varied to suit the construction provided that the number of ties per unit is maintained. Wall ties spaced not more than 300 mm apart vertically should be provided within 225 mm from the sides of all openings with unbonded jambs. Vertical twist type ties or ties of equivalent performance should be used in cavities wider than 75 mm.

Internal load bearing walls

The minimum thickness of all internal loadbearing walls in brickwork or blockwork (except compartment walls or separating walls) should have a thickness not less than that shown in Table 6.31.   A1/2 (1C10)

Table 6.31 Minimum thickness of external walls, compartment walls and separating walls

Height of wall	Length of wall	Minimum thickness of wall
Less than 3.5 m	Less than 12 m	80 mm for whole of its height
3.5 m to 9 m	Less than 9 m	80 mm for whole of its height
	9 m to 12 m	140 mm from the base for the height of one storey and 80 mm for the rest of its height
9 m to 12 m	Less than 9 m	140 mm from the base for the height of one storey and 80 mm for the rest of its height
	9 m to 12 m	140 mm from the base for the height of two storeys and 80 mm for the rest of its height

Parapet walls

The minimum thickness and maximum   A1/2 (1C11)
height of parapet walls should be as given
in Table 6.31.

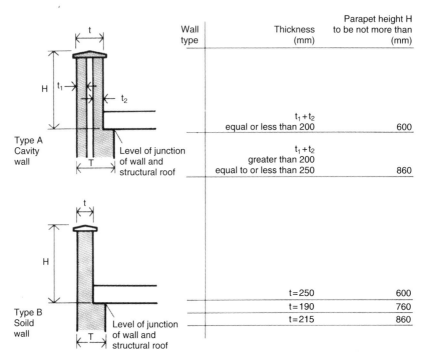

Wall type	Thickness (mm)	Parapet height H to be not more than (mm)
Type A Cavity wall	$t_1 + t_2$ equal or less than 200	600
	$t_1 + t_2$ greater than 200 equal to or less than 250	860
Type B Soild wall	$t = 250$	600
	$t = 190$	760
	$t = 215$	860

Figure 6.65 Height of parapet walls
📖 t should be less than or equal to T

Lateral support by walls

A wall in each storey of a building should   A1/2 (1C33)
extend to the full height of that storey,
and have horizontal lateral supports
to restrict movement of the wall at right angles
to its plane.

Table 6.32 Lateral support by floors

Type of wall	Wall length	Roof lateral support required	Floor lateral support required
External, compartment separating, solid or cavity wall	Any length	By every roof forming a junction with the supporting wall	
	Greater than 3 m		By every floor forming a junction with the supporting wall
Internal load bearing wall (not being a compartment or separating wall)	Any length	At the top of each storey	At the top of each storey

Loading on walls

The maximum span for any floor supported by a wall is 6 m measured centre to centre of bearing (see Figure 6.66).	A1/2 (1C24)
Vertical loading caused by concrete floor slabs, precast concrete floors, and timber floors on the walls, should be distributed.	A1/2 (1C25a)
Differences in the level of ground or other solid construction between one side of the wall and the other should be less than four times the thickness of the wall (see Figure 6.67).	A1/2 (1C25b)
The combined dead and imposed load should not exceed 70 kN/m at base of wall (see Figure 6.67).	A1/2 (1C25c)
Walls should **not** be subjected to lateral load other than from wind, and that caused by the differences in the level of the ground or other solid construction between one side of the wall and the other.	A1/2 (1C25d)

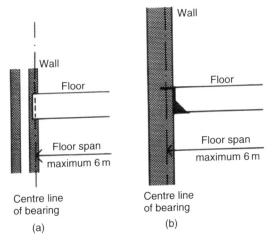

Figure 6.66 Maximum floor span. (a) Floor member bearing on wall. (b) Floor member bearing on joist hanger

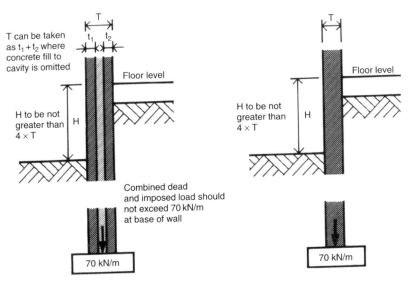

Figure 6.67 Differences in ground level

Lateral support by floors

Where contact between floors and walls on both sides of the wall is at intervals no greater than 2 m, floors should be at or about the same level on each side of the wall.   A1/2 (1C36)

Where lateral support is intermittent, the point   A1/2 (1C36)
of contact should be in line or nearly in line.

Walls should be strapped to floors above ground   A1/2 (1C36)
level, at intervals not exceeding 2 m by galvanized
mild steel or other durable metal straps which
have a minimum cross-section of 30 mm × 5 mm
as indicated in Figure 6.68.

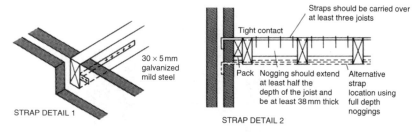

Figure 6.68 Lateral support by floors

Straps need not be provided in the longitudinal   A1/2 (1C36)
direction of joists in houses of not more than
two storeys:

1. If the joists are at not more than 1.2 m centres
    and have at least 90 mm bearing on the
    supported walls or 75 mm bearing on a timber
    wall-plate at each end.
2. If the joists are carried on the supported
    wall by joist hangers of the restraint type
    (described in BS 5628: Part 1 and shown
    in Figure 6.69 and are incorporated at not
    more than 2 m centres.
3. When a concrete floor has at least 90 mm   A1/2 (1C36)
    bearing on the supported wall (see Figure 6.70).
4. Where floors are at or about the same level on   A1/2 (1C36)
    each side of a supported wall, and contact
    between the floors and wall is either continuous
    or at intervals not exceeding 2 m.

 Where contact is intermittent, the points of
contact should be in line or nearly in line on plan.

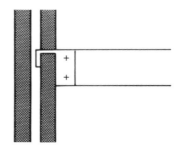

Figure 6.69 Restraint-type joist hanger

X to be not less than 90 mm

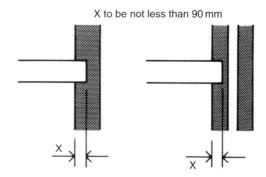

Figure 6.70 Restraint by concrete floor or roof

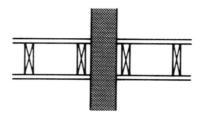

Figure 6.71 Restraint of internal walls

Wall ties

Wall ties should comply with BS 1243: 1978   A1/2 (1C20)
or be of other not less suitable type. In conditions
of severe exposure austenitic stainless steel
or suitable nonferrous ties should be used.

Brick and block construction

Walls should be properly bonded and solidly put
together with mortar and constructed of either:

A1/2 (1C21)

- clay bricks or blocks conforming to
  BS 3921: 1974 or BS 6649: 1985;
- calcium silicate bricks conforming to
  BS 187: 1978 or BS 6649: 1985;
- concrete bricks or blocks conforming to
  BS 6073: Part 1: 1981; or
- square dressed natural stone conforming to
  BS 5390: 1976 (1984).

Mortar

Mortar should be equivalent or of greater strength
(according to the masonry units and position of use)
to either the proportions given in BS 5628:
Part 1: 1978 for mortar designation or for Portland
cement – lime and fine aggregate measured by
volume of dry materials.

A1/2 (1C23)

End restraints

The ends of every wall (except single leaf walls
less than 2.5 m in storey height and length) in small
single storey non-residential buildings and annexes
should be bonded or otherwise securely tied
throughout their full height to a buttressing wall,
pier or chimney.

A1/2 (1C26)

Long walls may be provided with intermediate
support, dividing the wall into distinct lengths;
each distinct length is a supported wall for the
purposes of this section. The buttressing wall,
pier or chimney should provide support from
the base to the full height of the wall.

The buttressing wall should be bonded or securely
tied to the supported wall and at the other end to
a buttressing wall, pier or chimney (see Figure 6.72).

A1/2 (1C27)

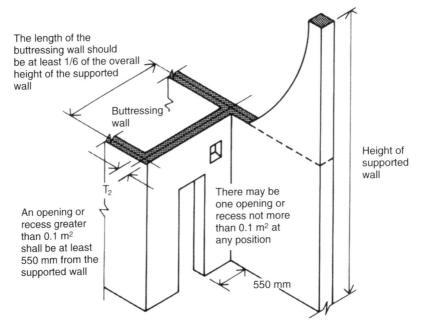

The length of the buttressing wall should be at least 1/6 of the overall height of the supported wall

Buttressing wall

T_2

An opening or recess greater than 0.1 m² shall be at least 550 mm from the supported wall

There may be one opening or recess not more than 0.1 m² at any position

550 mm

Height of supported wall

Figure 6.72 Openings in buttressing walls

If, the buttressed wall is not in itself a supported wall, then its thickness should not be less than:	A1/2 (1C27)

- 85 mm (i.e. half the thickness required for an external or separating wall of similar height and length, minus 5 mm);
- 75 mm if the wall forms part of a dwelling house (which does not exceed 6 m in total height and 10 m in length);
- 90 mm in any other cases.

Piers should measure at least three times the thickness of the supported wall and chimneys should be twice the thickness (measured at right angles to the wall).	A1/2 (1C28a)
Piers should have a minimum width of 190 mm (see Figure 6.73).	A1/2 (1C28a)
The sectional area on plan of chimneys (excluding openings for fireplaces and flues) should be not less than the area required for a pier in the same wall, and the overall thickness should not be less than twice the required thickness of the supported wall (see Figure 6.73).	A1/2 (1C28b)

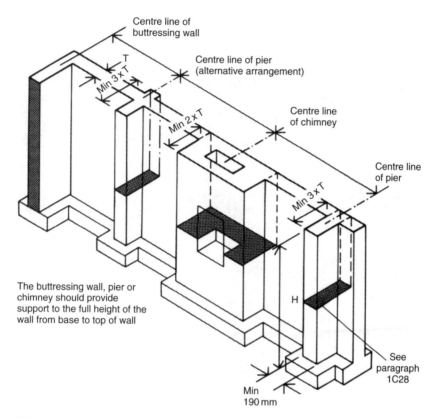

Figure 6.73 Buttressing

The buttressing wall, pier or chimney should provide support to the full height of the wall from base to top of the wall.	A1/2 (1C28)
The length of the buttressing wall should be at least 1/6 of the overall height of the supported wall.	A1/2 (1C28)
Openings or recesses in the buttressing wall should not impair the lateral support to be given by the buttressing wall.	A1/2 (1C27)
An opening or recess greater than 0.1 m² shall be at least 550 mm from the supported wall.	A1/2 (1C27)
Construction over openings and recesses should be adequately supported.	A1/2 (1C27)
There may be one opening or recess not more than 0.1 m² at any position.	A1/2 (1C27)
The buttressing wall, pier or chimney should provide support to the full height of the wall from base to top of the wall.	A1/2 (1C28)

Openings and recesses

The number, size and position of openings
and recesses should not impair the stability
of a wall or the lateral support afforded by
a buttressing wall to a supported wall.
Construction over openings and recesses
should be adequately supported.

A1/2 (1C29)

The size of openings and recesses is as
shown below with reference to Figure 6.74.
Requirements (using the value of factor x)
taken from Table 6.33.

1. $W_1 + W_2 + W_3$ should not exceed 3 m
2. W_1, W_2 or W_3 should not exceed 3 m
3. P_1 should be greater than or equal to W_1/x
4. P_2 should be greater than or equal to $W_1 + W_2/x$
5. P_3 should be greater than or equal to $W_2 + W_3/x$
6. P_4 should be greater than or equal to W_3/x
7. P_5 should be greater than or equal to W_4/x,
    but should not be less than 385 mm.

Table 6.33 Values of factor 'x'

Type of span	Maximum roof span	Minimum thickness of wall inner leaf	Span of floor that is parallel to the wall	Span of timber floor into wall		Span of concrete floor into wall	
				4.5 m max	6.0 m max	4.5 m max	6.0 m max
			Value of factor 'x'				
Roof spans parallel to the wall	Not applicable	100 m 90 m	6 6	6 6	6 6	6 6	6 5
Timber roof spans into the wall	9	100 m 90 m	6 6	6 4	5 4	4 3	3 3

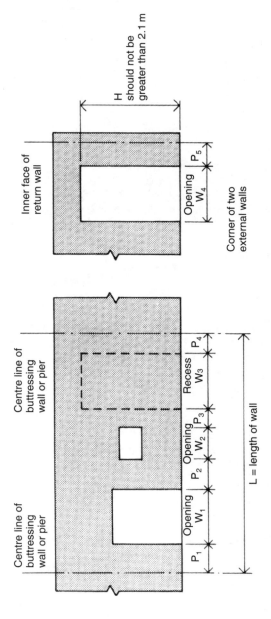

Figure 6.74 Size of openings and recesses

Chases

Vertical chases should not be deeper than 1/3 of the wall thickness or, in cavity walls, 1/3 of the thickness of the leaf.	A1/2 (1C31)
Horizontal chases should not be deeper than 1/6 of the thickness of the leaf or wall.	A1/2 (1C31)
Chases should not be so positioned as to impair the stability of the wall, particularly where hollow blocks are used.	A1/2 (1C31)

Overhangs

The amount of any projection should not impair the stability of the wall.	A1/2 (1C32)

Foundations

A wall shall be erected to prevent undue moisture from the ground reaching the inside of the building, and (if it is an outside wall) adequately resisting the penetration of rain and snow to the inside of the building (see Figure 6.75).	C3

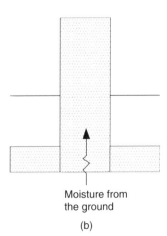

Figure 6.75 Wall, resistance to moisture. (a) External wall. (b) Internal wall

Resistance to the passage of moisture

All walls next to the ground (other than those of buildings used solely for storage) should:

• not be damaged by moisture from the ground, rain or snow and shall not carry that moisture to any part of the building that it would damage;	C3

☀ This damage can be avoided either by preventing moisture from getting to materials which would be damaged or by using materials that will not be damaged.

• resist the passage of ground moisture to the inside of the building (or any part which would be damaged by it);	C4 (4.2)
• not be damaged by moisture from the ground;	C4 (4.2, 4.8)
• resist the penetration of rain and snow to the inside of the building (or any part which would be damaged by it);	C4 (4.2, 4.7)
• not be damaged by rain and snow.	C4 (4.2, 4.7)
A damp-proof course (of bituminous material, engineering bricks or slates in cement, mortar or any other material that will prevent the passage of moisture) shall be continuous with any damp-proof membrane in the floors.	C4 (4.4, 4.8)
If the wall is an external wall, then the damp-proof course should be at least 150 mm above the level of the adjoining ground.	C4 (4.4, 4.8)
If the wall is a cavity wall then the cavity should be at least 150 mm below the level of the lowest damp-proof course.	C4 (4.4, 4.8, 4.11–4.13)
• All exposed walls shall be protected with a damp-proof course and a coping (see Figure 6.77).	C4 (4.8)
• All cladding (used to protect the building from rain or snow) shall be jointless or have sealed joints.	C4 (5.1–5.6)

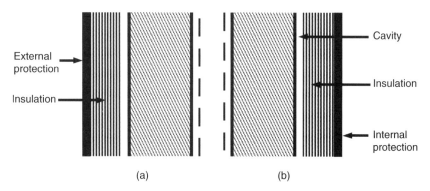

Figure 6.76 Insulated walls. (a) External insulation. (b) Internal insulation

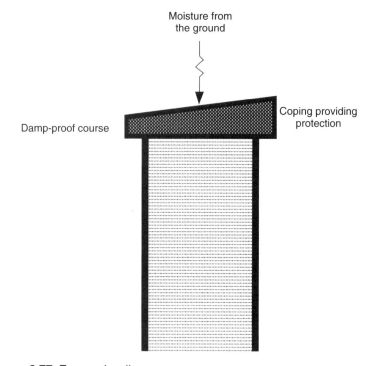

Figure 6.77 Exposed wall

Internal fire spread (linings)

The choice of materials for walls and ceilings can significantly affect the spread of a fire and its rate of growth, even though they are not likely to be the materials first ignited. Although furniture and fittings can have a major effect on fire spread it is not possible to control them through Building Regulations.

The surface linings of walls should meet the classifications shown in Table 6.34.

Table 6.34 Classification of linings

Location	Class*
Small rooms with an area of not more than 4 m² (in residential accommodation) or 30 m² (in non-residential accommodation)	3
Domestic garages not more than 40 m²	3
Other rooms (including garages)	1
Circulation spaces within buildings	1
Other circulation spaces (including the common area of flats and maisonettes)	0

*Classifications are based on tests as per BS 476 and as described in Appendix A of Approved Document B.

Air supported structures should comply with the recommendations given in BS 6661.	B2 (7.8)
Any flexible membrane covering a structure (other than an air supported structure) should comply with the recommendations given in Appendix A of BS 7157.	B2 (7.9)
The wall and any floor between the garage and the house shall have a 30 minute standard of fire resistance.	B2

Loadbearing elements of structure

All loadbearing elements of a structure shall have a minimum standard of fire resistance.	B3 (8.1)
Structural frames, beams, columns, loadbearing walls (internal and external), floor structures and gallery structures, should have at least the fire resistance given in Appendix A of Approved Document B.	B3 (8.2)

Compartmentation

To prevent the spread of fire within a building, whenever possible, the building should be sub-divided into compartments separated from one another by walls and/or floors of fire-resisting construction.	B3 (9.1)
A wall common to two or more buildings should be constructed as a compartment wall.	B3 (9.10)
Parts of a building that are occupied mainly for different purposes, should be separated from one another by compartment walls and/or compartment floors.	B3 (9.11)
Structural frames, beams, columns, loadbearing walls (internal and external), floor structures and gallery structures, should have at least the fire resistance given in Appendix A of Approved Document B.	B3 (8.2)
When altering an existing two-storey single-family dwellinghouse to provide additional storeys, the floor(s), both old and new, shall have the full 30 minute standard of fire resistance.	B3 (8.7)
If an existing house or other building is converted, the means of escape shall be adequately protected and there shall be a 30 minute fire resistance standard.	B3 (8.10–8.11)
There should be continuity at the junctions of the fire resisting elements enclosing a compartment.	B3 (9.6)
Spaces that connect compartments, such as stairways and service shafts, need to be protected to restrict fire spread between the compartments.	B3 (9.7)
Every place that is a potential fire hazard should be enclosed with fire-resisting construction.	B3 (9.12)
Every wall separating semi-detached houses, or houses in terraces, should be constructed as a compartment wall, and the houses should be considered as separate buildings.	B3 (9.13)
If a domestic garage is attached to (or forms an integral part of) a house, the garage should be separated from the rest of the house, as shown in Figure 6.78.	B3 (9.14)

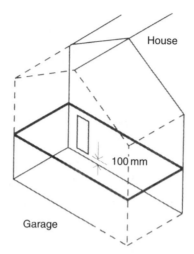

House

100 mm

Garage

Figure 6.78 Separation between garage and dwellinghouse

The wall and any floor between the garage and the house shall have a 30 minute standard of fire resistance. Any opening in the wall should be at least 100 mm above the garage floor level with an FD30 door.

In buildings containing flats or maisonettes, compartment walls or compartment floors shall be constructed between:	B3 (9.15)
<ul><li>every floor (unless it is within a maisonette);</li><li>one storey and another within one dwelling;</li><li>every wall separating a flat or maisonette from any other part of the building;</li><li>every wall enclosing a refuse storage chamber.</li></ul>	
Every compartment wall and compartment floor should:	B3 (9.22)
<ul><li>form a complete barrier to fire between the compartments they separate; and</li><li>have the appropriate fire resistance as indicated in Appendix A, Tables A6.1 and A6.2.</li></ul>	
Timber beams, joists, purlins and rafters may be built into or carried through a masonry or concrete compartment wall if the openings for them are kept as small as practicable and then fire-stopped.	B3 (9.22)
If trussed rafters bridge the wall, they should be designed so that failure of any part of the truss due to a fire in one compartment will not cause failure of any part of the truss in another compartment.	B3 (9.22)

Compartment walls that are common to two or more buildings should run the full height of the building in a continuous vertical plane so that the adjoining buildings are separated by walls, not floors.	B3 (9.23)
Compartment walls (used to form a separated part of a building) should run the full height of the building in a continuous vertical plane.	B3 (9.24)
Compartment walls in a top storey beneath a roof should be continued through the roof space.	B3 (9.26)
Where a compartment wall or compartment floor meets another compartment wall, or an external wall, the junction should maintain the fire resistance of the compartmentation.	B3 (9.27)
When a compartment wall meets the underside of the roof covering or deck, the wall/roof junction shall maintain continuity of fire resistance.	B3 (9.28)
Double skinned insulated roof sheeting should incorporate a band of material of limited combustibility.	B3 (9.29)
Any openings in a compartment wall which is common to two or more buildings should be provided with an escape door in case of fire.	B3 (9.33)
Openings in compartment walls should have the appropriate fire resistance and be limited to those for:	B3 (9.35)

- doors that have the appropriate fire resistance;
- the passage of pipes, ventilation ducts, chimneys, appliance ventilation ducts or ducts encasing one or more flue pipes;
- refuse chutes of non-combustible construction;
- protected shafts.

Any stairway or other shaft passing directly from one compartment to another should be enclosed in a protected shaft so as to delay or prevent the spread of fire between compartments.	B3 (9.36)

Protection of openings for pipes

Pipes that pass through a compartment wall or compartment floor (unless the pipe is in a protected shaft), or through a cavity barrier, should conform to one of the alternatives shown in Figure 6.79.

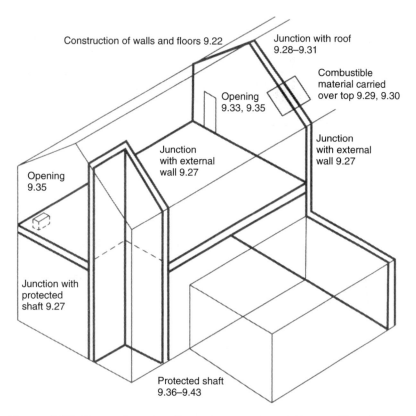

Construction of walls and floors 9.22

Junction with roof 9.28–9.31

Combustible material carried over top 9.29, 9.30

Opening 9.33, 9.35

Junction with external wall 9.27

Junction with external wall 9.27

Opening 9.35

Junction with protected shaft 9.27

Protected shaft 9.36–9.43

Figure 6.79 Compartment walls and compartment floors with reference to the relevant paragraphs in Approved Document B

Flue walls

Flue walls should have a fire resistance of at least one half of that required for the compartment wall or floor and be of non-combustible construction.

B3 (11.11)

If a flue, or duct containing flues or appliance ventilation duct(s), passes through a compartment wall or compartment floor, or is built into a compartment wall, each wall of the flue or duct should have a fire resistance of at least half that of the wall or floor in order to prevent the by-passing of the compartmentation (see Figure 6.80).

B3 (11.11)

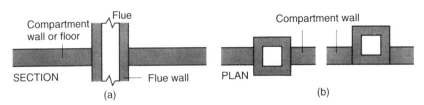

Figure 6.80 Flues penetrating compartment walls or floors. (a) Flue passing through compartment wall or floor. (b) Flue built into compartment wall

Fire resistance

Proprietary fire-stopping and sealing systems (including those designed for service penetrations) which have been shown by test to maintain the fire resistance of the wall or other element, are available and may be used. Other fire-stopping materials include:

- cement mortar,
- gypsum-based plaster,
- cement or gypsum-based vermiculite/perlite mixes,
- glass fibre, crushed rock, blast furnace slag or ceramic-based products (with or without resin binders), and
- intumescent mastics (B3 11.14).

Joints between fire separating elements should be fire-stopped.	B3 (11.12a)
All openings for pipes, ducts, conduits or cables to pass through any part of a fire separating element should be: • kept as few in number as possible; • kept as small as practicable; • fire-stopped (which in the case of a pipe or duct, should allow thermal movement).	B3 (11.12b)
To prevent displacement, materials used for fire-stopping should be reinforced with (or supported by) materials of limited combustibility.	B3 (11.13)

Construction of an external wall

Where a portal framed building is near a relevant boundary, the external wall near the boundary may need fire resistance to restrict the spread of fire between buildings.   B4 (13.4)

In cases where the external wall of the building cannot be wholly unprotected, the rafter members of the frame, as well as the column members, may need to be fire protected.   B4 (13.4)

The external surfaces of walls should meet the provisions shown in Figure 6.81.   B4 (13.5)

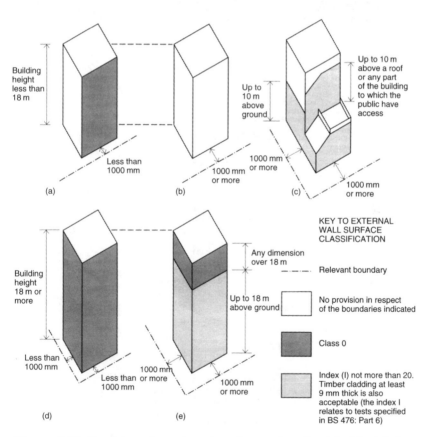

Figure 6.81 Provisions for external surfaces of walls. (a), (d), (e) Any building. (b) Any building other than (c). (c) Assembly or recreation building of more than one storey

It should be noted that the use of combustible materials for cladding framework, or the use of combustible thermal insulation as an overcladding may be risky in tall buildings, even though the provisions for external surfaces in Figure 6.81 may have been satisfied.

The external envelope of a building should not provide   B4 (13.7)
a medium for fire spread if it is likely to be a risk
to health or safety.

In a building with a storey 18 m or more above   B4 (13.7)
ground level, insulation material used in ventilated
cavities in the external wall construction
should be of limited combustibility
(this restriction does not apply to masonry
cavity wall construction).

Masonry wall construction

SECTION THROUGH CAVITY WALL

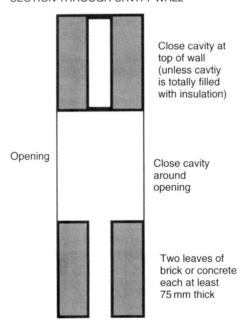

Close cavity at top of wall (unless cavtiy is totally filled with insulation)

Opening

Close cavity around opening

Two leaves of brick or concrete each at least 75 mm thick

Figure 6.82 Masonry cavity walls excluded from the provisions for cavity barriers

Combustible material should not be placed in or exposed to the cavity, except for:

- timber lintels, window or door frames, or the end stairway of timber joists;
- pipes, conduits or cables;
- DPC, flashing, cavity closer or wall ties;
- fire-resisting thermal insulating material;
- a domestic meter cupboard.

Cavity insulation

The outer leaf of the wall should be built of masonry or concrete.	D1 (1.1–1.2)
The inner leaf of the wall should be built of masonry (i.e. bricks or blocks).	D1 (1.1–1.2)
The wall being insulated with UF shall be assessed (in accordance with BS 8208) for suitability before any work commences.	D1 (1.1–1.2)
The person carrying out the work needs to hold (or operate under) a current BSI Certificate of Registration of Assessed Capability for the work he is doing.	D1 (1.1–1.2)
The installation shall be in accordance with BS 5618: 1985.	D1 (1.1–1.2)
The material shall be in accordance with the relevant recommendations of BS 5617: 1985.	D1 (1.1–1.2)

Airborne sound

Walls (see Figure 6.83) that separate: • a dwelling from another building • a dwelling from another dwelling • a habitable room or kitchen within a dwelling from another part of the same building which is not used exclusively as part of the dwelling shall resist the transmission of airborne sound.	E1
The flow of sound energy through walls and floors should be restricted.	E (0.6)

Walls should reduce the level of airborne sound.	E (0.7)
Floors should reduce airborne sound and also, if they are above a dwelling, impact sound.	E (0.9)
Air paths, including those due to shrinkage, must be avoided – porous materials and gaps at joints in the structure must be sealed.	E (0.10)
The possibility of resonance in parts of the structure (such as a dry lining) should be avoided.	E (0.10)
Flanking transmission (i.e. the indirect transmission of sound from one side of a wall or floor to the other side) should be minimized.	E (0.11–0.13)

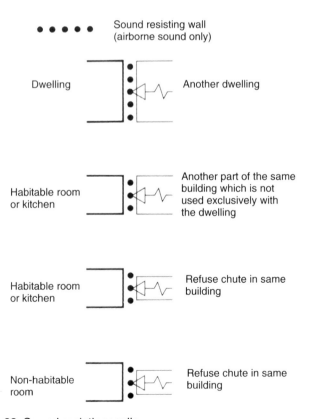

Figure 6.83 Sound resisting walls

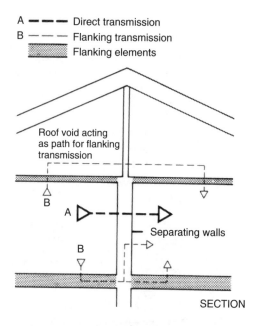

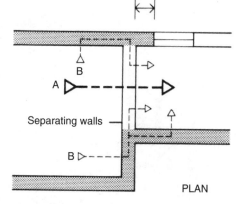

Figure 6.84 Direct and flanking transmission
For clarity not all flanking paths have been shown

Types of walls

As shown in Table 6.35, there are four types of separating wall.

The resistance to airborne sound depends mainly on the mass of the wall.

Table 6.35 Types of wall

Solid masonry (Wall type 1)	Type 1 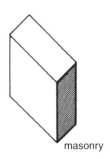 masonry	The resistance to airborne sound depends mainly on the mass of the wall.
Cavity masonry (Wall type 2)	Type 2 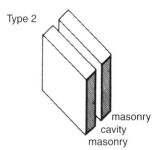 masonry cavity masonry	The resistance to airborne sound depends mainly on the mass of the leaves and the degree of isolation achieved.
Masonry between isolated panels (Wall type 3)	Type 3 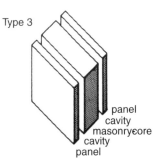 panel cavity masonry core cavity panel	The resistance to airborne sound depends partly on the mass and type of core and partly on the isolation and mass of the panels.
Timber frames with absorbent material (Wall type 4)	Type 4 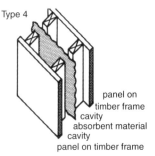 panel on timber frame cavity absorbent material cavity panel on timber frame	The resistance to airborne sound depends on the mass of leaves, isolation of the frames, plus absorption in the air space between.

Partitions

> There are no restrictions on partition walls meeting   E (p9)
> a Type 1 (masonry) separating wall.

Cavity masonry walls (Wall type 2)

With a cavity masonry wall, the resistance to airborne sound mainly depends on the mass of the leaves and the degree of isolation achieved. This can be achieved by:

> Filing all masonry joints with mortar and laying bricks   E (p10)
> frog upwards.
>
> Where plaster or plasterboard is specified, using a wall   E (p10)
> lining laminate of plasterboard and mineral wool.
>
> Maintaining the cavity up to the underside of the roof.   E (p10)
>
> Connecting the leaves only where necessary by butterfly   E (p10)
> pattern ties (spaced as per BS 5628).
>
> If external walls are to be filled with an insulating   E (p10)
> material other than unbonded particles or fibres,
> ensuring that the insulating material is prevented
> from entering the cavity in the separating wall
> by a flexible closer (e.g. mineral wool).

There are three types of wall constructions that provide a reasonable resistance to direct transmission (see A, B and C in Table 6.37).

Two other wall constructions (D and E in Table 6.38) are suitable between dwellings provided that a step in elevation and/or a stagger in the plan is incorporated at the separating wall.

Table 6.37 Construction of cavity walls (Wall type 2)

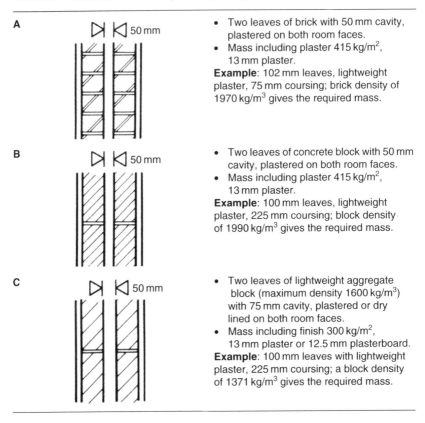

A   ▷◁ 50 mm
- Two leaves of brick with 50 mm cavity, plastered on both room faces.
- Mass including plaster 415 kg/m², 13 mm plaster.

Example: 102 mm leaves, lightweight plaster, 75 mm coursing; brick density of 1970 kg/m³ gives the required mass.

B   ▷◁ 50 mm
- Two leaves of concrete block with 50 mm cavity, plastered on both room faces.
- Mass including plaster 415 kg/m², 13 mm plaster.

Example: 100 mm leaves, lightweight plaster, 225 mm coursing; block density of 1990 kg/m³ gives the required mass.

C   ▷◁ 50 mm
- Two leaves of lightweight aggregate block (maximum density 1600 kg/m³) with 75 mm cavity, plastered or dry lined on both room faces.
- Mass including finish 300 kg/m², 13 mm plaster or 12.5 mm plasterboard.

Example: 100 mm leaves with lightweight plaster, 225 mm coursing; a block density of 1371 kg/m³ gives the required mass.

Table 6.38 Construction of cavity walls where a step and/or stagger of at least 300 mm is used

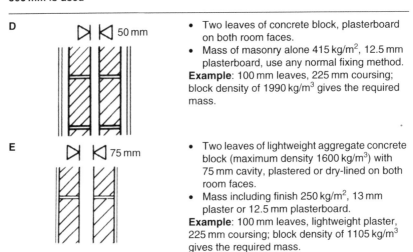

D   ▷◁ 50 mm
- Two leaves of concrete block, plasterboard on both room faces.
- Mass of masonry alone 415 kg/m², 12.5 mm plasterboard, use any normal fixing method.

Example: 100 mm leaves, 225 mm coursing; block density of 1990 kg/m³ gives the required mass.

E   ▷◁ 75 mm
- Two leaves of lightweight aggregate concrete block (maximum density 1600 kg/m³) with 75 mm cavity, plastered or dry-lined on both room faces.
- Mass including finish 250 kg/m², 13 mm plaster or 12.5 mm plasterboard.

Example: 100 mm leaves, lightweight plaster, 225 mm coursing; block density of 1105 kg/m³ gives the required mass.

Ceiling and roof space junctions

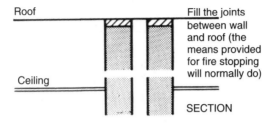

Figure 6.88 Roof junctions – cavity masonry (Wall type 2)

Heavy ceilings with sealed joints (e.g. 2.5 mm plasterboard) may reduce the mass of the wall above the ceiling to 150 kg/m². The cavity should still be maintained.	E (p11)
If lightweight aggregate blocks of density less than 1200 kg/m³ are used, seal one side with cement paint or plaster skim.	E (p11)

Floor junctions – intermediate and ground floors

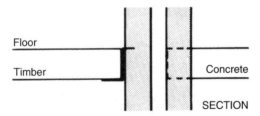

Figure 6.89 Floor junctions (Wall type 2)

Floor joists may be supported on hangers or built in (care should be taken to ensure that there are no airpaths through the wall).	E (p11)
A suspended concrete intermediate or ground floor should be carried through to the cavity face of each leaf.	E (p11)
A concrete slab on the ground may be continuous.	E (p11)

External wall junctions

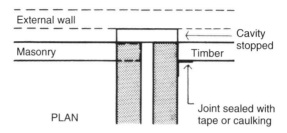

Figure 6.90 External wall junctions (Wall type 2)

The outer leaf of a cavity wall may be of any construction.	E (p11)
Where a cavity wall has an inner leaf of masonry:	E (p11)

- the masonry of the walls should be bonded together; or
- the masonry of the external wall should abut the separating wall and be tied to it with ties at no more than 300 mm centres vertically, to create a homogeneous unit; and
- the masonry should have a mass of 120 kg/m^2 except where separating wall type B is used when there is no minimum required mass.

Where a cavity wall has an inner leaf of timber construction it should:	E (p11)

- abut the separating wall;
- be tied to it with ties at no more than 300 mm centres vertically; and
- have the joints sealed with tape or caulking. Where the external wall has a cavity, the cavity should be stopped with a flexible closer.

The cavity in the separating wall should not be stopped by any material that connects the leaves together rigidly. Mineral wool is acceptable.	E (p11)

Partitions

There are no restrictions on partition walls meeting a type 2 (cavity masonry) separating wall.	E (p9)

Masonry walls between isolated panels (Wall type 3)

With a Type 3 wall, resistance to airborne sound depends partly on the mass and type of core and partly on the isolation and mass of the panels.

This is achieved by:

Filling masonry joints and sealing up all joints with mortar.	E (p12)
Avoiding air paths and laying bricks frog upwards.	E (p12)
Not fixing or tying panels to the masonry core but supporting them from the floor and ceiling so as to maintain isolation.	E (p12)

There are four masonry cores (A, B, C and D) and two panels (E and F), which, in any combination with core plus panels, should give reasonable resistance to direct transmission.

Table 6.39 Types of masonry cores available for Wall type 3

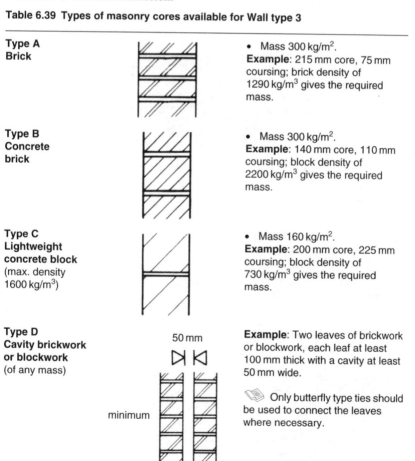

Type A Brick		• Mass 300 kg/m². **Example**: 215 mm core, 75 mm coursing; brick density of 1290 kg/m³ gives the required mass.
Type B Concrete brick		• Mass 300 kg/m². **Example**: 140 mm core, 110 mm coursing; block density of 2200 kg/m³ gives the required mass.
Type C Lightweight concrete block (max. density 1600 kg/m³)		• Mass 160 kg/m². **Example**: 200 mm core, 225 mm coursing; block density of 730 kg/m³ gives the required mass.
Type D Cavity brickwork or blockwork (of any mass)	50 mm minimum	**Example**: Two leaves of brickwork or blockwork, each leaf at least 100 mm thick with a cavity at least 50 mm wide. Only butterfly type ties should be used to connect the leaves where necessary.

Table 6.40 Types of panels used with Wall type 3

Type E Two sheets of plasterboard joined by a cellular core	• Mass including plaster finish if used 18 kg/m².
	Fix to ceiling and floor only and tape joints between panels.
Type F Two sheets of plasterboard with joints staggered	• Thickness of each sheet 12.5 mm if a supporting framework is used. • Total thickness of at least 30 mm if no framework is used.

Ceiling and roof space junctions

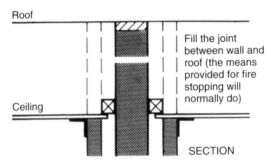

Figure 6.91 Roof junctions – masonry walls (Wall type 3)

Heavy ceilings should have with sealed joints (12.5 mm plasterboard or equivalent).	E (p13)
The freestanding panels may be omitted in the roof space and the mass of the core above the ceiling may be reduced to 150 kg/m² (**but** if core Type D is used the cavity should be maintained).	E (p13)
If open textured lightweight aggregate blocks less than 1200 kg/m³ are used to reduce mass, seal one side with cement paint or plaster skim.	E (p13)
The junction between ceiling and freestanding panels should be sealed with tape or caulking.	E (p13)

Type 3 floor junctions – intermediate and ground floors

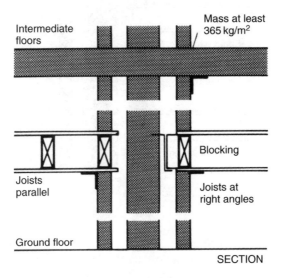

Figure 6.92 Floor junctions (Wall type 3)

Timber

With a timber intermediate floor:

• use joist hangers for any joists supported on the wall.	E (p13)
• seal the spaces between joists with full depth timber blocking.	E (p13)

External wall junctions

Joists

The outer leaf of a cavity wall may be of any construction.	E (p13)
With core Type C the inner leaf of a cavity external wall should have an internal finish of isolated panels.	E (p13)
Where the separating wall has core A, B or D, plaster or dry-lining with joints sealed with tape or caulking may be used.	E (p13)

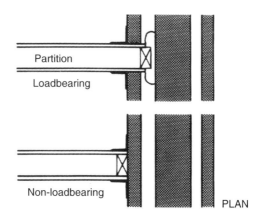

Figure 6.93 External wall junctions (Wall type 3)

A layer of insulation may be added to a dry-lining internal finish provided the 25 mm and 110 mm gaps shown in Figure 6.93 are maintained.	E (p13)
The inner leaf may be of any construction if it is lined with isolated panels.	E (p13)
If the inner leaf is plastered or dry-lined it should have a total mass of 120 kg/m^2 and be butt-jointed to the separating wall core with ties at no more than 300 mm centres, vertically.	E (p13)

Partitions

There are no restrictions on partition walls meeting a Type 3 separating wall.	E (p13)
Other loadbearing partitions should be fixed to the masonry core through a continuous pad of mineral wool.	E (p13)
Non-loadbearing partitions should be tight-butted to the isolated panels.	E (p13)
All joints between partitions and panels should be sealed with tape or caulking.	E (p13)

Timber framed walls (Wall type 4)

With a wall constructed with a timber frame and absorbent material, the resistance to airborne sound depends on the isolation of the frames plus absorption in the air space between.

With this type of construction:

Connect frames only if really necessary for structural reasons (using as few ties as possible not more than 14–15 gauge (40 mm × 3 mm) metal straps fixed at or just below ceiling level 1.2 m apart).	E (p14)
Power points may be set in the linings provided there is a similar thickness of cladding behind the socket box.	E (p14)
Power points should not be placed back to back across the wall.	E (p14)
Where fire-stops are needed in the cavity between frames they should either be flexible or fixed to only one frame.	E (p14)

There are two types of approved construction (A and B), which, with the appropriate lining (C) and absorbent material (D) will provide the resistance to direct transmission as required.

Table 6.41 Types of timber frames available for Wall type 4

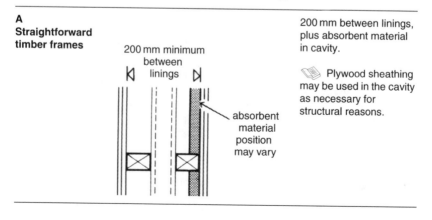

A
Straightforward timber frames

200 mm minimum between linings

absorbent material position may vary

200 mm between linings, plus absorbent material in cavity.

Plywood sheathing may be used in the cavity as necessary for structural reasons.

Table 6.41 Types of timber frames available for Wall type 4 (*Continued*)

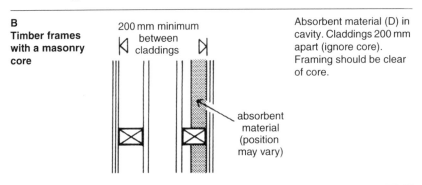

| **B**
Timber frames
with a masonry
core | 200 mm minimum between claddings | Absorbent material (D) in cavity. Claddings 200 mm apart (ignore core). Framing should be clear of core. |

Note: A masonry core does not normally improve sound resistance but may be useful for support and in stepped or staggered situations. There are no restrictions on type but the core should be connected to only one frame.

Table 6.42 Lining

| **C**
Lining on each side | Two or more layers of plasterboard with a combined thickness of 30 mm.
Joints staggered to avoid air paths. |

Table 6.43 Absorbent material

| **D**
Unfaced mineral fibre
batts or quilt (which
may be wire reinforced) | • Density at least 10 kg/m², thickness 25 mm if suspended in the cavity between frames, 50 mm if fixed to one frame, or 25 mm per quilt if one fixed to each frame. |

Timber framed walls – roof junctions

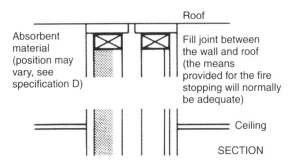

Figure 6.93(a) Roof junctions (Wall type 4)

There should be a fire-stop between the junction of the wall and roof (see Approved Document B, Fire safety).   E (p14)

Between the ceiling level and the underside of the roof either: E (p14)

- carry both frames through – the cladding on each frame may be reduced to not less than 25 mm; or
- close the cavity at ceiling level without connecting the two frames rigidly together and use one frame with at least 25 mm cladding on both sides.

In each case seal the space between frame and roof finish.

Floor junctions – intermediate and ground floors

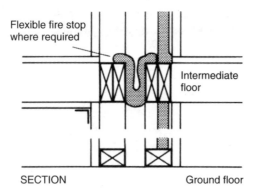

Figure 6.94 Intermediate and ground floor junctions (Wall type 4)

Block all air paths to the wall cavity either by carrying the cladding through the floor or by using a solid timber edge to the floor.   E (p15)

Where the joists are at right angles to the wall, seal spaces between joists with full depth timber blocking.   E (p15)

External wall junctions

There are no restrictions on a traditional timber-framed wall but if the wall is of cavity construction, then:

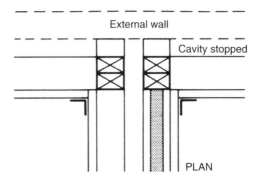

Figure 6.95 External wall junctions (Wall type 4)

The cavity should be sealed between the ends of the separating wall and the outer leaf to prevent sound paths.	E (p15)
The internal finish should be 12.5 mm plasterboard or other equally heavy material and resilient layers for thermal insulation may be incorporated (if desired).	E (p15)

Partitions

There are no restrictions on partition walls meeting a Type 4 (timber-framed) separating wall.	E (p159)

Junctions for concrete-based (Type 1) floors

External walls or cavity separating walls

The mass of the wall leaf adjoining the floor should be at least 120 kg/m² (including any finish), unless it is an external wall having openings of at least 20% of its area in each room, in which case there is no minimum requirement.	E (p19)
The mass of the wall leaf adjoining the floor should be at least 120 kg/m² (including any finish), unless it is an external wall having openings of at least 20% of its area in each room, in which case there is no minimum requirement.	E (p19)
The floor base (excluding any screed) should pass through the leaf whether spanning parallel to or at right angles to the wall.	E (p19)

The cavity should not be bridged.   E (p19)

In Type C or D floor bases, where the beams are parallel   E (p19)
to the wall, the first joint should be 300 mm from the
cavity face of the wall leaf.

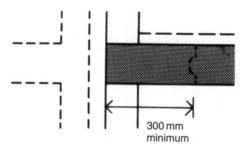

300 mm
minimum

Figure 6.96 External walls or cavity separating walls (Floor type 1)

Internal wall or solid separating wall

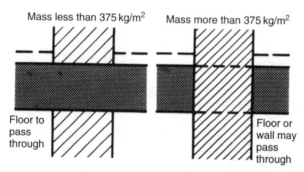

Mass less than 375 kg/m²   Mass more than 375 kg/m²

Floor to
pass
through

Floor or
wall may
pass
through

Figure 6.97 Internal walls or solid separating walls (Floor type 1)

If the wall mass is less than 375 kg/m² (including any   E (p19)
plaster or plasterboard) then the floor base excluding any
screed should pass through.

If the wall mass is more than 375 kg/m² (including any   E (p19)
plaster or plasterboard) either the wall excluding any finishes
or the floor base excluding any screed may pass through.

Where the wall does pass through, tie the floor base to the   E (p19)
wall and grout the joint.

Floor penetrations (excluding gas pipes)

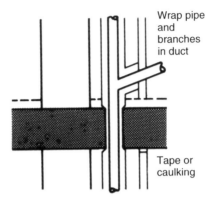

Figure 6.98 External wall junctions (Floor type 1)

Ducts or pipes which penetrate a floor separating habitable rooms should be in an enclosure (both above and below the floor). In all cases, the:

• enclosure material should have a mass of 15 kg/m²;	E (p19)
• enclosure should either be lined or the duct (or pipe) wrapped in 25 mm unfaced mineral wool.	E (p19)
Penetrations through a separating floor by ducts and pipes should have fire protection in accordance with Approved Document B, Fire safety.	E (p19)
Fire stopping should be flexible and prevent rigid contact betweenthe pipe and floor.	E (p19)

In the Gas Safety Regulations 1972 (S1, 197211178) and the Gas Safety (installation and use) Regulations 1984 (S1, 198411358) there are requirements for ventilation of ducts at each floor where they contain gas pipes. Gas pipes may be contained in a separate ventilated duct or they can remain unducted.

Junctions for concrete based (Type 2) floor junctions

External walls or cavity separating walls

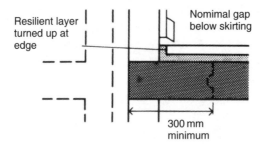

Figure 6.99 External walls or cavity separating walls (Floor type 2)

The mass of the wall leaf adjoining the floor should be at least 120 kg/m² (including any finish), unless it is an external wall having openings of at least 20% of its area in each room, in which case there is no minimum requirement.	E (p21)
The floor base (excluding any screed) should pass through the leaf whether spanning parallel to or at right angles to the wall.	E (p21)
The cavity should not be bridged.	E (p21)
If the floor base is Type C or D, where the beams are parallel to the wall the first joint should be 300 mm from the cavity face of the wall leaf.	E (p21)

Internal wall or solid separating wall

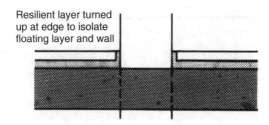

Figure 6.100 Internal walls or solid separating walls (Floor type 2)

If the wall mass is less than 375 kg/m² (including any plaster   E (p21)
or plasterboard) then the floor base excluding any screed
should pass through.

If the wall mass is more than 375 kg/m² (including any   E (p21)
plaster or plasterboard) either the wall excluding any finishes
or the floor base excluding any screed may pass through.

Where the wall does pass through, tie the floor base to the   E (p21)
wall and grout the joint.

Floor penetrations (excluding gas pipes)

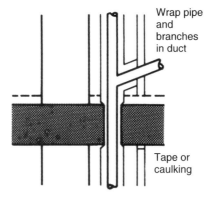

Figure 6.101 External wall junctions (Floor type 2)

Ducts or pipes which penetrate a floor separating habitable rooms should be in
an enclosure (both above and below the floor). In all cases, the:

- enclosure material should have a mass of 15 kg/m²;   E (p21)
- enclosure should either be lined or the duct (or pipe)   E (p21)
  wrapped in 25 mm unfaced mineral wool.

Penetrations through a separating floor by ducts and pipes   E (p21)
should have fire protection in accordance with Approved
Document B, Fire safety.

Fire stopping should be flexible and prevent rigid contact   E (p21)
between the pipe and floor.

In the Gas Safety Regulations 1972 (S1, 197211178) and the Gas Safety
(installation and use) Regulations 1984 (S1, 198411358) there are requirements
for ventilation of ducts at each floor where they contain gas pipes. Gas pipes
may be contained in a separate ventilated duct or they can remain unducted.

Junctions for timber-based (Type 3) floors

Timber framed walls

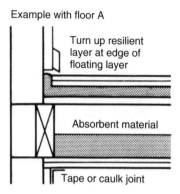

Example with floor A

Turn up resilient layer at edge of floating layer

Absorbent material

Tape or caulk joint

Figure 6.102 Timber framed walls (Floor type 3)

Seal the gap between wall and floating layer with a resilient strip glued to the wall.	E (p23)
Leave a 3 mm gap between skirting and floating layer (a seal is not necessary but if used it should be flexible).	E (p23)
Block air paths between the floor base and the wall, including the space between joists.	E (p23)
When joists are at right angles to the wall seal the junction of ceiling and wall lining with tape or caulking.	E (p23)

Heavy masonry leaf

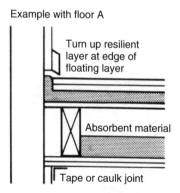

Example with floor A

Turn up resilient layer at edge of floating layer

Absorbent material

Tape or caulk joint

Figure 6.103 Heavy masonry leaf – Floor type 3

Building Regulations in Brief

Mass of leaf (including any finish) 375 kg/m², both above and below floor.	E (p23)
Seal the gap between wall and floating layer with a resilient strip.	E (p23)
Leave a 3 mm gap between skirting and floating layer (a seal is not necessary but if used it should be flexible).	E (p23)
Use any normal method of connecting floor base to wall.	E (p23)
Seal the junction of ceiling and wall lining with tape or caulking.	E (p23)

Light masonry leaf

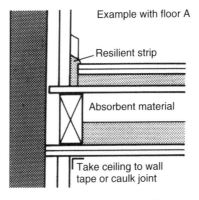

Figure 6.104 Light masonry leaf (Floor type 3)

If the mass including any plaster or plasterboard is less than 375 kg/m² a free-standing panel as specified in Wall type 3 should be used.	E (p23)
Seal the gap between panel and floating layer with a resilient strip.	E (p23)
Leave a 3 mm gap between skirting and floating layer (a seal is not necessary but if used it should be flexible).	E (p23)
Use any normal method of connecting floor base to wall but block air paths between floor and wall cavities.	E (p23)
Take ceiling through to masonry, seal junction with freestanding panel with tape or caulking.	E (p23)

Floor penetrations

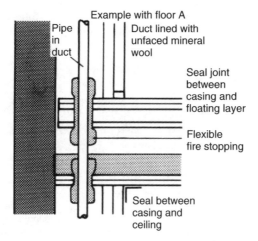

Figure 6.105 External wall junctions (Floor type 3)

Ducts or pipes which penetrate a floor separating habitable rooms should be in an enclosure (both above and below the floor). In all cases, the:

• enclosure material should have a mass of 15 kg/m^2.	
• enclosure should either be lined or the duct (or pipe) wrapped in 25 mm unfaced mineral wool.	E (p19)
Leave a 3 mm gap between enclosure and floating layer, seal with acrylic caulking or neoprene.	E (p24)
The enclosure may go down to the floor base if specification A is used, but in these cases, the enclosure needs to be isolated from the floating layer.	E (p24)
Penetrations through a separating floor by ducts and pipes should have fire protection in accordance with Approved Document B, Fire safety.	E (p24)
Fire stopping should be flexible and prevent rigid contact between the pipe and floor.	E (p24)

In the Gas Safety Regulations 1972 (S1, 197211178) and the Gas Safety (installation and use) Regulations 1984 (S1, 198411358) there are requirements for ventilation of ducts at each floor where they contain gas pipes. Gas pipes may be contained in a separate ventilated duct or they can remain unducted.

Remedial work and conversions

Where it cannot be shown that the existing construction (i.e. wall, floor or stair) meets the sound insulation requirements one of the following treatments shall be used in order to improve the level of sound insulation.

> As the implementation of upgrading measures will impose   E (5.3)
> additional loads on the existing structure, it (i.e. the existing
> structure) should be assessed to ensure that the additional
> loading can be carried safely with appropriate strengthening
> where necessary.

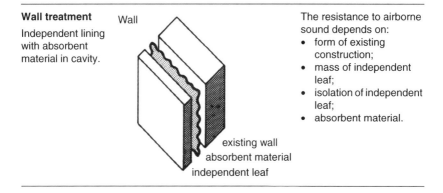

Wall treatment

Independent lining with absorbent material in cavity.

The resistance to airborne sound depends on:
- form of existing construction;
- mass of independent leaf;
- isolation of independent leaf;
- absorbent material.

Wall treatment (independent lining with absorbent material in cavity)

The construction may be used on one side of the existing wall only where the existing wall is masonry, has a thickness of at least 100 mm and is plastered on

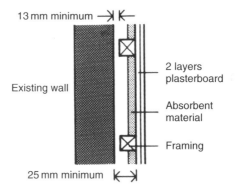

Figure 6.106 Non-contact with existing wall

both faces. With other types of existing wall the construction should be built on both sides.

The construction may be used on one side of the existing wall only where the existing wall is masonry, has a thickness of at least 100 mm and is plastered on both faces. With other types of existing wall the construction should be built on both sides.	E (p29)
Ensure that the independent leaf and its supporting frame are not in contact with the existing wall.	E (p29)
The absorbent material may bridge the cavity but should not be tightly compressed. A gap of 13 mm minimum should be left between the face of the wall and the framing.	E (p29)
The independent leaf should be two layers of plasterboard each at least 12.5 mm thick fixed to any type of framing or a plasterboard lining with a total thickness of 30 mm if no framework is used.	E (p29)

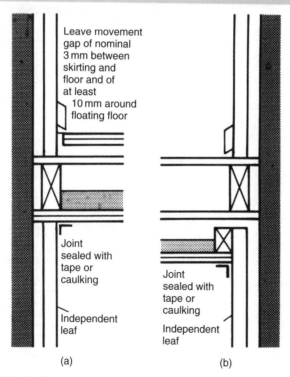

(a)   (b)

Figure 6.107 Wall treatment 1 – typical junction details. (a) With floating floor and existing or replacement ceiling. (b) With existing floor and independent ceiling

The plasterboard joints should be staggered, and a gap of   E (p29)
at least 25 mm left between the inside face of the plasterboard
and the face of the existing wall, and of at least 13 mm
between the framing and the face of the existing wall.

The perimeter of the independent leaf should be sealed   E (p29)
with tape or mastic.

The absorbent material should be mineral wool at least 25 mm   E (p29)
thick with a density of at least 10 kg/m³.

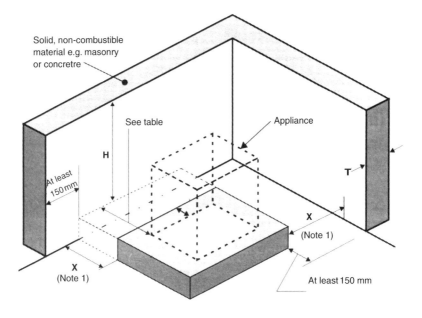

Location of hearth or appliance	Solid, non-combustibe material	
	Thickness (T)	Height (H)
Where the hearth abuts a wall and the appliance is not more than 50 mm from the wall	200 mm	At least 300 mm above the appliance and 1.2 m above the hearth
Where the hearth abuts a wall and the appliance is more than 50 mm but not more than 300 mm from the wall	75 mm	At least 300 mm above the appliance and 1.2 m above the hearth
Where the hearth does not abut a wall and is no more than 150 mm from the wall (see Note 1)	75 mm	At least 1.2 m above the hearth
Note 1: There is no requirement for protection of the wall where X is more than 150 mm		

Figure 6.108 Walls adjacent to hearths

Walls adjacent to hearths

> Walls that are not part of a fireplace recess or
> a prefabricated appliance chamber but are adjacent
> to hearths or appliances also need to protect the
> building from catching fire. A way of achieving
> the requirement is shown in Figure 6.108.

J (2.31)

6.8 Ceilings

6.8.1 The requirement

As a fire precaution, all materials used for internal linings of a building should have a low rate of surface flame spread and (in some cases) a low rate of heat release.

(Approved Document B2)

6.8.2 Meeting the requirements

Suspended ceilings

Table 6.44 sets out criteria appropriate to the suspended ceilings that can be accepted as contributing to the fire resistance of a floor.

Table 6.44 Limitations on fire protected suspended ceilings

Height of building or separated part	Type of floor	Provision for fire resistance of floor	Description of suspended ceiling
<18 m	Not compartment	60 mins or less	Type A, B, C or D
	Compartment	<60 mins	Type A, B, C or D
		60 mins	Type B, C or D
18 m or more	Any	60 mins or less	Type C or D
No limit	Any	60 mins	Type D

6.9 Roofs

The roof of a brick-built house is normally an aitched (sloping) roof comprising rafters fixed to a ridge board, braced by purlins, struts and ties and fixed to wall plates bedded on top of the walls. They are then usually clad with slates or tiles to keep the rain out.

Timber-framed houses usually have trussed roofs – prefabricated triangulated frames that combine the rafters and ceiling joists – which are lifted into place and supported by the ails. The trusses are joined together with horizontal and diagonal ties. A ridge board is not fitted, nor are purlins required. Roofing felt battens and tiling are applied in the usual way.

6.9.1 Requirements

As a fire precaution, all materials used for internal linings of a building should have a low rate of surface flame spread and (in some cases) a low rate of heat release.

(Approved Document B2)

💡 It is not necessary, however, to ventilate warm deck roofs or inverted roofs, i.e. those roofs where the moisture from the building cannot permeate the insulation.

External fire spread

- *The roof shall be constructed so that the risk of spread of flame and/or fire penetration from an external fire source is restricted.*
- *The risk of a fire spreading from the building to a building beyond the boundary, or vice versa shall be limited.*

(Approved Document B4)

Internal fire spread (structure)

- *Ideally the building should be sub-divided by elements of fire-resisting construction into compartments.*
- *All openings in fire-separating elements shall be suitably protected in order to maintain the integrity of the continuity of the fire separation.*
- *Any hidden voids in the construction shall be sealed and sub-divided to inhibit the unseen spread of fire and products of combustion, in order to reduce the risk of structural failure, and the spread of fire.*

(Approved Document B3)

Airborne and impact sound

Dwellings shall be designed and built so that the noise from normal domestic activity in an adjoining dwelling (or other building) is kept down to a level that:

- *does not affect the health of the occupants of the dwelling*
- *will allow them to sleep, rest and engage in their normal domestic activities without disruption.*

(Approved Document E)

6.9.2 Meeting the requirements

Environment

> Roofs shall not be damaged by moisture from rain or snow   C3
> and shall not carry that moisture to any part of the building
> which it would damage.
>
> 💡 This damage can be avoided either by preventing
> moisture from getting to materials that would be damaged
> or by using materials that will not be damaged.
>
> The roof of the building shall be resistant to the penetration   C3/4
> of moisture from rain or snow to the inside of the building
> (see Figure 6.109).

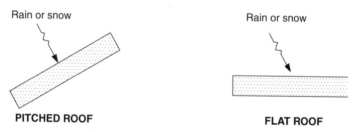

Figure 6.109 Roof, resistance to moisture

Trussed rafter roofs

> Trussed rafter roofs should be braced to the   A1/2 (1A2)
> recommendations of BS 5268.

Roof joists

> No notches or holes should be cut in roof rafters, other than   A1/2 (1B6)
> at supports where the rafter may be 'birds mouthed' to
> a depth not exceeding 0.33 times the rafter depth. In all cases:
>
> - notches should be no deeper than 0.125 times the   A1/2 (1B6)
>   depth of a joist and should not be cut closer to the
>   support than 0.07 of the span, nor further away than
>   0.25 times the span.

- holes should be no greater diameter than 0.25 times   A1/2 (1B6)
  the depth of the joist; should be drilled at the neutral
  axis; and should be not less than 3 diameters (centre to
  centre) apart; and should be located between 0.25 and
  0.4 times the span from the support.
- bearing areas and workmanship should comply with   A1/2 (1B7)
  the relevant requirements of BS 5268: Part 2 1991.

Table 6.45 Strutting of joists

Joist span (m)	Number of strutting rows
Less than 2.5 m	None
2.5–4.5 m	1 at mid span
More than 4.5 m	2 at one-third span positions

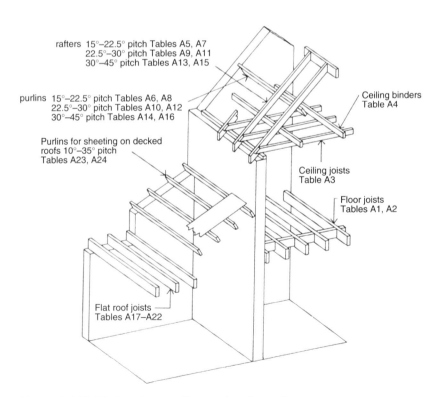

rafters 15°–22.5° pitch Tables A5, A7
         22.5°–30° pitch Tables A9, A11
         30°–45° pitch Tables A13, A15

purlins 15°–22.5° pitch Tables A6, A8
         22.5°–30° pitch Tables A10, A12
         30°–45° pitch Tables A14, A16

Purlins for sheeting on decked
roofs 10°–35° pitch
Tables A23, A24

Ceiling binders
Table A4

Ceiling joists
Table A3

Floor joists
Tables A1, A2

Flat roof joists
Tables A17–A22

Figure 6.110 Timber floor, ceiling and roof members

Maximum height of domestic buildings

> The maximum height of a building shall not exceed   A1/2 (1C17)
> the heights given in Table 6.46 with regard to the
> relevant wind speed.

Table 6.46 Maximum height of building on normal or slightly sloping sites

Location		Normal or slightly sloping sites							Steeply sloping sites, hills, cliffs and escarpments						
	Wind speed	36	38	40	42	44	46	48	36	38	40	42	44	46	48
Unprotected site, open countryside with no obstructions		15	15	15	15	15	11	9	8	6	4	3	0	0	0
Open countryside with scattered windbreaks		15	15	15	15	15	15	13	11	9	7.5	6	5	4	3
Country with many windbreaks, small towns, outskirts of large cities		15	15	15	15	15	15	15	15	15	14	12	10	8	6.5
Protected sites and country centres		15	15	15	15	15	15	15	15	15	15	15	15	15	14

Imposed loads on roofs

> The imposed loads on roofs shall not exceed those   A1/2 (1C16)
> shown in Table 6.47.

Table 6.47 Imposed loads

Element	Distributed loads	Concentrated load
Roofs	1.00 kN/m^2 for spans not exceeding 12 m. 1.50 kN/m^2 for spans not exceeding 6 m.	

Lateral support by roofs and floors

> Floors and roofs should act to transfer lateral forces (see   A1/2 (1C34)
> Table 6.48) from walls to buttressing walls, piers or
> chimneys and be secured to the supported wall as shown.

Table 6.48 Lateral support by floors

Type of wall	Wall length	Roof lateral support required	Floor lateral support required
External, compartment separating, solid or cavity wall	Any length	By every roof forming a junction with the supporting wall	
	Greater than 3 m		By every floor forming a junction with the supporting wall
Internal load bearing wall (not being a compartment or separating wall)	Any length	At the top of each storey	At the top of each storey

Rafters

When the spans of rafters or purlins are unequal, the section sizes should be determined for each span or by the longest span.	A1/2 Appendix A
No notches or holes should be cut in purlins unless checked by a competent person.	A1/2 Appendix A
The minimum bearing length at supports should be 35 mm for rafters and 50 mm for purlins.	A1/2 Appendix A

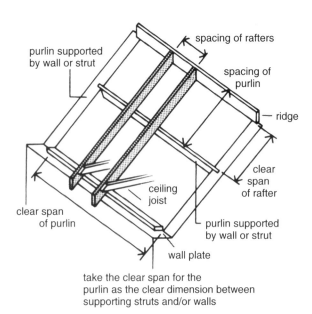

take the clear span for the purlin as the clear dimension between supporting struts and/or walls

Figure 6.111 Common (or jack rafters) and their supporting purlins

Tables 6.49 and 6.50 are an example of range of available choice for two sizes of rafters for two different strengths. For full details, see Tables A5, A9 and A13 of Appendix A to Approved Document A.

For purlins, see Tables A6, A10 and A14.

Table 6.49 Examples of imposed loading – SC3

Pitch	Size of rafter (mm)	Maximum clear span of rafters								
		Dead load less than 0.50 kN/m^3			Dead load between 0.50 and 0.75 kN/m^3			Dead load between 0.75 and 1.00 kN/m^3		
		400	450	600	400	450	600	400	450	600
		Spacing of rafters (mm)								
Greater than 15° but less than 22.5° with access only for maintenance or repair	38 × 100	2.1	2.05	1.93	1.93	1.88	1.75	1.80	1.75	1.61
Greater than 22.5° but less than 30° with access only for maintenance or repair	38 × 100	2.18	2.13	2.01	2.01	1.96	1.82	1.88	1.82	1.68
Greater than 30° but less than 45° with access only for maintenance or repair	38 × 100	2.28	2.23	2.10	2.10	2.05	1.91	1.96	1.91	1.76

Table 6.50 Examples of imposed loading – SC4

Pitch	Size of rafter (mm)	Maximum clear span of rafters								
		Dead load less than 0.50 kN/m^3			Dead load between 0.50 and 0.75 kN/m^3			Dead load between 0.75 and 1.00 kN/m^3		
		400	450	600	400	450	600	400	450	600
		Spacing of rafters (mm)								
Greater than 15° but less than 22.5° with access only for maintenance or repair	38 × 100	2.42	2.33	2.11	2.28	2.19	1.99	2.16	2.08	1.88
Greater than 22.5° but less than 30° with access only for maintenance or repair	38 × 100	2.48	2.38	2.17	2.33	2.24	2.03	2.21	2.12	1.93
Greater than 30° but less than 45° with access only for maintenance or repair	38 × 100	2.56	2.47	2.24	2.40	2.31	2.10	2.28	2.19	1.99

Joists (for flat roofs)

The minimum bearing length at supports for roof joists should be 35 mm.	A1/2 Appendix A

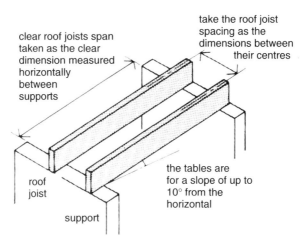

clear roof joists span taken as the clear dimension measured horizontally between supports

take the roof joist spacing as the dimensions between their centres

roof joist

support

the tables are for a slope of up to 10° from the horizontal

Figure 6.112 Joists for flat roofs

Table 6.51 Maximum clear span for floor joists

Timber strength class	Size of joist (mm)	Maximum clear span of rafters								
		Dead load less than 0.50 kN/m³			Dead load between 0.50 and 0.75 kN/m³			Dead load between 0.75 and 1.00 kN/m³		
		400	450	600	400	450	600	400	450	600
		Spacing of rafters (mm)								
SC3	38×97	1.74	1.72	1.67	1.67	1.64	1.58	1.61	1.58	1.51
SC4	38×97	1.84	1.82	1.76	1.76	1.73	1.66	1.69	1.66	1.59

Table 6.51 shows the range of choice available for two sizes of joist of two different strengths. For full details, see Tables A17 to A22 of Appendix A to Approved Document A.

Purlins supporting sheeting or decking for roofs

The minimum bearing length at supports for   A1/2 Appendix A
purlins should be 50 mm.

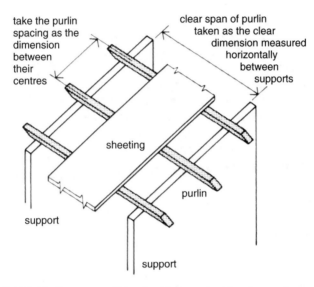

take the purlin
spacing as the
dimension
between
their
centres

clear span of purlin
taken as the clear
dimension measured
horizontally
between
supports

sheeting

purlin

support

support

Figure 6.113 Purlins supporting sheeting or decking for roofs

Table 6.52 shows the range of choice available for two sizes of joist of two
different strengths. For full details, see Tables A23 and A24 of Appendix A to
Approved Document A.

Table 6.52 Maximum clear span for floor joists

Timber strength class	Size of purlin (mm)	Maximum clear span of rafters								
		Dead load not more than 0.25 kN/m³			Dead load between 0.25 and 0.50 kN/m³			Dead load between 0.50 and 0.75 kN/m³		
		900	1500	2100	900	1500	2100	900	1500	2100
		Spacing of purlins (mm)								
SC3	50 × 100	1.68	1.51	1.34	1.55	1.40	1.24	1.45	1.31	1.16
SC4	50 × 100	1.79	1.58	1.40	1.64	1.46	1.30	1.53	1.37	1.21

Interruption of lateral support

Where an opening in a roof for a stairway or the like   A1/2 (1C38)
adjoins a supported wall and interrupts the continuity
of lateral support, then:

(a)   the maximum permitted length of the opening
      is to be 3 m, measured parallel to the supported
      wall, and
(b)   where a connection is provided by means
      other than by anchor, this should be provided
      throughout the length of each portion of the wall
      situated on each side of the opening, and
(c)   where connection is provided by mild steel anchors,
      these should be spaced closer than 2 m on each
      side of the opening to provide the same number
      of anchors as if there were no opening, and
(d)   there should be no other interruption of lateral
      support.

Means of escape

A flat roof being used as a means of escape should: C1

- be part of the same building from which escape is being made;
- should lead to a storey exit or external escape route;
- should provide 30 minutes' fire resistance.

Where a balcony or flat roof is provided for escape purposes
guarding may be needed (see also Approved Document K,
Protection from falling, collision and impact).

Compartmentation

To prevent the spread of fire within a building, whenever   B3 (9.1)
possible, the building should be sub-divided into
compartments separated from one another by walls
and/or floors of fire-resisting construction.

Compartment walls in a top storey beneath a roof should be continued through the roof space.	B3 (9.26)
When a compartment wall meets the underside of the roof covering or deck, the wall/roof junction shall maintain continuity of fire resistance.	B3 (9.28)
Double skinned insulated roof sheeting should incorporate a band of material of limited combustibility.	B3 (9.29)

Roof coverings

Roof covering (i.e. constructions which consist of one or more layers of material, but do not refer to the roof structure as a whole) are required to meet the following requirements:

Plastic rooflights should have a minimum of class 3 lower surface.	B4 (15.6)
When used in rooflights, unwired glass shall be at least 4 mm thick and shall be AA designated (see Table 6.52).	B4 (15.8)
Thatch and wood shingles should be regarded as having an AD/BD/CD designation (see Table 6.53).	B4 (15.9)

Table 6.53 Limitations on roof coverings

Designation of roof covering	Minimum distance from any point on relevant boundary			
	Less than 6 m	At least 6 m	At least 12 m	At least 20 m
AA, AB or AC	Acceptable	Acceptable	Acceptable	Acceptable
BA, BB or BC	Not acceptable	Acceptable	Acceptable	Acceptable
CA, CB or CC	Not acceptable	Acceptable	Acceptable	Acceptable
AD, BD or CD	Not acceptable	Acceptable	Acceptable	Acceptable
DA, DB, DC or DD	Not acceptable	Not acceptable	Not acceptable	Acceptable

Pitched roofs covered with slates or tiles

Covering material	Supporting structure	Designation
1. Natural slates 2. Fibre reinforced cement slates 3. Clay tiles 4. Concrete tiles	Timber rafters with or without underfelt, sarking, boarding, woodwool slabs, compressed straw slabs, plywood, wood chipboard, or fibre insulating board.	AA

📖 Although the table does not include guidance for pitched roofs covered with bitumen felt, it should be noted that there is a wide range of materials on the market and information on specific products is readily available from manufacturers.

Pitched roofs covered with self-supporting sheet

Roof covering material	Construction	Supporting structure	Designation
Profiled sheet of galvanized steel, aluminium, fibre reinforced cement, or pre-painted (coil coated) steel or aluminium with a PVC or PVF2 coating	Single skin without underlay, or with underlay or plasterboard, or woodwool slab	Structure of timber, steel or concrete	AA
Profiled sheet of galvanized steel, aluminium, fibre reinforced cement, or pre-painted (coil coated) steel or aluminium with a PVC or PVF2 coating	Double skin without interlayer, or with interlayer of resin bonded or concrete glass fibre, mineral wool slab, polystyrene, or polyurethane	Structure of timber, steel or concrete	AA

Flat roofs with bitumen felt

A flat roof consisting of bitumen felt should (irrespective of the felt specification) be deemed to be of designation AA if the felt is laid on a deck constructed of 6 mm plywood, 12.5 mm wood chipboard, 16 mm (finished) plain-edged timber boarding, compressed straw slab, screeded woodwool slab, profiled fibre reinforced cement or steel deck (single or double skin) with or without fibre insulating board overlay, profiled aluminium deck (single or double skin) with or without fibre insulating board overlay, or concrete or clay pot slab (in situ or pre-cast), and has a surface finish of:

(a)   bitumen-bedded stone chippings covering the whole surface to a depth of at least 12.5 mm;
(b)   bitumen-bedded tiles of a non-combustible material;
(c)   sand and cement screed; or
(d)   tarmacadam.

Pitched or flat roofs covered with fully supported materials

Covering material	Supporting structure	Designation
1. Aluminium sheet 2. Copper sheet 3. Zinc sheet	Timber joists and tongued and grooved boarding, or plain-edged boarding	AA
4. Lead sheet 5. Mastic asphalt 6. Vitreous enamelled steel 7. Lead/tin alloy-coated steel sheet 8. Zinc/aluminium alloy-coated steel sheet	Steel or timber joists with deck of woodwool slabs, compressed straw slab, wood chipboard, fibre insulating board, or 9.5 mm plywood	AA
9. Pre-painted (coil coated) steel sheet including liquid-applied PVC coatings	Concrete or clay pot slab (in situ or pre-cast) or non-combustible deck of steel, aluminium, or fibre cement (with or without insulation)	AA

Lead sheet supported by timber joists and plain edged boarding should be regarded as having a BA designation.

Rating of material and products

Table 6.54 Typical performance ratings of some generic materials and products

Rating	Material or product
Class 0	1. Any non-combustible material or material of limited combustibility. 2. Brickwork, blockwork, concrete and ceramic tiles. 3. Plasterboard (painted or not with a PVC facing not more than 0.5 mm thick) with or without an air gap or fibrous or cellular insulating material behind. 4. Woodwool cement slabs. 5. Mineral fibre tiles or sheets with cement or resin binding.
Class 3	6. Timber or plywood with a density more than 400 kg/ml, painted or unpainted. 7. Wood particle board or hardboard, either untreated or painted. 8. Standard glass reinforced polyesters.

Cavity masonry walls – ceiling and roof space junctions

Heavy ceilings with sealed joints (e.g. 2.5 mm plasterboard) may reduce the mass of the wall above the ceiling to 150 kg/m^2. The cavity should still be maintained.   E (p11)

If lightweight aggregate blocks density less than 1200 kg/m^3 are used, seal one side with cement paint or plaster skim.   E (p11)

Heavy ceilings should have sealed joints (12.5 mm plasterboard or equivalent).   E (p13)

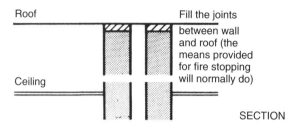

Figure 6.114 Roof junctions – cavity masonry

The freestanding panels may be omitted in the roof space and the mass of the core above the ceiling may be reduced to 150 kg/m^2 (**but** if core Type D is used the cavity should be maintained).	E (p13)
If open textured lightweight aggregate blocks less than 1200 kg/m^3 are used to reduce mass, seal one side with cement paint or plaster skim.	E (p13)
The junction between ceiling and freestanding panels should be sealed with tape or caulking.	E (p13)

Timber framed walls – roof junctions

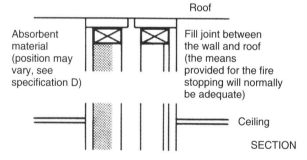

Figure 6.115 Roof junctions

There should be a fire-stop between the junction of the wall and roof (see Approved Document B, Fire safety)	E (p14)
Between the ceiling level and the underside of the roof either:	E (p14)

- carry both frames through – the cladding on each frame may be reduced to not less than 25 mm; or
- close the cavity at ceiling level without connecting the two frames rigidly together and use one frame with at least 25 mm cladding on both sides.

In each case seal the space between frame and roof finish.

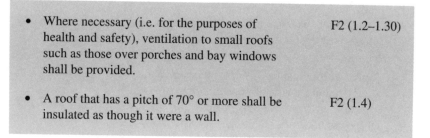

- Where necessary (i.e. for the purposes of   F2 (1.2–1.30)
  health and safety), ventilation to small roofs
  such as those over porches and bay windows
  shall be provided.

- A roof that has a pitch of 70° or more shall be   F2 (1.4)
  insulated as though it were a wall.

If the ceiling of a room follows the pitch of the roof, ventilation should be provided as if it were a flat roof.

Roof with a pitch of 15° or more

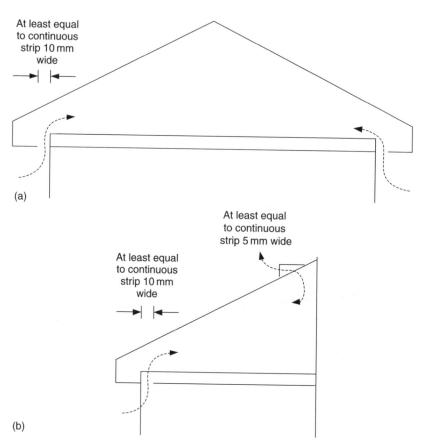

Figure 6.116 Ventilating roof voids. (a) Pitched roof. (b) Lean-to roof

- Pitched roof spaces should have ventilation   F2 (1.2–1.30)
  openings at least 10 mm wide at eaves level
  to promote cross-ventilation.
- A pitched roof that has a single slope and abuts   F2 (1.4)
  a wall should have ventilation openings at eaves
  level at least 10 mm wide and at high level (i.e. at the
  junction of the roof and the wall) at least 5 mm wide.

Roof with a pitch of less than 15°

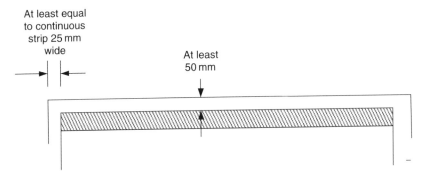

At least equal
to continuous
strip 25 mm
wide

At least
50 mm

Figure 6.117 Ventilating roof void – flat roof

Roof spaces should have ventilation openings at least 25 mm   F2 (2.2)
wide in two opposite sides to promote cross ventilation.

The void should have a free air space of at least 50 mm   F2 (2.4)
between the roof deck and the insulation.

Pitched roofs where the insulation follows the pitch of the   F2 (2.5)
roof need ventilation at the ridge at least 5 mm wide.

Where the edges of the roof abut a wall or other obstruction   F2 (2.6)
in such a way that free air paths cannot be formed to promote
cross ventilation or the movement of air outside any
ventilation openings would be restricted, an alternative
form of roof construction should be adopted.

💡 Roofs with a span exceeding 10 m may require more ventilation, totalling 0.6% of the roof area.

💡 Ventilation openings may be continuous or distributed along the full length and may be fitted with a screen, facia, baffle, etc.

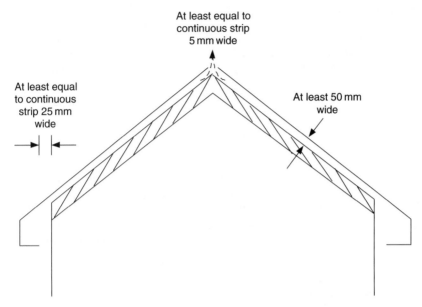

At least equal to
continuous strip
5 mm wide

At least equal
to continuous
strip 25 mm
wide

At least 50 mm
wide

Figure 6.118 Ventilating roof void – ceiling following pitch of roof

6.10 Chimneys

6.10.1 The requirement (Building Act 1984 Section 73)

If a person erects or raises a building that is (or is going to be) taller than the chimneys and/or flues from an adjoining building that is either joined by a party wall or less than six feet away from the taller building, then the local authority may:

- if reasonably practical, require that person to build up those chimneys and flues, so that their top is of the same height as the top of the chimneys of the taller building or the top of the taller building, whichever is the higher;
- require the owner or occupier of the adjoining building to allow the person erecting or raising the building, access to the adjacent building so that he can carry out such work as may be necessary to comply with the notice served on him.

💡 The owner or occupier of the adjacent building is entitled to complete the work himself by (within fourteen days) serving a 'counter-notice' that he has elected to carry out the work himself.

Protection of building

Combustion appliances and fluepipes shall be so installed, and fireplaces and chimneys shall be so constructed and installed, as to reduce to a reasonable level the risk of people suffering burns or the building catching fire in consequence of their use.

(Approved Document J3)

Provision of information

Where a hearth, fireplace, flue or chimney is provided or extended, a durable notice containing information on the performance capabilities of the hearth, fireplace, flue or chimney shall be affixed in a suitable place in the building for the purpose of enabling combustion appliances to be safely installed.

(Approved Document J4)

6.10.2 Meeting the requirement

End restraints

The ends of every wall (except single leaf walls less than 2.5 m in storey height and length) in small single storey non-residential buildings and annexes should be bonded or otherwise securely tied throughout their full height to a buttressing wall, pier or chimney. Long walls may be provided with intermediate support, dividing the wall into distinct lengths; each distinct length is a supported wall for the purposes of this section. The buttressing wall, pier or chimney should provide support from the base to the full height of the wall.	1C26
The buttressing wall should be bonded or securely tied to the supported wall and at the other end to a buttressing wall, pier or chimney (see Figure 6.72).	1C27
Piers should measure at least three times the thickness of the supported wall and chimneys should be twice the thickness (measured at right angles to the wall).	1C28a
The sectional area on plan of chimneys (excluding openings for fireplaces and flues) should be not less than the area required for a pier in the same wall, and the overall thickness should not be less than twice the required thickness of the supported wall (see Figure 6.73).	1C28b

The buttressing wall, pier or chimney should provide support to the full height of the wall from base to top.	1C28
Floors and roofs should act to transfer lateral forces (see Table 6.16) from walls to buttressing walls, piers or chimneys and be secured to the supported wall as shown.	1C34

Masonry chimneys

Where a chimney is not adequately supported by ties or securely restrained in any way, its height H (measured from the highest point of any chimney pot or other flue terminal) should not exceed 4.5 times the width W (the least horizontal dimension of the chimney measured at the same point of intersection) – provided that the density of the masonry is greater than $1500\,kg/m^3$.	1D1

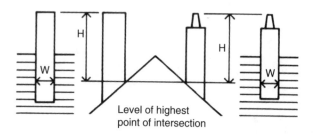

Figure 6.119 Proportions for masonry chimneys

The foundation of piers, buttresses and chimneys should project as indicated in Figure 6.120 and the projection X should never be less than P.	1E2g

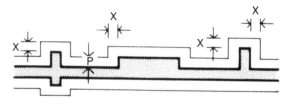

Projection X should not be less than P

Figure 6.120 Piers and chimneys

Flues, etc.

Flue walls should have a fire resistance of at least one half of that required for the compartment wall or floor and be of non-combustible construction.	B3 (11.11)
If a flue, or duct containing flues or appliance ventilation duct(s), passes through a compartment wall or compartment floor, or is built into a compartment wall, each wall of the flue or duct should have a fire resistance of at least half that of the wall or floor in order to prevent the by-passing of the compartmentation.	B3 (11.11)

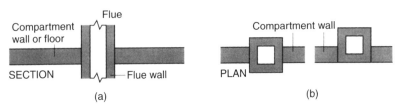

Figure 6.121 Flues penetrating compartment walls or floors. (a) Flue passing through compartment wall or floor. (b) Flue built into compartment wall

Fire stopping

Proprietary fire-stopping and sealing systems (including those designed for service penetrations) which have been shown by test to maintain the fire resistance of the wall or other element, are available and may be used. Other fire-stopping materials include:

- cement mortar,
- gypsum-based plaster,
- cement or gypsum-based vermiculite/perlite mixes,
- glass fibre, crushed rock, blast furnace slag or ceramic based products (with or without resin binders), and
- intumescent mastics (B3 11.14).

Joints between fire separating elements should be fire-stopped.	B3 (11.12a)
All openings for pipes, ducts, conduits or cables to pass through any part of a fire separating element should be:	B3 (11.12b)

- kept as few in number as possible;
- kept as small as practicable;
- fire-stopped (which in the case of a pipe or duct, should allow thermal movement).

To prevent displacement, materials used for fire-stopping should be reinforced with (or supported by) materials of limited combustibility.	B3 (11.13)

Chimney construction

Chimneys shall consist of a wall or walls enclosing one or more flues (see Figure 6.122).	J (0.4–7)
📖 In the gas industry, the chimney for a gas appliance is commonly called the flue.	
Down-draughts that could interfere with the combustion performance of an open-flued appliance (see Figure 6.123) shall be minimized.	J (0.4–11)

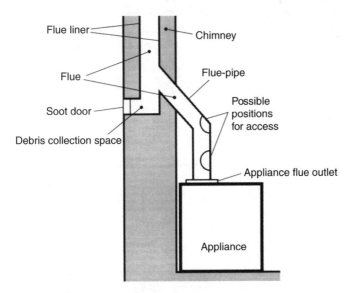

Figure 6.122 Chimneys and flues

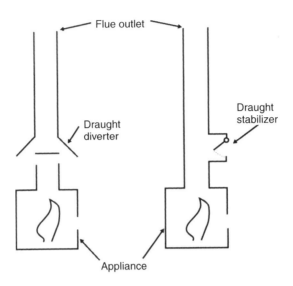

Figure 6.123 Draught diverters and draught stabilizers

Fireplaces

Fireplace recesses (sometimes called a builder's opening) shall be formed in a wall or in a chimney breast, from which a chimney leads and which has a hearth at its base.   J (0.4–16)

Simple recesses are suitable for closed appliances such as roomheaters, stoves, cookers or boilers. They are **not** suitable for an open fire without a canopy.

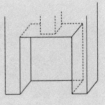

Fireplace recesses are used for accommodating open fires and freestanding fire baskets.

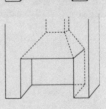

Fireplace recesses are often lined with firebacks to accommodate inset open fires.

📖 Lining components and decorative treatments fitted around openings reduce the opening area. It is the finished fireplace opening area that determines the size of flue required for an open fire in such a recess.

Hearths

> A hearth shall safely isolate a combustion appliance from   J (0.4–26)
> people, combustible parts of the building fabric and soft
> furnishings.

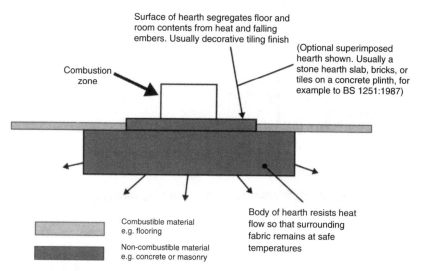

Surface of hearth segregates floor and room contents from heat and falling embers. Usually decorative tiling finish

(Optional superimposed hearth shown. Usually a stone hearth slab, bricks, or tiles on a concrete plinth, for example to BS 1251:1987)

Combustion zone

Body of hearth resists heat flow so that surrounding fabric remains at safe temperatures

Combustible material e.g. flooring

Non-combustible material e.g. concrete or masonry

Figure 6.124 The functions of a hearth

Flueblock chimneys

> Flueblock chimneys should be constructed of factory-made   J (1.29)
> components suitable for the intended application installed in
> accordance with manufacturer's instructions.
>
> Joints should be sealed in accordance with the flueblock   J (1.30)
> manufacturer's instructions.
>
> Bends and offsets should only be formed with matching   J (1.30)
> factory-made components.

Masonry chimneys (change of use)

> Where a building is to be altered for different use (e.g. it is   J (1.31)
> being converted into flats) the fire resistance of walls of
> existing masonry chimneys may need to be improved.

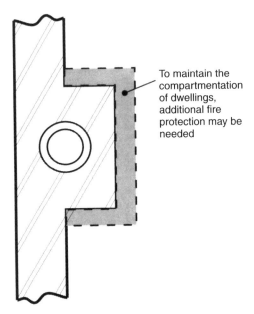

To maintain the compartmentation of dwellings, additional fire protection may be needed

Figure 6.125 Fire protection of chimneys passing through other dwellings

Connecting fluepipes

Whenever possible, fluepipes should be manufactured from:	J (1.32)

- cast iron (BS41: 1973 (1998));
- mild steel fluepipes (BS1449, Part 1: 1991, with a flue wall thickness of at least 3 mm);
- stainless steel (BS EN 10088-1: 1995 grades 1.4401, 1.4404, 1.4432 or 1.4436 with a flue wall thickness of at least 1 mm);
- vitreous enamelled steel (BS 6999: 1989 (1996)).

Fluepipes with spigot and socket joints should be fitted with the socket facing upwards to contain moisture and other condensates in the flue.	J (1.33)
Joints should be made gas-tight.	J (1.33)

Repair of flues

If renovation, refurbishment or repair amounts to or involves the provision of a new or replacement flue liner, it is considered 'building work' within the meaning of Regulation 3 of the Building Regulations and must, therefore, **not** be undertaken without prior notification to the local authority. Examples of work that would need to be notified include: J (1.34–1.35)

- relining work comprising the creation of new flue walls by the insertion of new linings such as rigid or flexible prefabricated components;
- a cast in situ liner that significantly alters the flue's internal dimensions.

If you are in doubt you should consult the building control department of your local authority, or an approved inspector.

Re-use of existing flues

Where it is proposed to bring a flue in an existing chimney back into use (or to re-use a flue with a different type or rating of appliance) the flue and the chimney should be checked and, if necessary, altered to ensure that they satisfy the requirements for the proposed use. J (1.36)

Oversize flues can be unsafe. A flue may, however, be lined to reduce the flue area to suit the intended appliance. J (1.38)

Relining

If a chimney has been relined in the past using a metal lining system and the appliance is being replaced, the metal liner should also be replaced unless the metal liner can be proven to be recently installed and can be seen to be in good condition. J (1.39)

🔔 In certain circumstances, relining is considered 'building work' within the meaning of Regulation 3 of the Building Regulations and must, therefore, **not** be undertaken without prior notification to the local authority. If you are in doubt you should consult the building control department of your local authority, or an approved inspector.

Flexible flue liners should only be used to reline a chimney and should not be used as the primary liner of a new chimney.   J (1.40)

Plastic fluepipe systems can be acceptable in some cases, for example with condensing boiler installations, where the fluepipes are supplied by or specified by the appliance manufacturer.   J (1.41)

Factory-made metal chimneys

Where a factory-made metal chimney passes through a wall, sleeves should be provided to prevent damage to the flue or building through thermal expansion.   J (1.43)

To facilitate the checking of gas-tightness, joints between chimney sections should not be concealed within ceiling joist spaces or within the thicknesses of walls.   J (1.43)

When providing a factory-made metal chimney, provision should be made to withdraw the appliance without the need to dismantle the chimney.   J (1.44)

Factory-made metal chimneys should be kept a suitable distance away from combustible materials.   J (1.45)

📖 One way of meeting this requirement is by locating the chimney not less than distance 'X' from combustible material, where 'X' is defined in BS 4543-1: 1990 (1996) as shown in Figure 6.126.

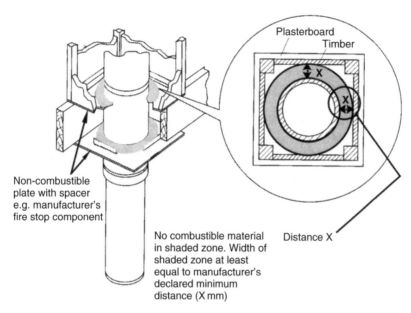

Plasterboard
Timber

Non-combustible
plate with spacer
e.g. manufacturer's
fire stop component

No combustible material
in shaded zone. Width of
shaded zone at least
equal to manufacturer's
declared minimum
distance (X mm)

Distance X

Figure 6.126 The separation of combustible material from a factory-made metal chimney meeting BS 4543, Part 1 (1990)

Flue systems

Flue systems should offer least resistance to the passage of flue gases by minimizing changes in direction or horizontal length.	J (1.47)
Wherever possible flues should be built so that they are straight and vertical except for the connections to combustion appliances with rear outlets where the horizontal section should not exceed 150 mm. Where bends are essential, they should be angled at no more than 45° to the vertical.	J (1.47)
Provisions should be made to enable flues to be swept and inspected (see Figure 6.127).	J (1.48)
A flue should **not** have openings into more than one room or space except for the purposes of:	J (1.49)

- inspection or cleaning; or
- fitting an explosion door, draught break, draught stabilizer or draught diverter.

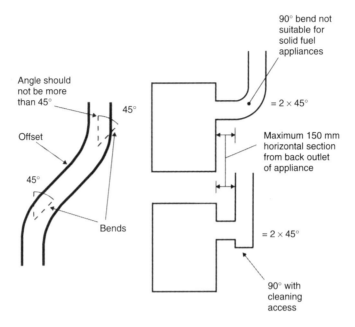

Figure 6.127 Bends in flues

Openings for inspection and cleaning should be formed using purpose factory-made components compatible with the flue system, having an access cover that has the same level of gas-tightness as the flue system and an equal level of thermal insulation.	J (1.50)
After the appliance has been installed, it should be possible to sweep the whole flue.	J (1.50)

Dry lining around fireplace openings

Where a decorative treatment, such as a fireplace surround, masonry cladding or dry lining is provided around a fireplace opening, any gaps that could allow flue gases to escape from the fireplace opening into the void behind the decorative treatment, should be sealed to prevent such leakage.	J (1.52)
The sealing material should be capable of remaining in place despite any relative movement between the decorative treatment and the fireplace recess.	J (1.53)

◉ **Notice plates for hearths and flues (Requirement J4)**

Where a hearth, fireplace (including a flue box), flue or chimney is provided or extended (including cases where a flue is provided as part of the refurbishment work), information essential to the correct application and use of these facilities should be permanently posted in the building. A way of meeting this requirement would be to provide a notice plate conveying the following information:

- the location of the hearth, fireplace (or flue box) or the location of the beginning of the flue;
- the category of the flue and generic types of appliances that can be safely accommodated;
- the type and size of the flue (or its liner if it has been relined) and the manufacturer's name;
- the installation date.

Additional provisions for appliances burning solid fuel (with a rated output up to 50 kW)

Any room or space containing an appliance should have a permanent air vent opening of at least the size shown in Figure 6.128.   J (2.2)

Open fire with no throat (e.g. a fire under a canopy)

Permanently open air vent(s) with a total free area of at least 50% of the cross-sectional area of the flue.

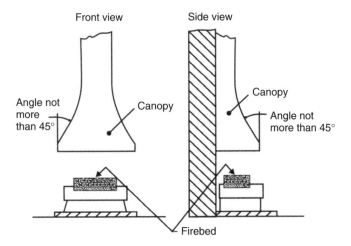

Figure 6.128 Canopy for an open solid fuel fire

Open fire with a throat and gather

Permanently open air vent(s) with a total free area of at least 50% of the throat opening area.

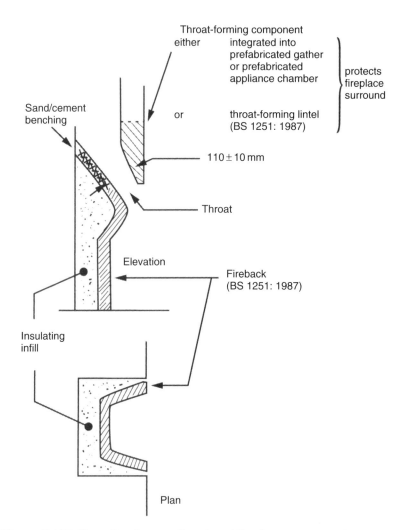

Figure 6.129 Open fireplaces – throat and fireplace components

Other appliance (such as a stove, cooker or boiler)

Permanently open air vent(s).

Size of flues

Fluepipes should have the same diameter or equivalent cross-sectional area as that of the appliance flue outlet.	J (2.4)
Flues should be not less than the size of the appliance flue outlet or that recommended by the appliance manufacturer.	J (2.5)

Flues in chimneys

Table 6.55 Size of flues in chimneys

Fireplace with an opening up to 500 mm × 550 mm.	200 mm diameter or rectangular/ square flues having the same cross-sectional dimension not less than 175 mm.
Fireplace with an opening in excess of 500 mm × 550 mm or a fireplace exposed on two or more sides.	If rectangular/square flues are used the minimum dimension should not be less than 200 mm.
Closed appliance up to 20 kW rated output which: • burns smokeless or low volatile fuel; or • is an appliance that meets the requirements of the Clean Air Act when burning an appropriate bitumous coal (these appliances are known as 'exempted fireplaces').	125 mm diameter or rectangular/ square flues having the same cross-sectional area and a minimum dimension not less than 100 mm for straight flues or 125 mm for flues with bends or offsets.
Other closed appliance of up to 35 kW rated output burning any fuel.	150 mm diameter or rectangular/ square flues having the same cross-sectional area and a minimum dimension not less than 125 mm.
Closed appliance of up to 30 kW and up to 50 kW rated output burning any fuel.	175 mm diameter or rectangular/ square flues having the same cross-sectional area and a minimum dimension not less than 150 mm.
For fireplaces with openings larger than 500 mm × 550 mm or fireplaces exposed on two or more sides (such as a fireplace under a canopy or open on both sides of a central chimney breast) a way of showing compliance would be to provide a flue with a cross-sectional area equal to 15% of the total face area of the fireplace opening(s) using the formula:	J (2.7)

Fireplace opening area (mm^2) = Total horizontal length of fireplace opening L (mm) × Height of fireplace opening H (mm)

Examples of L and H for large and unusual fireplace openings are shown in Figure 6.130

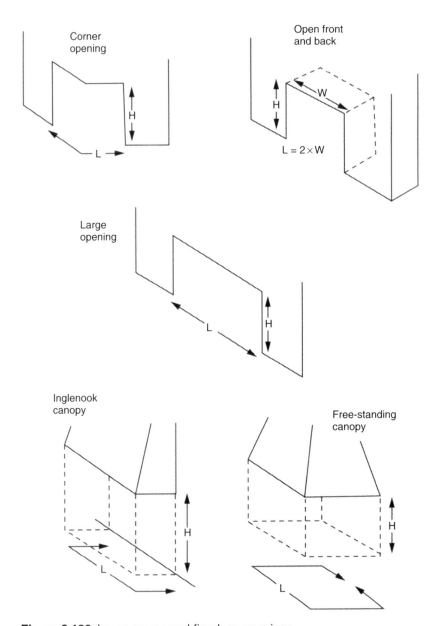

Figure 6.130 Large or unusual fireplace openings

Height of flues

Flues should be high enough (normally 4.5 m is sufficient) to ensure sufficient draught to clear the products of combustion.   J (2.8)

The outlet from a flue should be above the roof of the building in a position where the products of combustion can discharge freely and will not present a fire hazard, whatever the wind conditions (see Figure 6.131 and Table 6.56).   J (2.10)

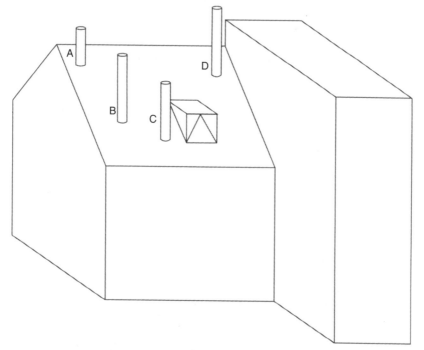

Figure 6.131 Flue outlet positions for solid fuel appliances

Flue outlet clearances – thatched or shingled roof

The clearances to flue outlets which discharge on, or are in close proximity to, roofs with surfaces which are readily ignitable (e.g. covered in thatch or shingles) should be increased to those shown in Figure 6.132.   J (2.12)

Table 6.56 Flue outlet positions

Point where flue passes through weather surfaces (e.g. roof, tiles or external walls)		Clearance to flue outlet
A	At or within 600 mm of the ridge.	At least 600 mm above the ridge.
B	Elsewhere on a roof (whether pitched or flat).	At least 2300 mm horizontally from the nearest point on the weather surface and: • at least 1000 mm above the highest point of intersection of the chimney and the weather surface; or • at least as high as the ridge.
C	Below (on a pitched roof) or within 2300 mm horizontally to an openable rooflight, dormer window or other opening.	At least 1000 mm above the top of the opening.
D	Within 2300 mm of an adjoining or adjacent building whether or not beyond the measurements is the boundary.	At least 600 mm above the adjacent building.

Connecting fluepipes

Connecting fluepipes should not pass through any roof space, partition, internal wall or floor, except to pass directly into a chimney through either a wall of the chimney or a floor supporting the chimney.	J (2.14)
Connecting fluepipes should be guarded if they could be at risk of damage or if the burn hazard they present to people is not immediately apparent.	J (2.14)
Connecting fluepipes should be located so as to avoid igniting combustible material by minimizing horizontal and sloping runs and separation of the fluepipe from combustible material.	J (2.15 and 1.45)

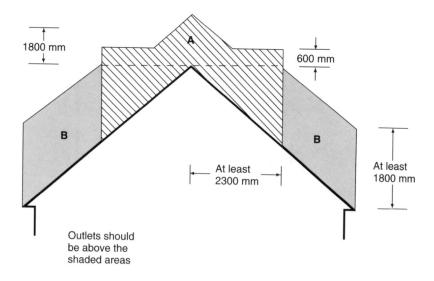

Figure 6.132 Flue outlet positions for solid fuel appliances – discharging near easily ignited roof coverings

Area	Location of flue outlet
A	At least 1800 mm vertically above the weather surface, and at least 600 mm above the ridge.
B	At least 1800 mm vertically above the weather surface, and at least 2300 mm horizontally from the weather surface.

Masonry and flueblock chimneys

The thickness of the walls around the flues, excluding the thickness of any flue liners, should be in accordance with Figure 6.134.	J (2.17)
Combustible material should not be located where it could be ignited by the heat dissipating through the walls of fireplaces or flues.	J (2.18)

Construction of fireplace gathers

To minimize resistance to the proper working of flues, tapered gathers should be provided in fireplaces for open fires (J (2.21)), or corbelling of masonry,

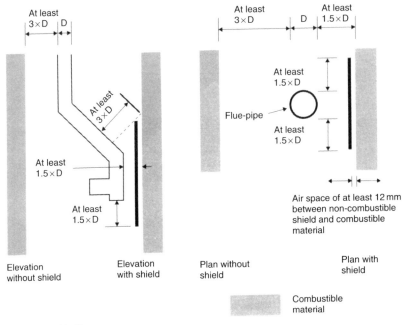

Shields should either:
a) extend beyond the fluepipe by at least 1.5×D; or
b) make any path between fluepipes and combustible material at least 3×D long

Figure 6.133 Protecting combustible material from uninsulated fluepipes for solid fuel appliances

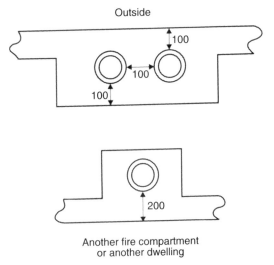

Figure 6.134 Wall thickness for masonry and flueblock chimneys. Dimensions in mm

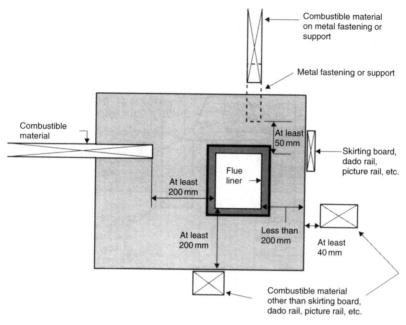

Figure 6.135 Minimum separation distances for combustible material in or near a chimney

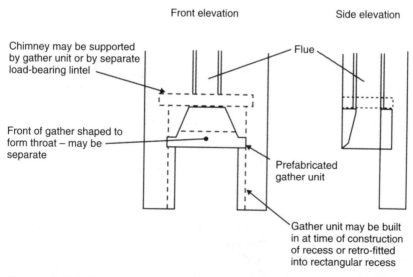

Figure 6.136 Construction of fireplace gathers – using prefabricated components

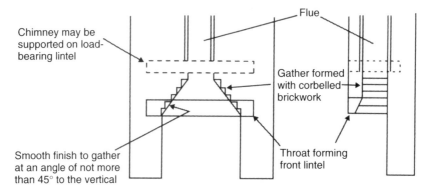

Figure 6.137 Construction of fireplace gathers – using masonry

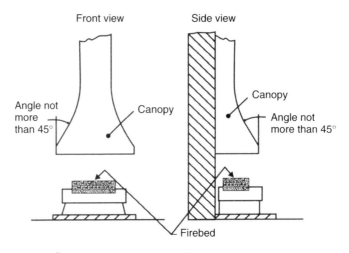

Figure 6.138 Canopy for an open fuel fire

as shown in Figure 6.137. Alternatively a suitable canopy (as shown in Figure 6.138) or a prefabricated appliance chamber incorporating a gather may be used.

This can be achieved by using prefabricated gather components built into a fireplace recess, as shown in Figure 6.136.

Construction of hearths

> Hearths should be constructed of suitably robust materials   J (2.22)
> and to appropriate dimensions such that, in normal use,
> they prevent combustion appliances setting fire to the
> building fabric and furnishings, and they limit the risk
> of people being accidentally burnt.

The hearth should be able to accommodate the weight of the appliance and its chimney if the chimney is not independently supported.	J (2.22)
Appliances should stand wholly above either hearths made of non-combustible board/sheet material, tiles at least 12 mm thick or constructional hearths.	J (2.23)
Constructional hearths should have plan dimensions as shown in Figure 6.139.	J (2.24a)
Constructional hearths should be made of solid, non-combustible material, such as concrete or masonry, at least 125 mm thick, including the thickness of any non-combustible floor and/or decorative surface.	J (2.24b)
Combustible material should not be placed beneath constructional hearths unless there is an air-space of at least 50 mm between the underside of the hearth and the combustible material, or the combustible material is at least 250 mm below the top of the hearth (see Figure 6.140).	J (2.25)
An appliance should be located on a hearth so that it is surrounded by a surface free of combustible material (as shown in Figure 6.141) or it may be the surface of a superimposed hearth laid wholly or partly upon a constructional hearth.	J (2.26)
The edges of this surface should be marked to provide a warning to the building occupants and to discourage combustible floor finishes such as carpet from being laid too close to the appliance. A way of achieving this would be to provide a change in level.	J (2.26)

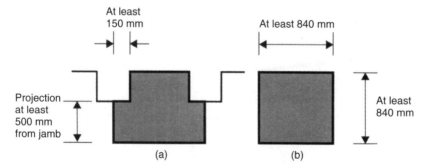

Figure 6.139 Constructional hearth suitable for solid fuel appliances (including open fires) – plan. (a) Fireplace recess. (b) Freestanding

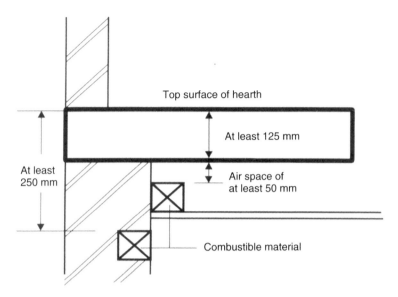

Figure 6.140 Constructional hearth suitable for solid fuel appliances (including open fires) – section

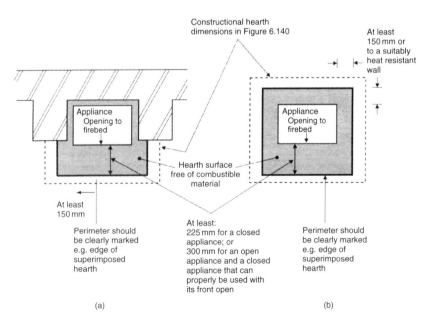

Figure 6.141 Non-combustible hearth surface surrounding a solid fuel appliance. (a) Fireplace recess. (b) Freestanding

Fireplace recesses

Fireplaces need to be constructed such that they adequately protect the building fabric from catching fire.

J (2.29)

Fireplace recesses can be from masonry or concrete as shown in Figure 6.141.

J (2.29a)

Fireplace recesses can also be prefabricated factory-made appliance chambers using components that are made of insulating concrete having a density of between 1200 and 1700 kg/m³ and with the minimum thickness as shown in Table 6.55.

J (2.29b)

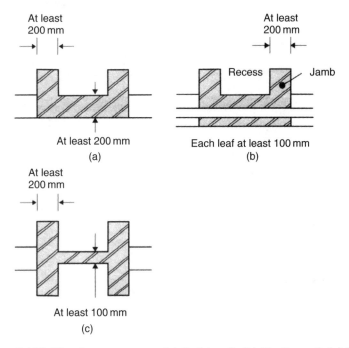

Figure 6.142 Fireplace recesses. (a) Solid wall. (b) Cavity wall. (c) Back-to-back (within the same dwelling)

Fireplace lining components

Fireplace recesses containing inset, open fires that need to be heat protected should either be lined with suitable firebricks or lining components as shown in Table 6.57.

J (2.30)

Table 6.57 Prefabricated appliance chambers: minimum thickness

Component	Minimum thickness (mm)
Base	50
Side section, forming wall on either side of chamber	75
Back section, forming rear of chamber	100
Top slab, lintel or gather, forming top of chamber	100

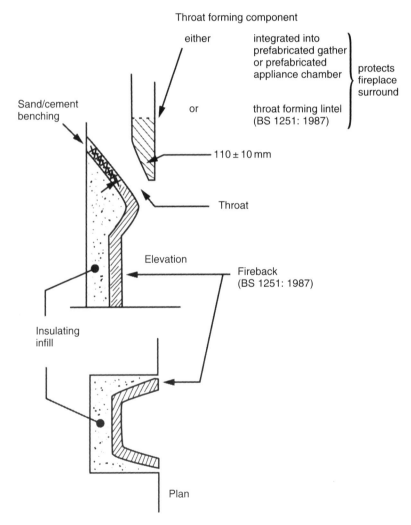

Figure 6.143 Open fireplaces – throat and fireplace components

Walls adjacent to hearths

Walls that are not part of a fireplace recess or a prefabricated   J (2.31)
appliance chamber but are adjacent to hearths or appliances
also need to protect the building from catching fire.
A way of achieving the requirement is shown in
Figure 6.108.

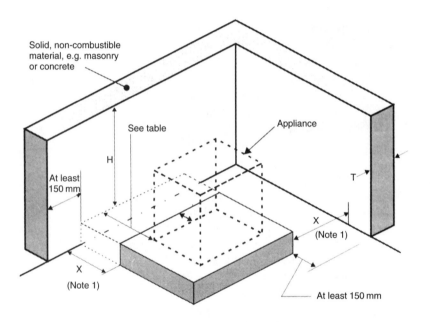

Location of hearth or appliance	Solid, non-combustible material	
	Thickness (T)	Height (H)
Where the hearth abuts a wall and the appliance is not more than 50 mm from the wall	200 mm	At least 300 mm above the appliance and 1.2 m above the hearth
Where the hearth abuts a wall and the appliance is more than 50 mm but not more than 300 mm from the wall	75 mm	At least 300 mm above the appliance and 1.2 m above the hearth
Where the hearth does not abut a wall and is no more than 150 mm from the wall (see Note 1)	75 mm	At least 1.2 m above the hearth
Note 1: There is no requirement for protection of the wall where X is more than 150 mm		

Figure 6.144 Walls adjacent to hearths

Additional provisions for gas burning devices

The Gas Safety (Installation and Use) Regulations require that (a) gas fittings, appliances and gas storage vessels must only be installed by a person with the required competence and (b) any person having control to any extent of gas work must ensure that the person carrying out that work has the required competence and (c) any person carrying out gas installation, whether an employee or self-employed, must be a member of a class of persons approved by the HSE; for the time being this means they must be registered with CORGI, the Council for Registered Gas Installers.

Important elements of the Regulations include that:

(a) any appliance installed in a room used or intended to   J (3.5a)
be used as a bath or shower room must be of
the room-sealed type

(b) a gas fire, other gas space heater or gas water heater   J (3.5b)
of more than 14 kW (gross) heat input (12.7 kW (net)
heat input) must not be installed in a room used or
intended to be used as sleeping accommodation unless
the appliance is room-sealed

(c) a gas fire, other space heater or gas water heater of   J (3.5c)
up to 14 kW (gross) heat input (12.7 kW (net) heat
input) must not be installed in a room used or intended
to be used as sleeping accommodation unless it is
room-sealed or equipped with a device designed to
shut down the appliance before there is a build-up
of a dangerous quantity of the products of combustion
in the room concerned

The restrictions in (a)–(c) above also apply in respect of   J (3.5d)
any cupboard or compartment within the rooms concerned,
and to any cupboard, compartment or space adjacent to, and
with an air vent into such a room.

Instantaneous water heaters (installed in any room) must be   J (3.5e)
room-sealed or have fitted a safety device to shut down the
appliance as in (c) above.

Precautions must be taken to ensure that all installation   J (3.5f)
pipework, gas fittings, appliances and flues are installed
safely. When any gas appliance is installed, checks are
required for ensuring compliance with the Regulations,
including the effectiveness of the flue, the supply of
combustion air, the operating pressure or heat input
(or where necessary both), and the operation of the
appliance to ensure its safe functioning.

Any flue must be installed in a safe position.   J (3.5g)

No alteration is allowed to any premises in which a gas   J (3.5h)
fitting or gas storage vessel is fitted that would adversely
affect the safety of that fitting or vessel, causing it no
longer to comply with the Regulations.

LPG storage vessels and LPG-fired appliances fitted with   J (3.5i)
automatic ignition devices or pilot lights must not be
installed in cellars or basements.

Outlets from flues should be situated externally so
as to allow the products of combustion to dispel, and,
if a balanced flue, the intake of air – see Figure 6.145.

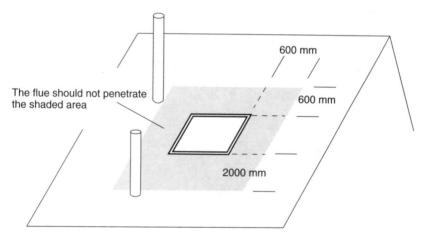

The flue should not penetrate
the shaded area

600 mm

600 mm

2000 mm

Terminals adjacent to windows
or openings on pitched and flat roofs

Figure 6.145 Location of outlets near roof windows from flues serving gas
appliances

Fireplaces – gas fires

Provided it can be shown to be safe, gas fires may be   J (3.7)
installed in fireplaces that have flues designed to serve solid
fuel appliances.

Bases for back boilers

> Back boilers should adequately protect the fabric of　　　J (3.39)
> the building from heat. (see example at Figure 4.146.)

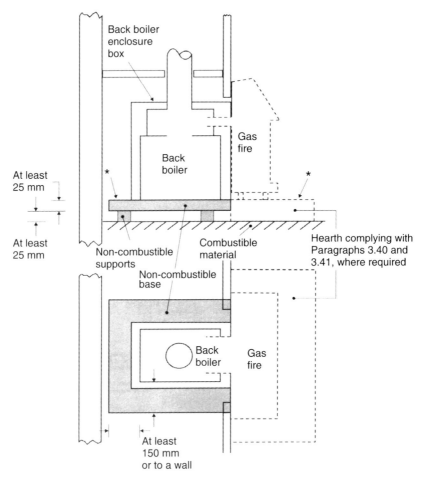

* Where the gas fire requires a hearth, the
back boiler base should be level with it

Figure 6.146 Bases for back boilers

Kerosene and gas oil burning appliances

Kerosene (Class C2) and gas oil (Class D) appliances have the following,
additional, requirements:

Open-fired oil appliances should not be installed in rooms such as bedrooms and bathrooms where there is an increased risk of carbon monoxide poisoning.	J (4.2)
The outlet from a flue should be so situated externally to ensure:	J (4.6)

- the correct operation of a natural draft flue;
- the intake of air if a balanced flue;
- dispersal of the products of combustion.

Figure 6.147 (and Table 6.58) indicates typical positioning to meet this requirement.

Flueblock chimneys

Flueblock chimneys should be installed with sealed joints in accordance with the flueblock manufacturer's installation instructions.	J (4.16)
Flueblocks that are not intended to be bonded into surrounding masonry should be supported and restrained in accordance with the manufacturer's installation instructions.	J (4.16)
Where a fluepipe or chimney penetrates a fire compartment wall or floor, it must not breach the fire separation requirements.	J (4.18) Approved Doc. B

Relining chimney flues (for oil appliances)

Flexible metal flue liners should be installed in one complete length without joints within the chimney.	J (4.22)
Other than for sealing at the top and the bottom, the space between the chimney and the liner should be left empty (unless this is contrary to the manufacturer's instructions).	J (4.22)
Flues that may be expected to serve appliances burning Class D oil (i.e. gas oil) should be made of materials that are resistant to acids.	J (4.23)

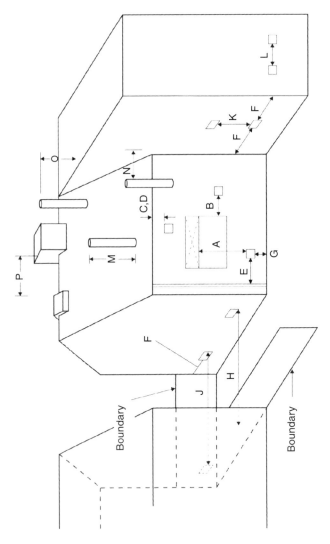

Figure 6.147 Location of outlets from flues serving oil-fired appliances

Table 6.58 Location of outlets from flues serving oil fired appliances

Minimum separation distances for terminals in mm

Location of outlet[1]	Appliance with pressure jet burner	Appliance with vaporizing burner
A Below an opening[2, 3]	600	should not be used
B Horizontally to an opening[2, 3]	600	should not be used
C Below a plastic/painted gutter, drainage pipe or eaves if combustible material protected[4]	75	should not be used
D Below a balcony or a plastic/painted gutter, drainage pipe or eaves without protection to combustible material	600	should not be used
E From vertical sanitary pipework	300	should not be used
F From an external or internal corner or from a surface or boundary alongside the terminal	300	should not be used
G Above ground or balcony level	300	should not be used
H From a surface or boundary facing the terminal	600	should not be used
J From a terminal facing the terminal	1200	should not be used
K Vertically from a terminal on the same wall	1500	should not be used
L Horizontally from a terminal on the same wall	750	should not be used
M Above the highest point of an intersection with the roof	600[6]	1000[5]
N From a vertical structure to the side of the terminal	750[6]	2300
O Above a vertical structure which is less than 750 mm (pressure jet burner) or 2300 mm (vaporizing burner) horizontally from the side of the terminal	600[6]	1000[5]
P From a ridge terminal to a vertical structure on the roof	1500	should not be used

Notes:
1. Terminals should only be positioned on walls where appliances have been approved for such configurations when tested in accordance with BS EN 303-1: 1999 or OFTEC standards OFS A100 or OFS A101.
2. An opening means an openable element, such as an openable window, or a permanent opening such as a permanently open air vent.
3. Notwithstanding the dimensions above, a terminal should be at least 300 mm from combustible material, e.g. a window frame.
4. A way of providing protection of combustible material would be to fit a heat shield at least 750 mm wide.
5. Where a terminal is used with a vaporizing burner, the terminal should be at least 2300 mm horizontally from the roof.
6. Outlets for vertical balanced flues in locations M, N and O should be in accordance with manufacturer's instructions.

Hearths for oil appliances

Hearths are needed to prevent the building catching fire and, whilst it is not a health and safety provision, it is customary to top them with a tray for collecting spilled fuel.   J (4.24)

6.11 Stairs

6.11.1 Requirements

The building shall be designed and constructed so that there are appropriate provisions for the early warning of fire, and appropriate means of escape in case of fire from the building to a place of safety outside the building capable of being safely and effectively used at all material times.

(Approved Document B)

For a typical one- or two-storey dwelling, the requirement is limited to the provision of smoke alarms and to the provision of openable windows for emergency exit (see B1.i).

Airborne and impact sound

Dwellings shall be designed and built so that the noise from normal domestic activity in an adjoining dwelling (or other building) is kept down to a level that:

- *does not affect the health of the occupants of the dwelling;*
- *will allow them to sleep, rest and engage in their normal domestic activities without disruption.*

(Approved Document E)

Stairs, ladders and ramps

All stairs, steps and ladders shall provide reasonable safety between levels in a building.

(Approved Document K1)

In a public building the standard of stair, ladder or ramp may be higher than in a dwelling, to reflect the lesser familiarity and greater number of users.

This requirement only applies to stairs, ladders and ramps that form part of the building.

Pedestrian guarding should be provided for any part of a floor, gallery, balcony, roof, or any other place to which people have access and any light well, basement area or similar sunken area next to a building.

(Approved Document K2)

Requirement K2 (a) applies only to stairs and ramps that form part of the building.

Access and facilities for disabled people

All precautions shall be taken to ensure that all new dwellings, other buildings and student living accommodation shall be reasonably safe and convenient for disabled people to:

- *gain access to and within, buildings other than dwellings and to use them.*
- *visit new dwellings and to use the principal storey.*

(Approved Document M)

6.11.2 Meeting the requirements

Except for kitchens, all habitable rooms in the upper storey(s) of a house served by only one stair should be provided with a window (or external door) that could be used as an emergency exit.	B1 (2.7)

Floors and/or stairs, which separate: E2

- a dwelling from another dwelling;
- a dwelling from another part of the same building that is not used exclusively as part of that dwelling;

shall resist the transmission of airborne sounds.

Floors and/or stairs above a dwelling: E3

- that separate it from another dwelling;
- or separate it from another part of the same building (which is not used exclusively as part of the dwelling);

shall resist the transmission of impact sound such as speech, musical instruments and loudspeakers and impact sources such as footsteps and furniture moving.

Remedial work and conversions

Where it cannot be shown that the existing construction (i.e. wall, floor or stair) meets the sound insulation requirements one of the following treatments shall be used in order to improve the level of sound insulation.

As the implementation of upgrading measures will impose additional loads on the existing structure, it (i.e. the existing structure) should be assessed to ensure that the additional loading can be carried safely with appropriate strengthening where necessary.   E (5.3)

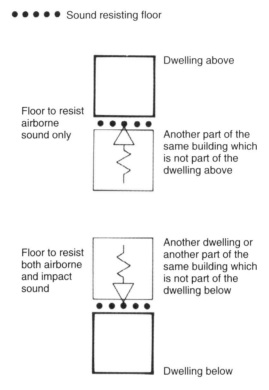

● ● ● ● ● Sound resisting floor

Dwelling above

Floor to resist airborne sound only

Another part of the same building which is not part of the dwelling above

Floor to resist both airborne and impact sound

Another dwelling or another part of the same building which is not part of the dwelling below

Dwelling below

Figure 6.148 Sound resisting floors

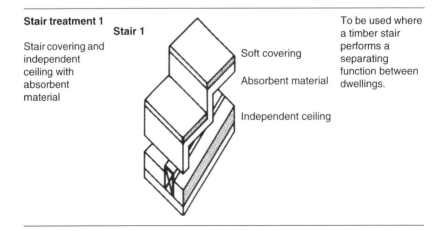

Stair treatment 1

Stair 1

Stair covering and independent ceiling with absorbent material

Soft covering

Absorbent material

Independent ceiling

To be used where a timber stair performs a separating function between dwellings.

Stair treatment 1 (stair covering and independent ceiling with absorbent material)

Timber stairs are subject to the same sound insulation requirements as floors where they perform a separating function. For fire resisting requirements see Approved Document B: Fire safety.

The sound insulation depends on the resilience of the stair covering, the mass of the stair, the mass and isolation of the independent ceiling or airtightness of the cupboard enclosure.

Lay soft covering (e.g. carpet) of at least 6 mm thickness over the stair treads.	E (p34)
If there is a cupboard under all, or part, of the stair:	E (p34)

- line the underside of the stair within the cupboard with plasterboard of at least 12.5 mm thickness with mineral wool within the space above the lining;
- build cupboard walls from two layers of plasterboard of at least 12.5 mm thickness or material of an equivalent mass/m²;
- use a small, heavy, well fitted door for the cupboard.

Where there is no cupboard under the stair, construct an independent ceiling below the stair (see Floor treatment 1).

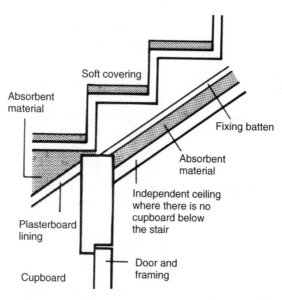

Figure 6.149 Stair covering and independent ceiling with absorbent material (Stair treatment 1)

Piped services

Piped services (excluding gas pipes) and ducts that pass through separating floors in conversions should be surrounded with sound absorbent material for their full height and enclosed in a duct above and below the floor.

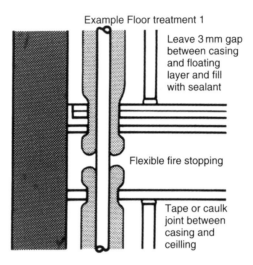

Figure 6.150 Piped services – casing and floor penetrations

Pipes and ducts that penetrate a floor separating habitable rooms in different dwellings should be enclosed above and below the floor.	E (p34)
The material used to form the enclosure should have a mass of at least 15 kg/m².	E (p34)
Either line the enclosure or wrap the pipe or duct within the enclosure with 25 mm of unfaced mineral wool.	E (p34)
Leave a 3 mm gap between the enclosure and floating layer of the floor, seal with caulking or neoprene.	E (p34)
The enclosure may go down to the floor base if Floor treatment 2 is used but ensure isolation from the floating layer.	E (p34)
Penetrations of a separating floor by ducts and pipes should have fire protection in accordance with Approved Document B: Fire safety.	E (p34)
The fire stopping should be flexible and should also prevent rigid contact between the pipe and the floor.	E (p34)

In the Gas Safety Regulations there are requirements for ventilation of ducts at each floor where they contain gas pipes. Gas pipes may be contained in a separate ventilated duct or they can remain unducted.

Stairs, ladders and ramps

The rise of a stair shall be between 155 mm and 220 mm with any going between 245 mm and 260 mm and a maximum pitch of 42°.	K1 (1.1–1.4) MK
The normal relationship between the dimensions of the rise and going is that twice the rise plus the going (2R+G) should be between 550 mm and 700 mm.	
Stairs with open risers that are likely to be used by children under 5 years should be constructed so that a 100 mm diameter sphere cannot pass through the open risers.	K1 (1.9)
Stairs which have more than 36 risers in consecutive flights should make at least one change of direction, between flights, of at least 30°.	K1 (1.14)
If a stair has straight and tapered treads, then the going of the tapered treads should not be less than the going of the straight tread.	K1 (1.20)
The going of tapered treads should measure at least 50 mm at the narrow end.	K1 (1.18)
The going should be uniform for consecutive tapered treads.	K1 (1.19) K1 (1.22–1.24)
Stairs should have a handrail on both sides if they are wider than 1 m and on at least one side if they are less than 1 m wide.	K1 (1.27) M
Handrail heights should be between 900 mm and 1000 mm measured to the top of the handrail from the pitch line or floor.	K1 (1.27) M
Spiral and helical stairs should be designed in accordance with BS 5395.	K1 (1.21)

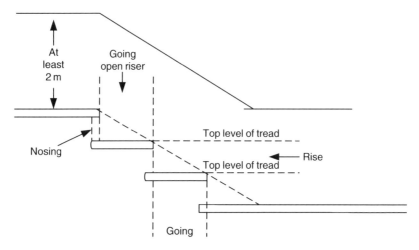

Figure 6.151 Rise and going plus headroom

Steps

Steps should have level treads.	K1 (1.8)
Steps may have open risers, but treads should then overlap each other by at least 16 mm.	K1 (1.8) M
Steps should be uniform with parallel nosings, the stair should have handrails on both sides and the treads should have slip resistant surfaces.	
The headroom on the access between levels should be no less than 2 m.	K1 (1.10)
Landings should be provided at the top and bottom of every flight.	K1 (1.15)
The width and length of every landing should be the same (or greater than) the smallest width of the flight.	K1 (1.15)
Landings should be clear of any permanent obstruction.	K1 (1.16)
Landings should be level.	K1 (1.17)
Any door (entrance, cupboard or duct) that swings across a landing at the top or bottom of a flight of stairs must leave a clear space of at least 400 mm across the full width of the flight.	K1 (1.16)

Flights and landings should be guarded at the sides   K1 (1.28–1.29)
when there is a drop of more than 600 mm.

⚰ For stairs that are likely to be used by children
under 5 years the construction of the guarding shall
be such that a 100 mm sphere cannot pass through
any openings in the guarding and children will not
easily climb the guarding.

• For loft conversions, a fixed ladder should have   K1 (1.25)
  fixed handrails on both sides.

💡 Whilst there are no recommendations for minimum stair widths, designers
should bear in mind the requirements of Approved Documents B (means of
escape) and M (access for disabled people).

Ramps

All ramps shall provide reasonable safety between levels in a building (where
the difference in level is more than 600 mm) and other buildings where the
change of level is more than 380 mm.

Ramps should be clear of permanent obstructions.   K1 (2.4)

The slope of a ramp shall be no more than 1:12.   K1 (2.1)

Ramps should have a handrail on both sides if they are   K1 (2.5)
wider than 1 m and on at least one side if they are   M
less than 1 m wide.

Handrail heights should be between 900 mm and 1000 mm   K1 (2.5)
measured to the top of the handrail from the pitch line   M
or floor.

All ramps should have landings.   K1 (2.6)

All ramps (and associated landings) should have a clear   K1 (2.2)
headroom throughout of at least 2 m.

Ramps and landings should be guarded at the sides when   K1 (2.7)
there is a drop of more than 600 mm.

⚰ For stairs that are likely to be used by children under
5 years the construction of the guarding shall be such that
a 100 mm sphere cannot pass through any openings in the
guarding and children will not easily climb the guarding.

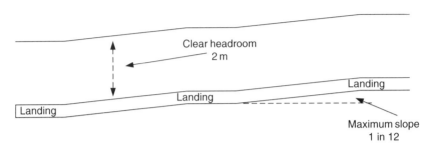

Figure 6.152 The recommended design of a ramp

Protection from falling

All stairs, landings, ramps and edges of internal floors shall have a wall, parapet, balustrade or similar guard at least 900 mm high.	K3 (3.2)
All guarding should be capable of resisting at least the horizontal force given in BS 6399: Part 1: 1996.	K3 (3.2)
If glazing is used as (or part of) the pedestrian guarding, see Approved Document N: Glazing – safety in relation to impact, opening and cleaning.	N
If a building is likely to be used by children under 5 years, the guarding should not have horizontal rails, should stop children from easily climbing it, and the construction should prevent a 100 mm sphere being able to pass through any opening of that guarding.	K3 (3.3)
All external balconies and edges of roofs shall have a wall, parapet, balustrade or similar guard at least 1100 mm high.	K3 (3.2)

💡 Requirement K2 (a) applies only to stairs and ramps that form part of the building.

Access and facilities for disabled people

Reasonable provision should be made to make sure that dwellings (including any purpose-built student living accommodation, other than a traditional hall of residence providing mainly bedrooms and not equipped as self-contained accommodation) ensure that:

A stair providing vertical circulation within the entrance   M2 (7.7a)
storey:

- should have flights whose clear widths are at least
  900 mm;

- should have a suitable continuous handrail   M2 (7.7b)
  on each side of the flight and any intermediate
  landings where the rise of the flight comprises
  three or more rises.

Requirements for common stairs

If a passenger lift is not installed, an internal stair may be used provided
that:

- all step nosings are distinguishable through contrasting   M2 (9.5a)
  brightness;

- there is a continuous handrail on each side of the   M2 (9.5f)
  flights (and the landings) if the overall rise of the stair
  is two or more risers;

- the rise of each step is uniform and not more than   M2 (9.5c)
  170 mm;

- the going of each step is uniform and not less than   M2 (9.5d)
  250 mm;

- the risers are not open.   M2 (9.5e)

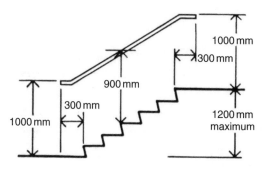

Figure 6.153 Common stairs in blocks of flats

6.12 Windows

6.12.1 Requirements

Protection from falling

Pedestrian guarding should be provided for any part of a floor (including the edge below an opening window) gallery, balcony, roof (including rooflight and other openings), any other place to which people have access and any light well, basement area or similar sunken area next to a building.

(Approved Document K2)

Conservation of fuel and power

Energy efficiency measures shall be provided which:

(a) • *limit the heat loss through the roof, wall, floor, windows and doors, etc. by suitable means of insulation;*
  • *where appropriate permit the benefits of solar heat gains and more efficient heating systems to be taken into account;*
  • *limit unnecessary ventilation heat loss by providing building fabric which is reasonably airtight;*
  • *limit the heat loss from hot water pipes and hot air ducts used for space heating;*
  • *limit the heat loss from hot water vessels and their primary and secondary hot water connections by applying suitable thicknesses of insulation (where such heat does not make an efficient contribution to the space heating).*

(b)   *provide space heating and hot water systems with reasonably efficient equipment such as heating appliances and hot water vessels where relevant, such that the heating and hot water systems can be operated effectively as regards the conservation of fuel and power.*

(c)   *provide lighting systems that utilize energy-efficient lamps with manual switching controls or, in the case of external lighting fixed to the building, automatic switching, or both manual and automatic switching controls as appropriate, such that the lighting systems can be operated effectively as regards the conservation of fuel and power.*

(d)   *provide information, in a suitably concise and understandable form (including results of performance tests carried out during the works), that shows building occupiers how the heating and hot water services can be operated and maintained.*

(Approved Document L1)

💡 Responsibility for achieving compliance with the requirements of Part L rests with the person carrying out the work. That person may be, for example, a developer, a main (or sub-) contractor, or a specialist firm directly engaged by a private client.

The person responsible for achieving compliance should either themselves provide a certificate, or obtain a certificate from the sub-contractor, that commissioning has been successfully carried out. The certificate should be made available to the client and the building control body.

Protection against impact

Glazing with which people are likely to come into contact whilst moving in or about the building, shall:

- *if broken on impact, break in a way which is unlikely to cause injury; or*
- *resist impact without breaking; or*
- *be shielded or protected from impact.*

(Approved Document N)

6.12.2 Meeting the requirement

Protection from falling

All stairs, landings, ramps and edges of internal floors shall have a wall, parapet, balustrade or similar guard at least 900 mm high.	K3 (3.2)
All guarding should be capable of resisting at least the horizontal force given in BS 6399: Part 1: 1996.	K3 (3.2)
If glazing is used as (or part of) the pedestrian guarding, see Approved Document N: Glazing – safety in relation to impact, opening and cleaning.	N
If a building is likely to be used by children under 5 years, the guarding should not have horizontal rails, should stop children from easily climbing it, and the construction should prevent a 100 mm sphere being able to pass through any opening of that guarding.	K3 (3.3)
All external balconies and edges of roofs shall have a wall, parapet, balustrade or similar guard at least 1100 mm high	K3 (3.2)

All windows, skylights, and ventilators shall be capable   K
of being left open without danger of people colliding with
them by:

- installing windows, etc. so that projecting parts are
  kept away from people moving in and around the
  building; or
- installing features that guide people moving in or about
  the building away from any open window, skylight or
  ventilator.

Parts of windows (skylights and ventilators) that project   K
either internally or externally more than about 1000 mm
horizontally into spaces used by people moving in or about
the building should not present a safety hazard.

Conservation of fuel and energy

Single-glazed panels can be acceptable in external doors   L1 (1.5)
provided that the heat loss through all the windows,
doors and rooflights matches the relevant figure in
Table 6.59 and the area of the windows, doors and
rooflights together does not exceed 25% of the total
floor area.

The average U-value of windows, doors and rooflights   L1 (1.7)
should match the relevant figure in Table 6.59 and
the area of the windows, doors and rooflights
together should not exceed 25% of the total floor
area.

💡 Examples of how the average U-value is calculated
are given in Appendix D to Approved Document L1.

For dwellings whose windows have metal frames   L1
(including thermally broken frames) the target U-value
can be increased by multiplying by a factor of 1.03, to
take account of the additional solar gain due to the
greater glazed proportion.

Table 6.59 Elemental method, U-values (W/m²K) for construction elements

Exposed element	U-value
Windows, doors and rooflights (area-weighted average), glazing in metal frames[1]	2.2
Windows, doors and (area-weighted average) rooflights[1], glazing in wood or PVC frames[2]	2.0

Notes:
1. Rooflights include roof windows.
2. The higher U-value for metal-framed windows allows for additional solar gain due to the greater glazed proportion.

Replacement of controlled services or fittings

> When windows, etc. are being upgraded they shall be replaced with new draught-proofed ones either with an average U-value not exceeding the appropriate entry in Table 6.59, or with a centre-pane U-value not exceeding 1.2 W/m²K.
>
> L1 (2.3a)
>
> This requirement does not apply to repair work on parts of these elements, such as replacing broken glass or sealed double-glazing units or replacing rotten framing members.

What about glazing?

Although the installation of replacement windows or glazing (e.g. by way of repair), is not considered as building work under Regulation 3 of the Building Regulations, on the other hand, glazing that:

- is installed in a location where there was none previously;
- is installed as part of an erection;
- is installed as part extension or material alteration of a building;

is subject to these requirements.

The existence of large uninterrupted areas of transparent glazing represents a significant risk of injury through collision. The risk is at its most severe between areas of a building or its surroundings that are essentially at the same level and where a person might reasonably assume direct access between locations that are separated by glazing.

The most likely places where people can sustain injuries are due to impacts with doors, door side panels (especially between waist and shoulder level)

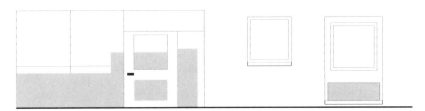

Figure 6.154 Shaded areas show critical locations in internal and external walls

when initial impact can be followed by a fall through the glazing resulting in additional injury to the face and body. Hands, wrists and arms are particularly vulnerable.

Apart from doors, walls and partitions are a low-level, high-risk area, particularly where children are concerned.

The existence of large uninterrupted areas of transparent glazing represents a significant risk of injury through collision. The risk is at its most severe between areas of a building or its surroundings that are essentially at the same level and where a person might reasonably assume direct access between locations that are separated by glazing.

💡 Approved Document B: Fire safety includes guidance on fire-resisting glazing and the reaction of glass to fire.

💡 Approved Document K: Protection from falling, collision and impact covers glazing that forms part of the protection from falling from one level to another, and that needs to ensure containment as well as limiting the risk of sustaining injury through contact.

Some glazing materials, such as annealed glass, gain strength through thickness; others such as polycarbonates or glass blocks are inherently strong. Some annealed glass is considered suitable for use in large areas forming fronts to shops, showrooms, offices, factories, and public buildings.

Protection against impact

Measures shall be taken to limit the risk of sustaining cutting and piercing injuries.	N1 (0.1)
In critical locations, if glazing is damaged the breakage should only result in small, relatively harmless particles.	N1 (0.2)
Glazing should be sufficiently robust to ensure that the risk of breakage is low.	N1 (0.4)

Steps should be taken to limit the risk of contact with the glazing.	N1 (0.5)
Glazing in critical locations should either be permanently protected, be in small panes or if it breaks, break safely (see BS 6206).	N1 (1.2)
Small panes should not exceed 250 mm and an area of 0.5 m².	N1 (1.6)

Transparent glazing

Transparent glazing with which people are likely to come into contact while moving in or about the building shall incorporate features that make it apparent.

The presence of glazing should be made more apparent or visible to people using the building.	N2 (0.8)
The presence of large uninterrupted areas of transparent glazing should be clearly indicated.	N2 (0.6, 2.1, 2.2)
In critical locations (i.e. large areas where the glazing forms part of internal or external walls and doors of shops, showrooms, transoms, offices, factories, public or other non-domestic buildings) the presence of large uninterrupted areas of transparent glazing should be clearly indicated by the use of broken or solid lines, patterns or company logos at appropriate heights and intervals.	N2 (2.4–2.5)

Safe opening and closing of windows

Windows, skylights and ventilators that can be opened by people should be capable of being opened, closed or adjusted safely.

Where controls can be reached without leaning over an obstruction they should not be more than 1.9 m above the floor. Where there is an obstruction, the control should be lower (e.g. not more than 1.7 m where there is a 600 mm deep obstruction).	N3 (3.2)

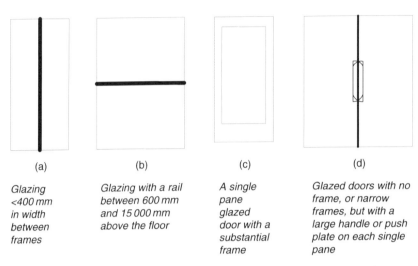

(a)	(b)	(c)	(d)
Glazing <400 mm in width between frames	*Glazing with a rail between 600 mm and 15 000 mm above the floor*	*A single pane glazed door with a substantial frame*	*Glazed doors with no frame, or narrow frames, but with a large handle or push plate on each single pane*

Figure 6.155 Examples of door height glazing not requiring identification further

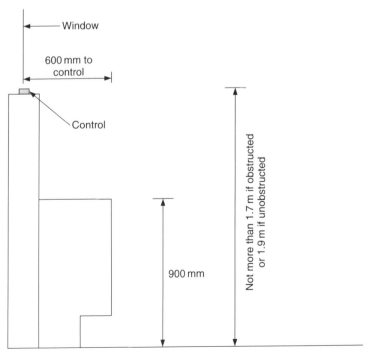

Figure 6.156 Height of controls

Where controls cannot be positioned within safe reach from a permanent stable surface, a safe means of remote operation, such as a manual or electrical system should be provided.

N3 (3.2)

Where there is a danger of the operator or other person falling through a window above ground floor level, suitable opening limiters should be fitted or guarding provided.

N3 (3.3)

Safe access for cleaning

All windows, skylights (or any transparent or translucent walls, ceilings or roofs) of a dwelling should be safely accessible for cleaning.

N

Where glazed surfaces cannot be cleaned safely by a person standing on the ground, the requirement for a floor, or other permanent stable surface, could be satisfied by provisions such as the following:

Safe means of access shall be provided for cleaning both sides of glazed surfaces where there is danger of falling more than 2 m.

N4

Where possible windows should be of a size and design that allow the outside surface to be cleaned safely from inside the building.

N4 (4.2)

Windows that reverse for cleaning should be fitted with a mechanism that holds the window in the reversed position (see BS 8213).

N4 (4.2)

For large buildings (e.g. office blocks) a firm, level surface shall be provided to enable portable ladders (not more than 9 m long) to be used and the use of suspended cradles, travelling ladders, abseiling equipment should also be considered.

N4 (4.2)

6.13 Doors

6.13.1 Requirements

Conservation of fuel and power

Energy efficiency measures shall be provided that limit the heat loss through the doors, etc. by suitable means of insulation.

💡 Responsibility for achieving compliance with the requirements of Part L rests with the person carrying out the work. That person may be, for example, a developer, a main (or sub-) contractor, or a specialist firm directly engaged by a private client.

The person responsible for achieving compliance should either themselves provide a certificate, or obtain a certificate from the sub-contractor, that commissioning has been successfully carried out. The certificate should be made available to the client and the building control body.

Access and facilities for disabled people

All precautions shall be taken to ensure that all new dwellings, other buildings and student living accommodation shall be reasonably safe and convenient for disabled people to:

- *gain access to and within, buildings other than dwellings and to use them.*
- *visit new dwellings and to use the principal storey.*

(Approved Document M)

6.13.2 Meeting the requirement

Doors and gates on main traffic routes (and those that can be pushed open from either side) should have vision panels unless they are low enough (e.g. 900 mm) to see over.	K5 (5.2a)
Sliding doors and gates should have a retaining rail to prevent them falling should the suspension system fail or the rollers leave the track.	K5 (5.2b)
Upward opening doors and gates should be fitted with a device to stop them falling in a way that could cause injury.	K5 (5.2c)
Power operated doors and gates should have safety features to prevent injury to people who are struck or trapped (such as a pressure sensitive door edge that operates the power switch).	K5 (5.2d)

Power operated doors and gates should have a readily identifiable and accessible stop switch.   K5 (5.2d)

Power operated doors and gates should be provided with a manual or automatic opening device in the event of a power failure where and when necessary for health or safety.   K5 (5.2d)

Single-glazed panels can be acceptable in external doors provided that the heat loss through all the windows, doors and rooflights matches the relevant figure in Table 6.57 and the area of the windows, doors and rooflights together does not exceed 25% of the total floor area.   L1 (1.5)

The average U-value of windows, doors and rooflights should match the relevant figure in Table 6.57 and the area of the windows, doors and rooflights together should not exceed 25% of the total floor area.   L1 (1.7)

Examples of how the average U-value is calculated are given in Appendix D to Approved Document L1.

Replacement of controlled services or fittings

When doors, etc. are being upgraded they shall be replaced with new draught-proofed ones either with an average U-value not exceeding the appropriate entry in Table 6.57, or with a centre-pane U-value not exceeding 1.2 W/m²K.   L1 (2.3a)

This requirement does not apply to repair work on parts of these elements, such as replacing broken glass or sealed double-glazing units or replacing rotten framing members.

Access and facilities for disabled people

Reasonable provision should be made to make sure that dwellings (including any purpose-built student living accommodation, other than traditional halls of residence providing mainly bedrooms and not equipped as self-contained accommodation) ensure that:

- an external door providing access for disabled people has a minimum clear opening width of 775 mm;   M2 (6.23)

- corridors and passageways in the entrance storey should be sufficiently wide to allow convenient circulation by a wheelchair user;   M2 (7.2)

- internal doors need to be of a suitable width to facilitate wheelchair manoeuvre;   M2 (7.4)

- a corridor or other access route in the entrance storey or principal storey serving habitable rooms and a room containing a WC (which may be a bathroom) on that level, has an unobstructed width in accordance with Table 6.60;   M2 (7.5a)

- any local permanent obstruction in a corridor, such as a radiator, should not be longer than 2 m and wider than 750 mm;   M2 (7.5b)

- doors to habitable rooms and/or WCs should have a minimum clear opening width as shown in Table 6.60.   M2 (7.5c)

Table 6.60 Minimum widths of corridors and passageways for a range of doorway widths

Doorway clear opening width (mm)	Corridor/passageway width (mm)
750 or wider	900 (when approach is head on)
750	1200 (when approach is not head on)
775	1050 (when approach is not head on)
800	800 (when approach is not head on)

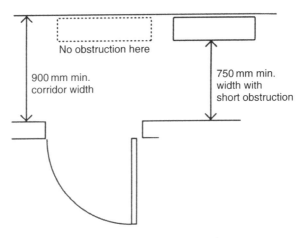

No obstruction here

900 mm min.
corridor width

750 mm min.
width with
short obstruction

Figure 6.157 Corridors, passages and internal doors

6.14 Water (and earth) closets

6.14.1 The requirement

Sanitary conveniences (provided in separate rooms or in bathrooms) shall be:

- *separated from places where food is prepared;*
- *provided with washbasins plumbed with hot and cold water;*
- *be designed and installed so as to allow effective cleaning.*

(Building Act 1984 Section 26)

All plans for buildings must include at least one (or more) water or earth closets **unless** the local authority are satisfied in the case of a particular building that one is not required (for example in a large garage separated from the house).

If you propose using an earth closet, the local authority cannot reject the plans unless they consider that there is insufficient water supply to that earth closet.

Ventilation

Ventilation (mechanical and/or air-conditioning systems designed for domestic buildings) shall be capable of restricting the accumulation of moisture and pollutants originating within a building.

(Approved Document F)

Sanitary conveniences

All dwellings (houses, flats or maisonettes) should have at least one closet and one washbasin which should:

- *be separated by a door from any space used for food preparation or where washing-up is done in washbasins;*
- *ideally, be located in the room containing the closet;*
- *have smooth, non-absorbent surfaces and be capable of being easily cleaned;*
- *be capable of being flushed effectively;*
- *only be connected to a flush pipe or discharge pipe;*
- *washbasins should have a supply of hot and cold water.*

(Approved Document G1)

Access and facilities for disabled people

All precautions shall be taken to ensure that all new dwellings, other buildings and student living accommodation shall be reasonably safe and convenient for disabled people to:

- *gain access to and within buildings other than dwellings and to use them;*
- *visit new dwellings and to use the principal storey.*

(Approved Document M)

If the proposed building is going to be used as a workplace or a factory in which persons of both sexes are going to be employed, then separate closet accommodation **must** be provided unless the local authority approve otherwise.

Business premises

There may be some additional requirements regarding numbers, types and siting of appliances in business premises. If this applies to you then you will need to look at:

- the Offices, Shops and Railway Premises Act 1963,
- the Factories Act 1961, or
- the Food Hygiene (General) Regulations 1970.

6.14.2 Meeting the requirement

All dwellings (houses, flats or maisonettes) should have at least one closet and one washbasin.	G1
Closets (and/or urinals) should be separated by a door from any space used for food preparation or where washing-up is done.	G1
Washbasins should, ideally, be located in the room containing the closet, or, if not, adjacent to the WC.	G1
The surfaces of a closet, urinal or washbasin should be smooth, non-absorbent and capable of being easily cleaned.	G1
Closets (and/or urinals) should be capable of being flushed effectively.	G1
Closets (and/or urinals) should only be connected to a flush pipe or discharge pipe.	G1
Washbasins should have a supply of hot and cold water.	G1
Closets fitted with flushing apparatus should discharge through a trap and discharge pipe into a discharge stack or a drain.	G1

Although the above are the minimum requirements for meeting sanitary conveniences and washing facilities, other local authority regulations may apply and it is worth seeking the advice of the local planning officer before proceeding.

Disabled access

Although there is no requirement to improve access or facilities for disabled people in business premises, if the existing building has been partially demolished (so that only the external walls remain) or the building is going to be a new building, then the following additional requirements apply for disabled people:

Wheelchair users should not have to travel more than one storey to reach a suitable closet.	M3
Closets should allow ease of access and use at any time.	
Sanitary accommodation for wheelchair users shall	M3
<ul><li>be provided on a 'unisex' or 'integral' basis (if unisex, then they should have a separate entrance from other sanitary accommodation);</li><li>enable wheelchair manoeuvrability;</li><li>allow for frontal, lateral, diagonal and backward transfer onto the closet;</li><li>have facilities for hand washing and hand drying;</li><li>be within reach from the closet, prior to transfer.</li></ul>	
Urinals fitted with flushing apparatus should discharge through a grating, a trap and a branch pipe to a discharge stack or a drain.	G1
Closets fitted with macerators and pump can be connected to a discharge pipe discharging to a discharge stack if the macerator and pump system is approved under the current European Technical Approval system.	G1
Washbasins should discharge through a grating, a trap and a branch discharge pipe to a discharge stack or (if it is a ground floor location) into a gully or directly into a drain.	G1
If there is no suitable water supply or means of disposing foul water, closets (and/or urinals) can use chemical treatment.	G1

The Building Act 1984 Sections 64–68

Under existing regulations, all buildings (except factories and buildings used as workplaces) shall be provided with sufficient closet accommodation (or privy) according to the intended use of that building and the amount of people using that building. The only exceptions are if the building (in the view of the local authority) has an insufficient water supply and a sewer is not available.

If a building already has a sufficient water supply and sewer available, the local authority have the authority to insist that the owner of the property replaces any other closet (e.g. an earth closet) with a water closet. In these cases the owner is entitled to claim 50% of the expense of doing this off the local authority.

If the local authority completes the work, then they are entitled to claim 50% back from the owner.

 The owner of the property has **no** right of appeal in these cases.

 In the Greater London area, a 'water closet' can **also** be taken to mean a urinal.

Workplace conveniences

If the building is a workplace used by both sexes, then sufficient and satisfactory accommodation is required for persons of each sex.

This requirement does not apply to premises to which the Offices, Shops and Railway Premises Act of 1963 applies.

Loan of temporary sanitary conveniences

If the local authority is maintaining, improving or repairing drainage systems and this requires the disconnection of existing buildings from these sanitary conveniences, then, on request from the occupier of the building, the local authority are required to supply (on temporary loan and at no charge) sanitary conveniences:

- if the disconnection is caused by a defect in a public sewer;
- if the local authority has order the replacement of earth closets (see above);
- for the first seven days of any disconnection.

Erection of public conveniences

You are not allowed to erect a public sanitary convenience in (or on) any location that is accessible from a street, without the consent of the local authority. Any person who contravenes this requirement is liable to a fine and can be made (at his own expense) to remove or permanently close it.

This requirement does not apply to sanitary conveniences erected by a railway company within their railway station, yard or approaches and by dock undertakers on land belonging to them.

Protection of openings for pipes

Pipes which pass through a compartment wall or compartment floor (unless the pipe is in a protected shaft), or through a cavity barrier, should conform to one of the following alternatives:

Proprietary seals (any pipe diameter) that maintain the fire resistance of the wall, floor or cavity barrier.	B3 (11.5–11.6)

Pipes with a restricted diameter where fire-stopping is used around the pipe, keeping the opening as small as possible.	B3 (11.5 and 11.7)
Sleeving – a pipe of lead, aluminium, aluminium alloy, fibre-cement or UPVC, with a maximum nominal internal diameter of 160 mm, may be used with a sleeving of non-combustible pipe as shown in Figure 6.158.	B3 (11.5 and 11.8)

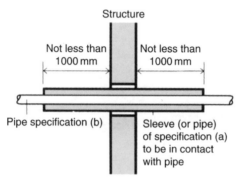

Figure 6.158 Pipes penetrating a structure

Make the opening in the structure as small as possible and provide fire-stopping between pipe and structure

Ventilation

Buildings (or spaces within buildings) other than those:

- into which people do not normally go; or
- that are used solely for storage; or
- that are garages used solely in connection with a single dwelling;

shall be provided with ventilation to:

• extract water vapour from non-habitable areas where it is produced in significant quantities (e.g. kitchens, utility rooms and bathrooms);	F1 (1.1–1.5)
• extract pollutants (which are a hazard to health) from areas where they are produced in significant quantities (e.g. rooms containing processes that generate harmful contaminants and rest rooms where smoking is permitted);	F1 (2.3–2.5)

- rapidly dilute (when necessary) pollutants and   F1 (1.1–1.4)
  water vapour produced in habitable rooms,   F1 (1.6–1.8)
  occupiable rooms and sanitary accommodation;
- provide a minimum supply of fresh air for occupants.   F1 (2.6–2.8)

Table 6.61 Ventilation of rooms containing openable windows (i.e. located on an external wall)

Room	Rapid ventilation (e.g. opening windows)	Background ventilation	Extract ventilation fan rates or passive stack (PSV)
Habitable room	1/20th of floor area	8000 mm²	
Kitchen	Opening window (no minimum size)	4000 mm²	30 litres/second adjacent to a hob or 60 litres/second elsewhere or PSV
Utility room	Opening window (no minimum size)	4000 mm²	30 litres/second or PSV
Bathroom (with or without WC)	Opening window (no minimum size)	4000 mm²	15 litres/second or PSV
Sanitary accommodation (separate from bathroom)	1/20th of floor area or mechanical extract at 6 litres/second.	4000 mm²	

Sanitary conveniences

All dwellings (houses, flats or maisonettes) should have at least one closet and one washbasin.	G1
Closets (and/or urinals) should be separated by a door from any space used for food preparation or where washing-up is done.	G1
Washbasins should, ideally, be located in the room containing the closet if it is not adjacent to the WC.	G1
The surfaces of a closet, urinal or washbasin should be smooth, non-absorbent and capable of being easily cleaned.	G1
Closets (and/or urinals) should be capable of being flushed effectively.	G1
Closets (and/or urinals) should only be connected to a flush pipe or discharge pipe.	G1
Washbasins should have a supply of hot and cold water.	G1
Closets fitted with flushing apparatus should discharge through a trap and discharge pipe into a discharge stack or a drain.	G1

Business premises

There may be some additional requirements regarding numbers, types and siting of appliances in business premises. If this applies to you then you will need to look at:

- the Offices, Shops and Railway Premises Act 1963,
- the Factories Act 1961, or
- the Food Hygiene (General) Regulations 1970.

Although there is no requirement to improve access or facilities for disabled people, if the existing building has been partially demolished (so that only the external walls remain) or the building is going to be a new building, then the following additional requirements apply for disabled people:

Wheelchair users should not have to travel more than one storey to reach a suitable closet.	M3
Closets should allow ease of access and use at any time.	
Sanitary accommodation for wheelchair users	M3
• can be provided on a 'unisex' or 'integral' basis (if unisex, then they should have a separate entrance from other sanitary accommodation); • enable wheelchair manoeuvrability; • allow for frontal, lateral, diagonal and backward transfer onto the closet; • have facilities for hand washing and hand drying; • be within reach from the closet, prior to transfer.	
Urinals fitted with flushing apparatus should discharge through a grating, a trap and a branch pipe to a discharge stack or a drain.	G1
Closets fitted with macerators and pump can be connected to a discharge pipe discharging to a discharge stack if the macerator and pump system is approved under the current European Technical Approval system.	G1
Washbasins should discharge through a grating, a trap and a branch discharge pipe to a discharge stack or (if it is a ground floor location) into a gully or directly into a drain.	G1
If there is no suitable water supply or means of disposing foul water, closets (and/or urinals) can use chemical treatment.	G1

Access and facilities for disabled people

Reasonable provision should be made to make sure that dwellings (including any purpose-built student living accommodation, other than traditional halls of residence providing mainly bedrooms and not equipped as self-contained accommodation) ensure that:

• sanitary accommodation on the principal storey is provided;	M (0.8ii c)
• whenever possible a WC should be provided in the entrance storey or the principal storey of a dwelling;	M2 (10.1–10.3)
• the door to the WC compartment should open outwards, and be positioned so as to enable wheelchair users to access the WC and for the WC to have a washbasin positioned so that it does not impede access;	M2 (10.3)
• to enable transfer, the wheelchair should be able to approach within 400 mm of the front of the WC.	M2 (10.3)

It is appreciated that in most circumstances, it will not always be practical for the wheelchair to be fully accommodated within the WC compartment.

6.15 Lighting

6.15.1 The requirement

Conservation of fuel and power

Energy efficiency measures shall be provided which:

- *provide lighting systems that utilize energy-efficient lamps with manual switching controls or, in the case of external lighting fixed to the building, automatic switching, or both manual and automatic switching controls as appropriate, such that the lighting systems can be operated effectively as regards the conservation of fuel and power.*
- *provide information, in a suitably concise and understandable form (including results of performance tests carried out during the works) that shows building occupiers how the heating and hot water services can be operated and maintained.*

(Approved Document L1)

Responsibility for achieving compliance with the requirements of Part L rests with the person carrying out the work. That person may be, for example, a developer, a main (or sub-) contractor, or a specialist firm directly engaged by a private client.

The person responsible for achieving compliance should either themselves provide a certificate, or obtain a certificate from the sub-contractor, that commissioning has been successfully carried out. The certificate should be made available to the client and the building control body.

Access and facilities for disabled people

All precautions shall be taken to ensure that all new dwellings, other buildings and student living accommodation shall be reasonably safe and convenient for disabled people to:

- *gain access to and within buildings other than dwellings and to use them;*
- *visit new dwellings and to use the principal storey.*

(Approved Document M)

6.15.2 Meeting the requirement

Internal lighting

Table 6.62 gives an indication of a recommended number of locations (excluding garages, lofts and outhouses) that need to be equipped with efficient lighting (L1 (1.55–1.56)).

Table 6.62

Number of rooms created (see Note 1)	Recommended minimum number of locations
1–3	1
4–6	2
7–9	3
10–12	4

Note 1: Hall, stairs and landing(s) count as one room as does a conservatory.

In locations where lighting can be expected to have most use, fixed lighting (e.g. fluorescent tubes and compact fluorescent lamps – but not GLS tungsten lamps with bayonet cap or Edison screw bases) with a luminous efficacy greater than 40 lumens per circuit-watt should be available.	L1 (1.54)

External lighting fixed to the building

External lighting (including lighting in porches, but not lighting in garages and carports) should:

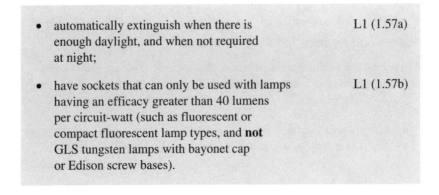

- automatically extinguish when there is enough daylight, and when not required at night;   L1 (1.57a)

- have sockets that can only be used with lamps having an efficacy greater than 40 lumens per circuit-watt (such as fluorescent or compact fluorescent lamp types, and **not** GLS tungsten lamps with bayonet cap or Edison screw bases).   L1 (1.57b)

Access and facilities for disabled people

Reasonable provision should be made to make sure that dwellings (including any purpose-built student living accommodation, other than traditional halls of residence providing mainly bedrooms and not equipped as self-contained accommodation) ensure that:

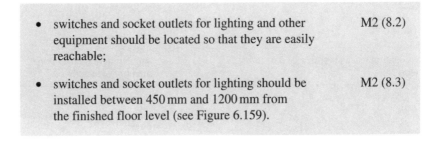

- switches and socket outlets for lighting and other equipment should be located so that they are easily reachable;   M2 (8.2)

- switches and socket outlets for lighting should be installed between 450 mm and 1200 mm from the finished floor level (see Figure 6.159).   M2 (8.3)

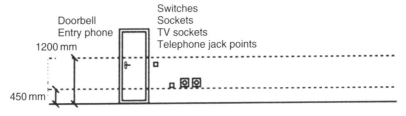

Figure 6.159 Heights for switches and sockets

6.16 Combustion appliances

6.16.1 The requirement

Protection of building

Combustion appliances and fluepipes shall be so installed, and fireplaces and chimneys shall be so constructed and installed, as to reduce to a reasonable level the risk of people suffering burns or the building catching fire in consequence of their use.

(Approved Document J3)

Provision of information

Where a hearth, fireplace, flue or chimney is provided or extended, a durable notice containing information on the performance capabilities of the hearth, fireplace, flue or chimney shall be affixed in a suitable place in the building for the purpose of enabling combustion appliances to be safely installed.

(Approved Document J4)

Conservation of fuel and power

Energy efficiency measures shall be provided which:

(a) • *limit the heat loss through the roof, wall, floor, windows and doors, etc. by suitable means of insulation;*
  • *limit unnecessary ventilation heat loss by providing building fabric which is reasonably airtight;*
  • *limit the heat loss from hot water pipes and hot air ducts used for space heating;*
  • *limit the heat loss from hot water vessels and their primary and secondary hot water connections by applying suitable thicknesses of insulation (where such heat does not make an efficient contribution to the space heating);*

(b) *provide space heating and hot water systems with reasonably efficient equipment such as heating appliances and hot water vessels where relevant, such that the heating and hot water systems can be operated effectively as regards the conservation of fuel and power.*

(c) *provide lighting systems that utilize energy-efficient lamps with manual switching controls or, in the case of external lighting fixed to the building, automatic switching, or both manual and automatic switching controls as appropriate, such that the lighting systems can be operated effectively as regards the conservation of fuel and power.*

*(d) provide information, in a suitably concise and understandable form
      (including results of performance tests carried out during the works) that
      shows building occupiers how the heating and hot water services can be
      operated and maintained.*

(Approved Document L1)

Responsibility for achieving compliance with the requirements of Part L
rests with the person carrying out the work. That person may be, for example,
a developer, a main (or sub-) contractor, or a specialist firm directly engaged
by a private client.

The person responsible for achieving compliance should either themselves
provide a certificate, or obtain a certificate from the sub-contractor, that
commissioning has been successfully carried out. The certificate should be
made available to the client and the building control body.

6.16.2 Meeting the requirement

Air supplies for combustion installations

A room containing an open-flued appliance may need permanently open air vents (Figure 6.160(a) and (c)).	J (1.4)
Appliance compartments that enclose open-flued combustion appliances should be provided with vents large enough to admit all of the air required by the appliance for combustion and proper flue operation, whether the compartment draws its air from a room or directly from outside (Figure 6.160(b) and (c)).	J (1.5)
Where appliances require cooling air, appliance compartments should be large enough to enable air to circulate and high and low level vents should be provided.	J (1.6)

In a flueless situation, air for combustion (and to carry away its products)
can be achieved as shown in Figure 6.162.

Where appliances are to be installed within balanced compartments, special provisions will be necessary.	J (1.7)
If an appliance is room-sealed but takes its combustion air from another space in the building (such as the roof void) or if a flue has a permanent opening to another space in the	J (1.8)

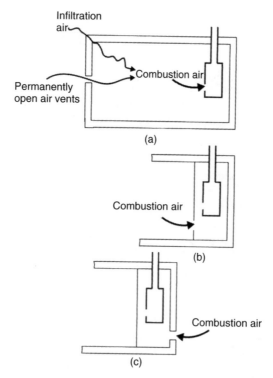

Figure 6.160 Air for combustion and operation of the flue (open flued). (a) Appliance in room. (b) Appliance in appliance compartment with internal vent. (c) Appliance in appliance compartment with external vent

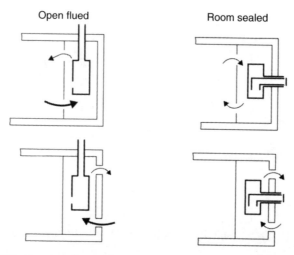

Figure 6.161 Combustion and operation requiring air cooling

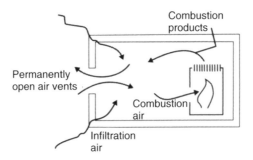

Figure 6.162 Air for combustion and operation of the flue (flueless)

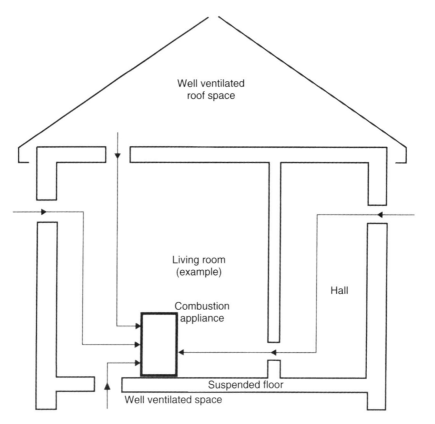

Figure 6.163 Locating permanent air vent openings (examples)

building (such as where it feeds a secondary flue in the roof void), that space should have ventilation openings directly to outside.

Where flued appliances are supplied with combustion   J (1.9)
air through air vents which open into adjoining rooms or
spaces, the adjoining rooms or spaces should have air vent
openings of at least the same size direct to the outside.
Air vents for flueless appliances however, should open
directly to the outside air.

Air vents

Permanently open air vents should be non-adjustable,   J (1.10)
sized to admit sufficient air for the purpose intended and
positioned where they are unlikely to become blocked.

Air vents should be sufficient for the appliances to be   J (1.11)
installed (taking account where necessary of obstructions
such as grilles and anti-vermin mesh).

Air vents should be sited outside fireplace recesses and   J (1.11a)
beyond the hearths of open fires so that dust or ash will not
be disturbed by draughts.

Air vents should be sited in a location unlikely to cause   J (1.11b)
discomfort from cold draughts.

Grilles or meshes protecting air vents from the entry   J (1.15)
of animals or birds should have aperture dimensions
no smaller than 5 mm.

In noisy areas, it may be necessary to install proprietary   J (1.16)
noise attenuated ventilators to limit the entry of noise
into the building.

Discomfort from cold draughts can be avoided by placing vents close to appliances (for example by using floor vents), by drawing air from intermediate spaces such as hallways or by ensuring good mixing of incoming cold air by placing air vents close to ceilings.

In buildings where it is intended to install open-flued   J (1.20)
combustion appliances and extract fans, the combustion
appliances should be able to operate safely whether or not
the fans are running.

For gas appliances where a kitchen contains an open-flued appliance, the extract rate of the kitchen extract fan should not exceed 20 litres/second (72 m³/hour).	J (1.20a)
When installing ventilation for solid fuel appliances avoid installing extract fans in the same room.	J (1.20c)

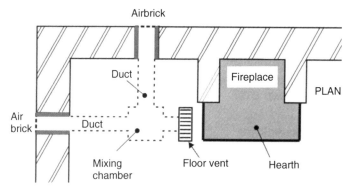

Figure 6.164 Air vent openings in a solid floor

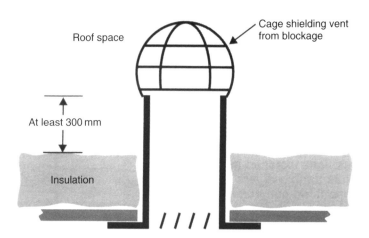

Figure 6.165 Ventilator used in a roof space (e.g. a loft)

Flues

Appliances other than flueless appliances should incorporate or be connected to suitable flues which discharge to the outside air.	J (1.24)
Chimneys and flues should provide satisfactory control of water condensation.	J (1.26)
New chimneys should be constructed with flue liners (clay, concrete or pre-manuafctured) and masonry (bricks, medium weight concrete blocks or stone) suitable for the intended application.	J (1.27)
Liners should be installed in accordance with their manufacturer's instructions.	J (1.28)
Liners need to be placed with the sockets or rebate ends uppermost to contain moisture and other condensates in the flue.	J (1.28)
Joints should be sealed with fire cement, refractory mortar or installed in accordance with their manufacturer's instructions.	J (1.28)
Spaces between the lining and the surrounding masonry should not be filled with ordinary mortar.	J (1.28)

Space heating system controls

Heating systems other than space heating provided by individual solid fuel, gas and electric fires or room heaters should be provided with:

- zone controls,
- timing controls, and
- boiler control interlocks.

For electric storage heaters the system should have an automatic charge control that detects the internal temperature and adjusts the charging of the heater accordingly.

Zone controls

Hot water central heating systems, fan-controlled electric storage heaters and electric panel heaters should control the temperatures independently in areas (such as separate sleeping and living areas) that have different heating needs.	L1 (1.38)

Room thermostats and/or thermostatic radiator valves (or   L1 (1.38)
any other suitable temperature sensing devices) should be
used where appropriate.

In most dwellings one timing zone divided into two   L1 (1.39)
temperature control sub-zones would be sufficient.

Large dwellings should be divided into zones with floor   L1 (1.39)
area no greater than 150 m².

Timing controls

For gas-fired and oil-fired systems and for systems   L1 (1.40)
with solid-fuel-fired boilers (where forced-draught fans
operate when heat is required), timing devices should be
provided to control the periods when the heating systems
operate.

Separate timing control should be provided for space   L1 (1.40)
heating and water heating, except for combination boilers
or solid fuel appliances.

Boiler control interlocks

Gas- and oil-fired hot water central heating system   L1 (1.41)
controls should switch the boiler off when no heat
is required whether control is by room thermostats
or by thermostatic radiator valves.

The boiler in systems controlled by thermostats should   L1 (1.41a)
operate only when a space heating or vessel thermostat
is calling for heat.

A room thermostat or flow switch should also be   L1 (1.41b)
provided to switch off the boiler when there is no
demand for heating or hot water.

Hot water systems

Systems incorporating integral or separate hot water storage vessels should:

Have an insulating vessel with a 35 mm thick, factory-applied coating of PU-foam having a minimum density of 30 kg/m³.	L1 (1.43)
Unvented hot water systems need additional insulation to control the heat losses through the safety fittings and pipework.	L1 (1.43)

Operating and maintenance instructions for heating and hot water systems

The building owner and/or occupier should be given information on the operation and maintenance of the heating and hot water systems.	L1 (1.51)
The instructions should be directly related to the system(s) in the dwelling and should explain to householders how to operate the systems so that they can perform efficiently, and what routine maintenance is advisable for the purposes of the conservation of fuel and power.	L1 (1.51)

Insulation of pipes and ducts

Pipes and ducts should be insulated to conserve heat and hence maintain the temperature of the water or air heating service.	L1 (1.52)
Space heating pipe work located outside the building fabric insulation layer(s) should be wrapped with insulation material having a thermal conductivity at 40 °C not exceeding 0.035 W/mK and a thickness equal to the outside diameter of the pipe up to a maximum of 40 mm.	L1 (1.52a)

Warm air ducts should be insulated in accordance with BS 5422: 2001.

L1 (1.52b)

Hot pipes connected to hot water storage vessels (including the vent pipe, and the primary flow and return to the heat exchanger, where fitted) should be insulated for at least 1 m from their points of connection (or up to the point where they become concealed).

L1 (1.52c)

To protect against freezing, central heating and hot water pipework in unheated areas may also need increased insulation thicknesses.

L1 (1.53)

Replacement of controlled services or fittings

Heating boilers: Replacement heating boilers in dwellings having a floor area greater than 50 m² shall be treated as if it were a new dwelling and in the case of:

L1 (2.3a)

- ordinary oil or gas boilers, shall be by a boiler with a SEDBUK not less than the appropriate entry in Table 2 of L1;

L1 (2.3a(1))

- back boilers, shall be by a boiler having a SEDBUK of not less than three percentage points lower than the appropriate entry in Table 2 of L1;

L1 (2.3a(2))

- solid fuel boilers, shall be by a boiler having an efficiency not less than that recommended for its type in the HETAS certification scheme.

L1 (2.3a(3))

Hot water vessels: Replacements shall all be new equipment as if for a new dwelling.

L1 (2.3c)

Boiler and hot water storage controls: the work may also need to include replacement of the time switch or programmer, room thermostat, and hot water vessel thermostat, and provision of a boiler interlock and fully pumped circulation.

L1 (2.3d)

6.17 Hot water storage

6.17.1 The requirement

Hot water storage

A hot water storage system shall:

- *be installed by a competent person*
- *not exceed 100 °C*
- *discharge safely*
- *not cause danger to persons in or about the building.*

(Approved Document G3)

 Requirement G3 does not apply to a storage system that:

- has a storage capacity of 15 litres or less;
- is used only for providing space heating only;
- heats or stores water for industrial process.

Conservation of fuel and power

Energy efficiency measures shall be provided which:

(a) • *limit the heat loss through the roof, wall, floor, windows and doors, etc. by suitable means of insulation;*
- *where appropriate permit the benefits of solar heat gains and more efficient heating systems to be taken into account;*
- *limit unnecessary ventilation heat loss by providing building fabric which is reasonably airtight;*
- *limit the heat loss from hot water pipes and hot air ducts used for space heating;*
- *limit the heat loss from hot water vessels and their primary and secondary hot water connections by applying suitable thicknesses of insulation (where such heat does not make an efficient contribution to the space heating).*

(b) *Provide space heating and hot water systems with reasonably efficient equipment such as heating appliances and hot water vessels where relevant, such that the heating and hot water systems can be operated effectively as regards the conservation of fuel and power.*

(c) *Provide information, in a suitably concise and understandable form (including results of performance tests carried out during the works) that shows building occupiers how the heating and hot water services can be operated and maintained.*

(Approved Document L1)

💡 Responsibility for achieving compliance with the requirements of Part L rests with the person carrying out the work. That person may be, for example, a developer, a main (or sub-) contractor, or a specialist firm directly engaged by a private client.

The person responsible for achieving compliance should either themselves provide a certificate, or obtain a certificate from the sub-contractor, that commissioning has been successfully carried out. The certificate should be made available to the client and the building control body.

6.17.2 Meeting the requirement

Insulation of pipes and ducts

Pipes and ducts should be insulated to conserve heat and hence maintain the temperature of the water or air heating service.	L1 (1.52)
Space heating pipework located outside the building fabric insulation layer(s) should be wrapped with insulation material having a thermal conductivity at 40 °C not exceeding 0.035 W/mK and a thickness equal to the outside diameter of the pipe up to a maximum of 40 mm.	L1 (1.52a)
Warm air ducts should be insulated in accordance with BS 5422: 2001.	L1 (1.52b)
Hot pipes connected to hot water storage vessels (including the vent pipe, and the primary flow and return to the heat exchanger, where fitted) should be insulated for at least 1 m from their points of connection (or up to the point where they become concealed).	L1 (1.52c)
To protect against freezing, central heating and hot water pipework in unheated areas may also need increased insulation thicknesses.	L1 (1.53)

Replacement of controlled services or fittings

Heating boilers: Replacement heating boilers in dwellings having a floor area greater than 50 m² shall be treated as if it were a new dwelling and in the case of:	L1 (2.3a)
• ordinary oil or gas boilers, shall be by a boiler with a SEDBUK not less than the appropriate entry in Table 2 of L1;	L1 (2.3a(1))
• back boilers, shall be by a boiler having a SEDBUK of not less than three percentage points lower than the appropriate entry in Table 2 of L1;	L1 (2.3a(2))
• solid fuel boilers, shall be by a boiler having an efficiency not less than that recommended for its type in the HETAS certification scheme.	L1 (2.3a(3))
Hot water vessels: Replacements shall all be new equipment as if for a new dwelling.	L1 (2.3c)
Boiler and hot water storage controls: the work may also need to include replacement of the time switch or programmer, room thermostat, and hot water vessel thermostat, and provision of a boiler interlock and fully pumped circulation.	L1 (2.3d)

6.18 Liquid fuel

6.18.1 Meeting the requirement

Storage and supply

Oil and LPG fuel storage installations (including the pipework connecting them to the combustion appliances in the buildings they serve) shall:

• be located and constructed so that they are reasonably protected from fires that may occur in buildings or beyond boundaries;	J (5.1a)
• be reasonably resistant to physical damage and corrosion;	J (5.1a)
• be designed and installed so as to minimize the risk of oil escaping during the filling or maintenance of the tank;	J (5.1bi)

- incorporate secondary containment when there is   J (5.1bii)
  a significant risk of pollution;
- contain labelled information on how to respond   J (5.1biii)
  to a leak.

Table 6.63 Fire protection for oil storage tanks

Location of tank	Protection usually satisfactory
Within a building	Locate tanks in a place of special fire hazard, which should be directly ventilated to outside. Without prejudice to the need for compliance with all the requirements in Schedule 1, the need to comply with Part B should particularly be taken into account.
Less than 1800 mm from any part of a building	(a) Make building walls imperforate:[1] (1) within 1800 mm of tanks with at least 30 minutes' fire resistance;[2] (2) to internal fire and construct eaves within 1800 mm of tanks and extending 300 mm beyond each side of tanks with at least 30 minutes' fire resistance to external fire and with non-combustible cladding; or (b) Provide a firewall[3] between the tank and any part of the building within 1800 mm of the tank and construct eaves as in (a) above. The firewall should extend at least 300 mm higher and wider than the affected parts of the tank.
Less than 760 mm from a boundary	Provide a firewall between the tank and the boundary or a boundary wall having at least 30 minutes' fire resistance on either side. The firewall or the boundary wall should extend at least 300 mm higher and wider than the top and sides of the tank.
At least 1800 mm from the building and at least 760 mm from a boundary	No further provisions necessary.

Notes:
1. Excluding small openings such as air bricks etc.
2. Fire resistance in terms of insulation, integrity and stability.
3. Firewalls are imperforate non-combustible walls or screens, such as masonry walls or steel screens.

Oil pollution

The Control of Pollution (Oil Storage) (England) Regulations 2001 (SI 2001/2954) came into force on 1 March 2002. They apply to a wide range of oil storage installations in England, but they do not apply to the storage of oil on any premises used wholly or mainly as one or more private dwellings, if the capacity of the tank is 3500 litres or less.

LPG storage

LPG installations are controlled by legislation enforced by the HSE which includes the following requirements applicable to dwellings:

The LPG tank should be installed outdoors and not within an open pit.	J (5.15)
The tank should be adequately separated from buildings, the boundary and any fixed sources of ignition to enable safe dispersal in the event of venting or leaks and in the event of fire to reduce the risk of fire spreading (see Figure 6.166).	J (5.15)
Firewalls may be free-standing built between the tank and the building, boundary and fixed source of ignition (see Figure 6.166(b)) or a part of the building or a fire resistance (insulation, integrity and stability) boundary wall belonging to the property.	J (5.16)
Where a firewall is part of the building or a boundary wall, it should be located in accordance with Figure 6.166(c).	J (5.16)
If the firewall is part of the building then it should be constructed as shown in Figure 6.166(d).	J (5.16)
Firewalls would be imperforate and of solid masonry, concrete or similar construction.	J (5.17)
Firewalls should have a fire resistance (insulation, integrity and stability) of at least 30 minutes.	J (5.17)
If firewalls are part of the building as shown in Figure 6.166(d), they should have a fire resistance (insulation, integrity and stability) of at least 60 minutes.	J (5.17)
To ensure good ventilation, firewalls should not normally be built on more than one side of a tank.	J (5.17)
A firewall should be at least as high as the pressure relief valve.	J (5.18)

Where an LPG storage installation consists of a set of cylinders, a way of meeting the requirements is as shown in Figure 6.167.

Cylinders to stand upright, secured by straps or chains against a wall outside the building in a well ventilated position at ground level.	J (5.20)
Cylinders should be provided with a firm, level base such as concrete at least 50 mm thick or paving slabs bedded on mortar.	J (5.20)

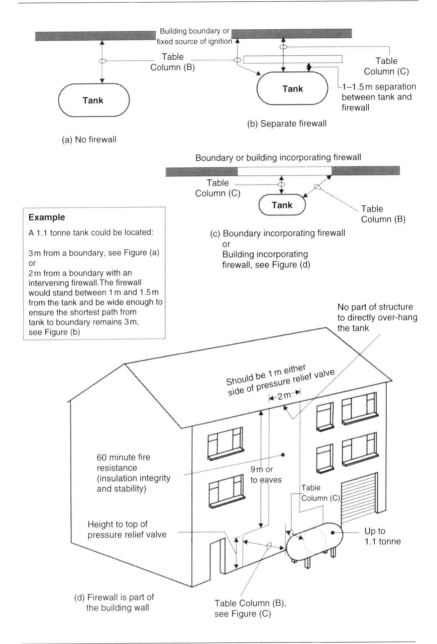

Example

A 1.1 tonne tank could be located:

3 m from a boundary, see Figure (a)
or
2 m from a boundary with an intervening firewall. The firewall would stand between 1 m and 1.5 m from the tank and be wide enough to ensure the shortest path from tank to boundary remains 3 m, see Figure (b)

A Capacity of tank (tonnes)	Minimum separation	
	B tank with no firewall	**C** tank shielded by firewall
0.25	2.5	0.3
1.1	3.0	1.5

Figure 6.166 Separation or shielding of LPG tanks from building, boundaries and fixed sources of ignition

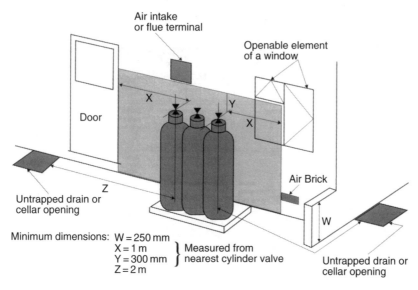

Figure 6.167 Location of LPG cylinders

6.19 Cavities and concealed spaces

6.19.1 The requirement

Internal fire spread (structure)

- *ideally the building should be sub-divided by elements of fire-resisting construction into compartments;*
- *any hidden voids in the construction shall be sealed and sub-divided to inhibit the unseen spread of fire and products of combustion, in order to reduce the risk of structural failure, and the spread of fire.*

(Approved Document B3)

6.19.2 Meeting the requirement

All concealed spaces and/or cavities formed in the construction of a building shall restrict smoke and flame spread.	B3 (10.1–10.4)

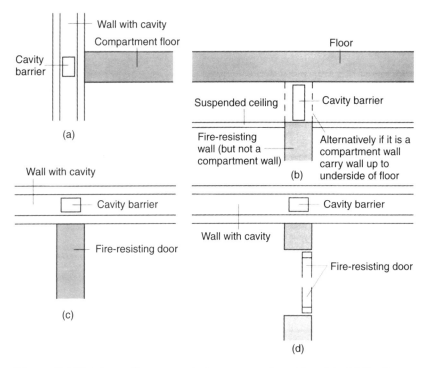

Figure 6.168 Interrupting concealed spaces and cavities. (a), (b) Sections. (c), (d) Plans

Cavities

Cavities should be constructed to provide at least 30 minutes' fire resistance.	B3 (10.6)
Cavity barriers should be tightly fitted to and mechanically fixed in position so that they are unlikely to be affected by building movement or failure.	B3 (10.8–10.9)
Any openings in a cavity barrier should be limited to those for:	B3 (10.14)

- doors that have at least 30 minutes' fire resistance;
- the passage of pipes;
- the passage of cables or conduits;
- openings fitted with a suitably mounted automatic fire damper;
- ducts that (unless they are fire-resisting) are fitted with a suitably mounted automatic fire damper where they pass through the cavity barrier.

Table 6.64 Provision of cavity barriers

Cavity barriers to be provided	Dwelling houses	Flats or maisonettes	Other residential or institutional	Non-residential (e.g. office, shop, storage)
At the junction between an external cavity wall and a compartment wall that separates buildings; and at the top of such an external cavity wall.	X	X	X	X
Above the enclosures to a protected stairway in a house with a floor more than 4.5 m above ground level.	X			
At the junction between an external cavity wall and every compartment floor and compartment wall.		X	X	X
At the junction between a cavity wall and every compartment floor, compartment wall, or other wall or door assembly that forms a fire-resisting barrier.		X	X	X
In a protected escape route, above and below any fire-resisting construction that is not carried full storey height, or (in the case of a top storey) to the underside of the roof covering.		X	X	X
Where the corridor should be sub-divided to prevent fire or smoke affecting two alternative escape routes simultaneously above any such corridor enclosures that are not carried full storey height, or (in the case of the top storey) to the underside of the roof covering.			X	X
Above any bedroom partitions that are not carried full storey height, or (in the case of the top storey) to the underside of the roof covering.			X	
To sub-divide any cavity (including any roof space but excluding any underfloor service void).			X	X
Within the void behind the external face of rainscreen cladding at every floor level, and on the line of compartment walls abutting the external wall, of buildings that have a floor 18 m or more above ground level.		X	X	
At the edges of cavities (including around openings).	X	X	X	X

6.20 Kitchens and utility rooms

6.20.1 The requirement

Ventilation

Ventilation (mechanical and/or air-conditioning systems designed for domestic buildings) shall be capable of restricting the accumulation of moisture and pollutants originating within a building.

(Approved Document F1)

6.20.2 Meeting the requirement

Buildings (or spaces within buildings) other than those:

- into which people do not normally go; or
- that are used solely for storage; or
- that are garages used solely in connection with a single dwelling;

shall be provided with ventilation to:

• extract water vapour from non-habitable areas where it is produced in significant quantities (e.g. kitchens, utility rooms and bathrooms);	F1 (1.1–1.5)
• extract pollutants (which are a hazard to health) from areas where they are produced in significant quantities (e.g. rooms containing processes that generate harmful contaminants and rest rooms where smoking is permitted);	F1 (2.3–2.5)
• rapidly dilute (when necessary) pollutants and water vapour produced in habitable rooms, occupiable rooms and sanitary accommodation;	F1 (1.1–1.4) F1 (1.6–1.8)
• provide a minimum supply of fresh air for occupants.	F1 (2.6–2.8)

Table 6.65 Ventilation of rooms containing openable windows (i.e. located on an external wall)

Room	Rapid ventilation (e.g. opening windows)	Background ventilation	Extract ventilation fan rates or passive stack (PSV)
Habitable room	1/20th of floor area	8000 mm^2	
Kitchen	Opening window (no minimum size)	4000 mm^2	30 litres/second adjacent to a hob or 60 litres/second elsewhere or PSV
Utility room	Opening window (no minimum size)	4000 mm^2	30 litres/second or PSV
Bathroom (with or without WC)	Opening window (no minimum size)	4000 mm^2	15 litres/second or PSV
Sanitary accommodation (separate from bathroom)	1/20th of floor area or mechanical extract at 6 litres/second	4000 mm^2	

6.21 Storage of food

6.21.1 The requirement (Building Act 1984 Sections 28 and 70)

All houses or buildings that have been converted into houses must provide sufficient and suitable accommodation for storing food or *'sufficient and suitable space for the provision of such accommodation by the occupier'*.

This could prove to be a problem when submitting plans for approval and you would be wise to consider the possibilities.

6.22 Refuse facilities

6.22.1 The requirement (Building Act 1984 Section 23)

Probably due to new EU agreements, local authorities have become more strict in seeing that the requirements contained in the Building Act for storage and collection of refuse are applied. This means that you have to ensure that the building is equipped with a satisfactory method for storing refuse (with a house this would normally be a simple dustbin; with a block of flats or a factory, however, the system for storage would have to be more sophisticated). The local council will also need to be able to collect and remove this refuse easily and so this should be borne in mind when siting the refuse collection point.

🔔 Under the Building Act 1984 it is **unlawful** for any person (except with the consent of the local authority) to close or obstruct the means of access by which refuse or faecal matter is removed from a building.

Airborne and impact sound

Dwellings shall be designed and built so that the noise from normal domestic activity in an adjoining dwelling (or other building) is kept down to a level that:

- *does not affect the health of the occupants of the dwelling;*
- *will allow them to sleep, rest and engage in their normal domestic activities without disruption.*

(Approved Document E)

6.22.2 Meeting the requirement

Refuse chutes

A wall separating a habitable room or kitchen and a refuse chute should have mass (including any finishes) of at least 1320 kg/m².	E (1.3)
A wall separating a non-habitable room, which is in a dwelling, from a refuse chute should have a mass (including any finishes) of at least 220 kg/m².	E (1.3)

6.23 Fire resistance

The overall aim of fire safety precautions is to ensure that:

- a satisfactory means of giving an alarm of fire is available;
- a satisfactory means of escape for persons in the event of fire in a building is available (see Approved Document B1);
- that fire spread over the internal linings of buildings is inhibited (see Approved Document B2);
- the stability of buildings is ensured in the event of fire;
- there is a sufficient degree of fire separation within buildings and between adjoining buildings;
- the unseen spread of fire and smoke in concealed spaces in buildings is inhibited (see Approved Document B3);
- external walls and roofs have adequate resistance to the spread of fire over the external envelope;
- the spread of fire from one building to another is restricted (see Approved Document B4);

- there is satisfactory access for fire appliances to buildings;
- there are facilities in buildings to assist fire fighters in the saving of life of people in and around buildings (see Approved Document B5).

Many of the requirements are, of course, closely interlinked. For example, there is a close link between the provisions for means of escape (B1) and those for the control of fire growth (B2), fire containment (B3), and facilities for the fire service (B5). Similarly there are links between B3 and the provisions for controlling external fire spread (B4), and between B3 and B5. Interaction between these different requirements should be recognized where variations in the standard of provision are being considered.

Factors that should be taken into account include:

- the anticipated probability of a fire occurring;
- the anticipated fire severity;
- the ability of a structure to resist the spread of fire and smoke;
- the consequential danger to people in and around the building.

Measures that could be incorporated include:

- the adequacy of means to prevent fire;
- early fire warning by an automatic detection and warning system;
- the standard of means of escape;
- provision of smoke control;
- control of the rate of growth of a fire;
- the adequacy of the structure to resist the effects of a fire;
- the degree of fire containment;
- fire separation between buildings or parts of buildings;
- the standard of active measures for fire extinguishment or control;
- facilities to assist the fire service;
- availability of powers to require staff training in fire safety and fire routines, e.g. under the Fire Precautions Act 1971, the Fire Precautions (Workplace) Regulations 1997, or registration or licensing procedures;
- consideration of the availability of any continuing control under other legislation that could ensure continued maintenance of such systems;
- management.

The design of fire safety in hospitals is covered by Health Technical Memorandum (HTM) 81 Fire precautions in new hospitals (revised 1996).

Building Regulations are intended to ensure that a reasonable standard of life safety is provided, in case of fire. The protection of property, including the building itself, may require additional measures, and insurers will in general seek their own higher standards, before accepting the insurance risk. Guidance is given in the LPC *Design guide for the fire protection of buildings*.

💡 Guidance for assisting protection in Civil and Defence Estates is given in the *Crown Fire Standards* published by the Property Advisers to the Civil Estate (PACE).

6.23.1 The requirement

The building shall be designed and constructed so that there are appropriate provisions for the early warning of fire, and appropriate means of escape in case of fire from the building to a place of safety outside the building capable of being safely and effectively used at all material times.

(Approved Document B1)

💡 For a typical one- or two-storey dwelling, the requirement is limited to the provision of smoke alarms and to the provision of openable windows for emergency exit (see B1.i).

As a fire precaution, all materials used for internal linings of a building should have a low rate of surface flame spread and (in some cases) a low rate of heat release.

(Approved Document B2)

Internal fire spread (structure)

- *all loadbearing elements of structure of the building shall be capable of withstanding the effects of fire for an appropriate period without loss of stability;*
- *ideally the building should be subdivided by elements of fire-resisting construction into compartments;*
- *all openings in fire-separating elements shall be suitably protected in order to maintain the integrity of the continuity of the fire separation;*
- *any hidden voids in the construction shall be sealed and subdivided to inhibit the unseen spread of fire and products of combustion, in order to reduce the risk of structural failure, and the spread of fire.*

(Approved Document B3)

External fire spread

- *external walls shall be constructed so that the risk of ignition from an external source, and the spread of fire over their surfaces, is restricted;*
- *the amount of unprotected area in the side of the building shall be restricted so as to limit the amount of thermal radiation that can pass through the wall;*
- *the roof shall be constructed so that the risk of spread of flame and/or fire penetration from an external fire source is restricted;*
- *the risk of a fire spreading from the building to a building beyond the boundary, or vice versa shall be limited.*

(Approved Document B4)

Access facilities for the fire service

The following should be available:

- *vehicle access for fire appliances;*
- *access for fire-fighting personnel;*
- *the provision of fire mains within the building (for non-domestic buildings);*
- *venting for heat and smoke from basement areas.*

(Approved Document B5)

6.23.2 Meeting the requirements

Access facilities for the fire service

The following should be available:

- *vehicle access for fire appliances;*
- *access for fire-fighting personnel;*
- *the provision of fire mains within the building (for non-domestic buildings);*
- *venting for heat and smoke from basement areas.*

(Approved Document B5)

For dwellings and other small buildings, it is usually only necessary to ensure that the building is sufficiently close to a point accessible to fire brigade vehicles. In more detail this includes:

There should be vehicle access for a pump appliance to small buildings (those of up to 2000 m² with a top storey up to 11 m above ground level) to either: B4 (17.2)

- 15% of the perimeter; or
- within 45 m of every point on the projected plan area (or 'footprint') of the building;

whichever is the less onerous.
Note: For single family dwelling houses, the 45 m may be measured to a door to the dwelling.

There should be vehicle access for a pump appliance to blocks of flats/maisonettes to within 45 m of every dwelling entrance door.   B4 (17.3)

Every elevation should have a suitable door, not less than 750 mm wide, giving access to the interior of the building.   B4 (17.5)

Fire mains installed in a building should be equipped with valves, etc. so that the fire service may connect hoses for water to fight fires inside the building.	B4 (16.1)
Buildings provided with firefighting shafts should be provided with fire mains in those shafts.	B4 (16.2)
There should be one fire main in every firefighting shaft.	B4 (16.4)

Fire resistance

Table 6.66 sets out the minimum periods of fire resistance for elements of structure for domestic residential buildings (see Approved Document B Table A2).

Table 6.66 Minimum periods of fire resistance (dwellings)

Purpose group of building	Minimum periods for elements for a structure in a:					
	Basement storey (depth of lowest basement)		Ground or upper storey (height of top floor above ground)			
	>10 m	<10 m	<5 m	<18 m	<30 m	>30 m
(a), (b) and (c) Flats and maisonettes	90 mins	60 mins	30 mins	60 mins	90 mins	120 mins
Dwellinghouses	N/A	30 mins	30 mins	60 mins	N/A	N/A

6.24 Means of escape

6.24.1 The requirement

Subject to Section 30(3) of the Fire Precautions Act 1971, if a building (or proposed building) exceeds two storeys in height and the floor of any upper storey is more than 20 ft above the surface of the street or ground on any side of the building and is:

- *let out as flats or tenement dwellings;*
- *used as an inn, hotel, boarding-house, hospital, nursing home, boarding-school, children's home or similar institution; or is*
- *used as a restaurant, shop, store or warehouse and has on an upper floor sleeping accommodation for persons employed on the premises.*

*then it must be equipped with adequate means of escape in case of fire, from
each storey.*

(Building Act 1984 Section 72)

*The building shall be designed and constructed so that there are appropriate
provisions for the early warning of fire, and appropriate means of escape in
case of fire from the building to a place of safety outside the building capable
of being safely and effectively used at all material times.*

(Approved Document B1)

For a typical one- or two-storey dwelling, the requirement is limited to
the provision of smoke alarms and to the provision of openable windows for
emergency exit (see B1.i).

6.24.2 Meeting the requirement

Fire detection and alarms

Dwellings of one or two storeys

Ideally dwellings should:

- be protected with an automatic fire detection   B1 (1.2–1.3)
  and alarm system in accordance with the relevant
  recommendations of BS 5839: Part 1 Fire
  detection and alarm systems for buildings,
  Code of practice for system design, installation
  and servicing, to at least an L3 standard,
  or BS 5839: Part 6 Code of practice for the
  design and installation of fire detection and
  alarm systems in dwellings, to at least a
  Grade E type LD3 standard;

- either be provided with a suitable number of   B1 (1.4)
  mains-operated smoke alarms (conforming to
  BS 5446 Components of automatic fire alarm
  systems for residential premises, Part 1
  Specification for self-contained smoke alarms
  and point-type smoke detectors) or a battery
  (either rechargeable or replaceable) device,
  or capacitor.

Large houses

A large house of more than three storeys (including basement storeys) with any of its storeys exceeding 200 m²:

• should either be fitted with an L2 system or a Grade B type LD3 as described in BS 5839: Part 1: 1988.	B1 (1.6 and 1.7)
Lofts in a one- or two-storey house that have been converted into habitable accommodation, should have an automatic smoke detection and alarm system linked to the main house alarm system.	B1 (1.8)

Flats and maisonettes

The same principles apply within flats and maisonettes as for houses, while noting that:

• a flat with accommodation on more than one level (i.e. a maisonette) should be treated in the same way as a house with more than one storey;	B1 (1.9)
• where groups of students share one flat with its own entrance door, an automatic detection system should be provided within each flat.	B1 (1.9)

Smoke alarms – dwellings

Smoke alarms should normally be positioned in the circulation spaces between sleeping spaces and places where fires are most likely to start (e.g. kitchens and living rooms) to pick up smoke in the early stages, while also being close enough to bedroom doors for the alarm to be effective when occupants are asleep.

Requirement

There should be at least one smoke alarm on every storey of a house (including bungalows).	B1 (1.12)
If more than one smoke alarm is installed in a dwelling then they should be linked so that if a unit detects smoke it will operate the alarm signal of all the smoke detectors.	B1 (1.13)

There should be a smoke alarm in the circulation
space within 7.5 m of the door to every habitable
room.

B1 (1.14a)

If a kitchen area is not separated from the
stairway or circulation space by a door, there
should also be an additional heat detector in the
kitchen, that is interlinked to the
other alarms.

B1 (1.14b)

Installation

Smoke alarms should ideally be ceiling mounted,
25 mm and 600 mm below the ceiling
(25–150 mm in the case of heat detectors)
and at least 300 mm from walls and light
fittings.

B1 (1.14c–d)

💡 Units designed for wall mounting may also
be used provided that the units are above the level
of doorways opening into the space.

Smoke alarms should not be fixed over a stair shaft
or any other opening between floors.

B1 (1.15)

Smoke alarms should not be fixed next
to or directly above heaters or air conditioning
outlets.

B1 (1.16)

Smoke alarms should not be fixed in bathrooms,
showers, cooking areas or garages, or any other
place where steam, condensation or fumes could
give false alarms.

B1 (1.16)

Smoke alarms should not be fitted in places that
get very hot (such as a boiler room), or very cold
(such as an unheated porch).

B1 (1.16)

Smoke alarms should not be fixed to surfaces
that are normally much warmer or colder than
the rest of the space.

B1 (1.16)

Power supplies

The power supply for a smoke alarm system:

• should be derived from the dwelling's mains electricity supply from a single independent circuit at the dwelling's main distribution board (consumer unit);	B1 (1.17)
• should include a stand-by power supply that will operate during mains failure;	B1 (1.17–18)
• should preferably not be protected by any residual current device (rcd).	B1 (1.20)

💡 Smoke alarms may be interconnected using radio-links, provided that this does not reduce the lifetime or duration of any stand-by power supply.

Smoke alarms – other than dwellings

The type of fire alarm/smoke detection system required for a particular building depends on the type of occupancy and the means of escape strategy (e.g. simultaneous, phased or progressive horizontal evacuation) – (B1 (1.23)).

Fire alarms

All fire detection and fire-warning systems shall be properly designed, installed and maintained (B1 (1.32)).

All buildings should have arrangements for detecting fire.	B1 (1.25)
All buildings should have the means of raising an alarm in case of fire (e.g. rotary gongs, handbells or shouting 'fire') or be fitted with a suitable electrically operated fire warning system (in compliance with BS 5839).	B1 (1.26)
The fire warning signal should be distinct from other signals that may be in general use.	B1 1.28)
In premises used by the general public, e.g. large shops and places of assembly, a staff alarm system (complying with BS 5839) may be used.	B1 (1.29)

Dwellings

In the event of fire, suitable means shall be provided for emergency egress from each storey.	B1 (2.1)
Floors more than 7.5 m above ground level (where the risk that the stairway will become impassable before occupants of the upper parts of the house have escaped is appreciable) will be provided with an alternative route.	B1 (2.1)
Except for kitchens, all habitable rooms in the upper storey(s) of a house served by only one stair should be provided with a window (or external door) that could be used as an emergency exit.	B1 (2.7)
Except for kitchens, all habitable rooms in the ground storey should either:	B1 (2.8)

- open directly onto a hall leading to the entrance or other suitable exit; or
- be provided with a window (or door).

Where a sleeping gallery is provided:	B1 (2.9a–c)

- the gallery should be not more than 4.5 m above ground level;
- the distance between the foot of the access stair to the gallery and the door to the room containing the gallery should not exceed 3 m;
- galleries longer than 7.5 m should be provided with a separate alternative exit.

Any cooking facilities within a room containing a gallery should either:	B1 (2.9d)

- be enclosed with fire-resisting construction; or
- be positioned so that they do not prejudice escape from the gallery.

Inner rooms

A room whose only escape route is through another room is termed an inner room and is at risk if a fire starts in that other room (access room). Such an arrangement is only acceptable where the inner room is:

- a kitchen;
- a laundry or utility room;

- a dressing room;
- a bathroom, WC, or shower room;
- any other room on a floor not more than 4.5 m above ground level.

Balconies and flat roofs

A flat roof being used as a means of escape should:

- be part of the same building from which escape is being made;
- should lead to a storey exit or external escape route;
- should provide 30 minutes' fire resistance.

💡 Where a balcony or flat roof is provided for escape purposes, guarding may be needed.

• all external balconies and edges of roofs shall have a wall, parapet, balustrade or similar guard at least 1100 mm high.	K3 (3.2)

Basements

If the basement storey contains a habitable room it should have either:

- an external door or window suitable for emergency egress; or
- a protected stairway leading from the basement to a final exit.

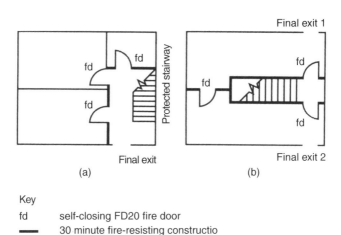

Key

fd   self-closing FD20 fire door
━━   30 minute fire-resisting constructio

Figure 6.169 Alternative arrangements for final exits. In loft conversions, existing doors need only be made self-closing

Emergency egress windows and external doors

any window provided for emergency egress purposes or external door provided for escape should have an unobstructed openable area of at least 0.33 m² that is a minimum of 450 mm high, 450 mm wide and not more than 1100 mm from the floor (see also Figure 6.170).	B1 (2.11)

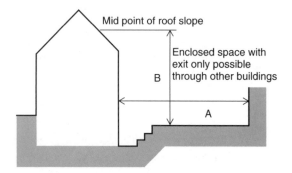

Depth of back garden A should exceed the height B of the house above ground level for it to be acceptable for an escape route from the ground or basement storey to open into the garden.

Figure 6.170 Ground or basement storey exit into an enclosed space

Houses with floors more than 4.5 m above ground level

The top storey should be separated from the lower storeys by fire-resisting construction and be provided with an alternative escape route leading to its own final exit.	B1 (2.13b)
If a house has two or more storeys with floors more than 4.5 m above ground level (typically a house of four or more storeys), then an alternative escape route should be provided from each storey or level situated 7.5 m or more above ground level.	B1 (2.14)

Means of escape

Loft conversions

Where an existing two-storey house is being enlarged by converting the existing roof space into habitable rooms then, provided the new second storey

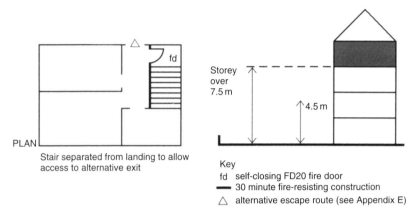

PLAN

Stair separated from landing to allow
access to alternative exit

Storey
over
7.5 m

4.5 m

Key
fd   self-closing FD20 fire door
━━ 30 minute fire-resisting construction
△   alternative escape route (see Appendix E)

Figure 6.171 Fire separation in houses with more than one floor over 4.5 m above ground level

does not exceed 50 m² in floor area, or contain more than two habitable rooms, then:

- the stairs in the ground and first storeys should be enclosed with walls and/or partitions that are fire-resisting;   B1 (2.18a)

- the enclosure should either: B1 (2.18b)

  – extend to a final exit, see Figure 6.169(a); or
  – give access to at least two escape routes at ground level

- every doorway within the enclosure to the existing stair should be fitted with a door, which (in the case of doors to habitable rooms) should be fitted with a self-closing device;   B1 (2.19)

- any new door to a habitable room should be a fire door;   B1 (2.19)

- any glazing (whether new or existing) in the enclosure to the existing stair, including all doors (whether or not they need to be fire doors), but excluding glazing to a bathroom or WC, should be fire-resisting and retained by a suitable glazing system and beads compatible with the type of glass;   B1 (2.20)

- the new storey should be served by a stair that is an extension of an existing stairway;   B1 (2.21)

- the new storey should be separated from the rest of the house by a fire-resistant door;   B1 (2.22)

- the room (or rooms) in any new storey should each have an openable window or rooflight that meets the relevant provisions in Figure 6.172. A door to a roof terrace is also acceptable.   B1 (2.24)

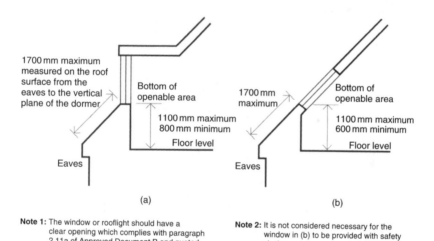

Note 1: The window or rooflight should have a clear opening which complies with paragraph 2.11a of Approved Document B and quoted on page 437

Note 2: It is not considered necessary for the window in (b) to be provided with safety glazing.

Figure 6.172 Position of dormer window or rooflight that is suitable for emergency exit from a loft conversion of a two-storey dwelling house. (a) Dormer window (the window may be in the end wall of the house, instead of the roof as shown). (b) Rooflight or roof window

Flats and maisonettes

The means of escape from a flat or maisonette 4.5 m above ground level are very similar, in most respects, to those for dwellinghouses. Emphasis is, however, made on the necessity to ensure that means are provided for early warning in the event of fire and that suitable means are available for emergency escape. For complete details of additional requirements, see Approved Document B Sections 3 and 6.

Buildings other than dwellings

The prime differences between the design of buildings other than dwellings, flats and/or maisonettes, is the necessity to address the availability of horizontal as well as vertical escape.

Horizontal escape

The general principle to be followed when designing facilities for means of escape is that any person confronted by an outbreak of fire within a building can turn away from it and make a safe escape. Full details of the requirements for horizontal escape are contained in B2 Section 4. The following are intended to provide an indication of these requirements.

The number of escape routes and exits to be provided depends on the number of occupants in the room, tier or storey in question.	B1 (4.2 and 4.7)
Separate means of escape should be provided from any storeys (or parts of storeys) used for residential or assembly and recreation purposes.	B1 (4.4)
In order to avoid occupants being trapped by fire or smoke, there should be alternative escape routes from all parts of the building.	B1 (4.5)
The occupant capacity of the inner room (i.e. a room from which the only escape route is through another room) should not exceed 60.	B1 (4.9a)
Buildings with more than one exit in a central core should be planned so that storey exits are remote from one another, and no two exits are approached from the same lift hall, common lobby or undivided corridor.	B1 (4.10)
All escape routes should have a clear headroom of not less than 2 m except in doorways.	B1 (4.15)
Whilst the width of escape routes and exits depends on the number of persons needing to use them, they should never be less than 750 mm wide.	B1 (4.16–18)
Every corridor more than 12 m long that connects two or more storey exits, should be sub-divided by self-closing fire doors.	B1 (4.23)

Unless the escape stairway(s) and corridors are protected by a pressurization system complying with BS 5588, every dead-end corridor exceeding 4.5 m in length should be separated by self-closing fire doors.	B1 (4.24)
Where an external escape route (other than a stair) is beside an external wall of the building, that part of the external wall within 1800 mm of the escape route should be of fire-resisting construction, up to a height of 1100 mm above the paving level of the route.	B1 (4.27)

Vertical escape

The availability of adequately sized and protected escape stairs as a means of escape in multi-storey buildings (other than dwellinghouses, flats and/or maisonettes) needs careful consideration. The actual number, width and protection of escape stairs needed in a building (or part of a building) will vary according to the constraints imposed by the design of horizontal escape routes, whether a single stair is acceptable and whether independent stairs are required in mixed occupancy buildings, etc. (for full details see B1 Sections 5 and 6).

Uninsulated glazed elements on escape routes

Table 6.67 sets out limitations on the use of uninsulated fire-resisting glazed elements. These limitations do not apply to the use of insulated fire-resisting glazed elements.

6.25 Bathrooms

6.25.1 The requirement

All dwellings (whether they are a house, flat or maisonette) should have at least one bathroom with a fixed bath or shower, and the bath or shower should be equipped with hot **and** cold water. This ruling applies to all plans for:

- new houses;
- new buildings, part of which are going to be used as a dwelling;
- existing buildings that are going to be converted, or partially converted into dwellings.

(Building Act 1984 Section 27)

Table 6.67 Limitations on the use of uninsulated glazed elements on escape routes

| Position of glazed element | Maximum total glazed area in parts of the building with access to | | | |
| | A single stairway | | More than one stairway | |
	Walls	Door leaf	Walls	Door leaf
Single family dwellinghouses				
1 (a) within the enclosures of (i) protected stairway	Fixed fanlights only	Unlimited	Fixed fanlights only	Unlimited
(ii) existing stair	Unlimited	Unlimited	Unlimited	Unlimited
(b) within fire resisting separation	100 mm from floor	100 mm from floor	100 mm from floor	100 mm from floor
(c) existing window between an attached/ integral garage and the house	Unlimited	N/A	Unlimited	N/A
Flats and maisonettes				
2 Within the enclosures of a protected entrance hall or protected landing	Fixed fanlights only	1100 mm from floor	Fixed fanlights only	1100 mm from floor

Ventilation

Ventilation (mechanical and/or air-conditioning systems designed for domestic buildings) shall be capable of restricting the accumulation of moisture and pollutants originating within a building.

(Approved Document F1)

Sanitary conveniences

All dwellings (houses, flats or maisonettes) should have at least one closet and one washbasin which should:

• *be separated by a door from any space used for food preparation or where washing-up is done in washbasins;*
• *ideally, be located in the room containing the closet;*
• *have smooth, non-absorbent surfaces and be capable of being easily cleaned;*
• *be capable of being flushed effectively;*
• *only be connected to a flush pipe or discharge pipe;*
• *washbasins should have a supply of hot and cold water.*

(Approved Document G1)

All dwellings (house, flat or maisonette) should have at least one bathroom with a fixed bath or shower.

(Approved Document G3)

6.25.2 Meeting the requirement

All dwellings (houses, flats or maisonettes) should have at least one bathroom with a fixed bath or shower and the bath or shower should:

• have a supply of hot and cold water;	G2
• discharge through a grating, a trap and branch discharge pipe to a discharge stack or (if on a ground floor) discharge into a gully or directly to a foul drain;	G2 G2 H1
• be connected to a macerator and pump (of an approved type) if there is no suitable water supply or means of disposing foul water.	G2

Bathrooms shall have either a fixed bath or shower bath that is provided with hot and cold water and connected to a foul water drainage system.

 Requirement G2 only applies to dwellings.

Ventilation

Buildings (or spaces within buildings) other than those:

* into which people do not normally go; or
* that are used solely for storage; or
* that are garages used solely in connection with a single dwelling;

shall be provided with ventilation to:

• extract water vapour from non-habitable areas where it is produced in significant quantities (e.g. kitchens, utility rooms and bathrooms);	F1 (1.1–1.5)
• extract pollutants (which are a hazard to health) from areas where they are produced in significant quantities (e.g. rooms containing processes that generate harmful contaminants and rest rooms where smoking is permitted);	F1 (2.3–2.5)
• rapidly dilute (when necessary) pollutants and water vapour produced in habitable rooms, occupiable rooms and sanitary accommodation;	F1 (1.1–1.4) F1 (1.6–1.8)
• provide a minimum supply of fresh air for occupants.	F1 (2.6–2.8)

Table 6.68 Ventilation of rooms containing openable windows (i.e. located on an external wall)

Room	Rapid ventilation (e.g. opening windows)	Background ventilation	Extract ventilation fan rates or passive stack (PSV)
Habitable room	1/20th of floor area	8000 mm^2	
Kitchen	Opening window (no minimum size)	4000 mm^2	30 litres/second adjacent to a hob or 60 litres/second elsewhere or PSV
Utility room	Opening window (no minimum size)	4000 mm^2	30 litres/second or PSV
Bathroom (with or without WC)	Opening window (no minimum size)	4000 mm^2	15 litres/second or PSV
Sanitary accommodation (separate from bathroom)	1/20th of floor area or mechanical extract at 6 litres/second	4000 mm^2	

Washbasins

Washbasins should have a supply of hot and cold water.	G1
Washbasins should discharge through a grating, a trap and a branch discharge pipe to a discharge stack or (if it is a ground floor location) into a gully or directly into a drain.	G1

6.26 Loft conversions

6.26.1 The requirements

Stairs, ladders and ramps

All stairs, steps and ladders shall provide reasonable safety between levels in a building.

(Approved Document K1)

Protection from falling

Pedestrian guarding should be provided for any part of a floor (including the edge below an opening window) gallery, balcony, roof (including rooflight

*and other openings), any other place to which people have access and any
light well, basement area or similar sunken area next to a building.*

(Approved Document K2)

💡 Requirement K2 (a) applies only to stairs and ramps that form part of the
building.

6.26.2 Meeting the requirements

Loft conversions

Where an existing two-storey house is being enlarged by converting the existing
roof space into habitable rooms then, provided the new second storey does not
exceed 50 m² in floor area, or contain more than two habitable rooms, then:

• the stairs in the ground and first storeys should be enclosed with walls and/or partitions which are fire-resisting;	B1 (2.18a)
• the enclosure should either: – extend to a final exit, see Figure 6.169(a); or – give access to at least two escape routes at ground level	B1 (2.18b)
• every doorway within the enclosure to the existing stair should be fitted with a door, which (in the case of doors to habitable rooms) should be fitted with a self-closing device;	B1 (2.19)
• any new door to a habitable room should be a fire door;	B1 (2.19)
• any glazing (whether new or existing) in the enclosure to the existing stair, including all doors (whether or not they need to be fire doors), but excluding glazing to a bathroom or WC, should be fire-resisting and retained by a suitable glazing system and beads compatible with the type of glass;	B1 (2.20)
• the new storey should be served by a stair that is an extension of an existing stairway;	B1 (2.21)
• the new storey should be separated from the rest of the house by a fire resistant door;	B1 (2.22)
• the room (or rooms) in any new storey should each have an openable window or rooflight that meets the relevant provisions in Figure 6.173. A door to a roof terrace is also acceptable.	B1 (2.24)

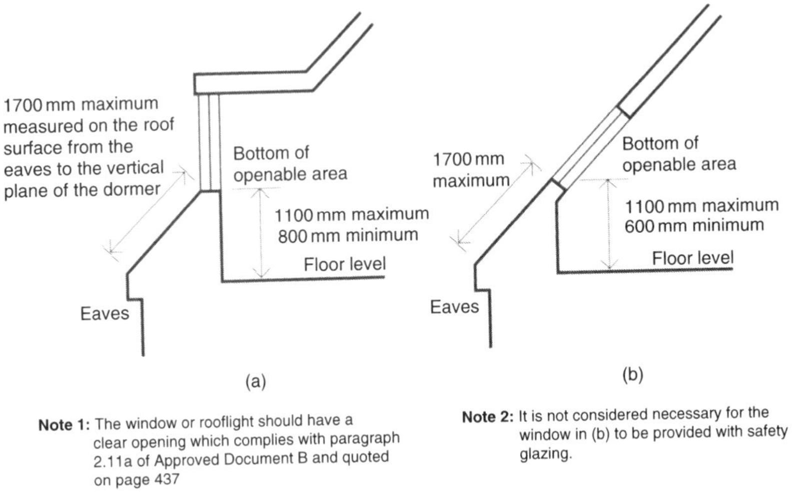

1700 mm maximum measured on the roof surface from the eaves to the vertical plane of the dormer

Bottom of openable area

1100 mm maximum
800 mm minimum

Floor level

Eaves

(a)

1700 mm maximum

Bottom of openable area

1100 mm maximum
600 mm minimum

Floor level

Eaves

(b)

Note 1: The window or rooflight should have a clear opening which complies with paragraph 2.11a of Approved Document B and quoted on page 437

Note 2: It is not considered necessary for the window in (b) to be provided with safety glazing.

Figure 6.173 Position of dormer window or rooflight that is suitable for emergency purposes from a loft conversion of a two-storey dwelling house. (a) Dormer window (the window may be in the end wall of the house, instead of the roof as shown). (b) Rooflight or roof window

Stairs, ladders and ramps

The rise of a stair shall be between 155 mm and 220 mm with any going between 245 mm and 260 mm and a maximum pitch of 42°.	K1 (1.1–1.4) MK
💡 The normal relationship between the dimensions of the rise and going is that twice the rise plus the going (2R+G) should be between 550 mm and 700 mm.	
Stairs with open risers that are likely to be used by children under five years should be constructed so that a 100 mm diameter sphere cannot pass through the open risers.	K1 (1.9)
Stairs that have more than 36 risers in consecutive flights should make at least one change of direction, between flights, of at least 30°.	K1 (1.14)
If a stair has straight and tapered treads, then the going of the tapered treads should not be less than the going of the straight tread.	K1 (1.20)
The going of tapered treads should measure at least 50 mm at the narrow end.	K1 (1.18)
The going should be uniform for consecutive tapered treads.	K1 (1.19) K1 (1.22–1.24)

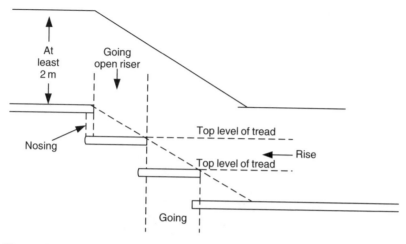

Figure 6.174 Rise and going plus headroom

Stairs should have a handrail on both sides if they are wider than 1 m and on at least one side if they are less than 1 m wide.	K1 (1.27) M
Handrail heights should be between 900 mm and 1000 mm measured to the top of the handrail from the pitch line or floor.	K1 (1.27) M
Spiral and helical stairs should be designed in accordance with BS 5395.	K1 (1.21)

Steps

• Steps should have level treads.	K1 (1.8)
• Steps may have open risers, but treads should then overlap each other by at least 16 mm.	K1 (1.8) M
Steps should be uniform with parallel nosings, the stair should have handrails on both sides and the treads should have slip-resistant surfaces.	
• The headroom on the access between levels should be no less than 2 m.	K1 (1.10)
• Landings should be provided at the top and bottom of every flight.	K1 (1.15)
• The width and length of every landing should be the same (or greater than) the smallest width of the flight.	K1 (1.15)

- Landings should be clear of any permanent obstruction. — K1 (1.16)

- Landings should be level. — K1 (1.17)

- Any door (entrance, cupboard or duct) that swings across a landing at the top or bottom of a flight of stairs must leave a clear space of at least 400 mm across the full width of the flight. — K1 (1.16)

- Flights and landings should be guarded at the sides when there is a drop of more than 600 mm. — K1 (1.28–1.29)

For stairs that are likely to be used by children under five years the construction of the guarding shall be such that a 100 mm sphere cannot pass through any openings in the guarding and children will not easily climb the guarding.

For loft conversions, a fixed ladder should have fixed handrails on both sides. — K1 (1.25)

Whilst there are no recommendations for minimum stair widths, designers should bear in mind the requirements of Approved Documents B (means of escape) and M (access for disabled people).

Protection from falling

All stairs, landings, ramps and edges of internal floors shall have a wall, parapet, balustrade or similar guard at least 900 mm high. — K3 (3.2)

All guarding should be capable of resisting at least the horizontal force given in BS 6399: Part 1: 1996. — K3 (3.2)

If glazing is used as (or part of) the pedestrian guarding, see Approved Document N: Glazing – safety in relation to impact, opening and cleaning. — N

If a building is likely to be used by children under five years, the guarding should not have horizontal rails, should stop children from easily climbing it, and the construction should prevent a 100 mm sphere being able to pass through any opening of that guarding. — K3 (3.3)

All external balconies and edges of roofs shall have a wall, parapet, balustrade or similar guard at least 1100 mm high. — K3 (3.2)

💡 Requirement K2 (a) applies only to stairs and ramps that form part of the building.

6.27 Entrances and courtyards

6.27.1 The requirement (Building Act 1984 Sections 24 and 71)

The Building Act is very specific about exits, passageways and gangways and local authorities are required to consult with the fire authority to ensure that proposed methods of ingress and egress are deemed satisfactory, depending on the type of building. The purpose for which the building is going to be used needs to be considered as each case can be different. In particular, is the building going to be:

- a theatre, hall or other public building that is used as a place of public resort;
- a restaurant, shop, store or warehouse to which members of the public are likely to be admitted;
- a club;
- a school;
- a church, chapel or other place of worship (erected or used after the Public Health Acts Amendment Acts 1890 came into force).

At all times, the means of ingress and egress and the passages and gangways, while persons are assembled in the building, are to be kept free and unobstructed.

💡 Private houses are not restricted by the actual Building Act of 1984 provided that members of the public are only admitted occasionally or exceptionally.

(Building Act 1984 Section 85)

All entrances to courts and yards must allow the free circulation of air and are not allowed to be closed, narrowed, reduced in height or in any other way altered so that it can impede the free circulation of air through the entrance.

Vehicle barriers and loading bays

- *Vehicle barriers should be provided that are capable of resisting or deflecting the impact of vehicles.*
- *Loading bays shall be provided with an adequate number of exits (or refuges) to enable people to avoid being crushed by vehicles.*

(Approved Document K3)

Access and facilities for disabled people

All precautions shall be taken to ensure that all new dwellings, other buildings and student living accommodation shall be reasonably safe and convenient for disabled people to:

- *gain access to and within buildings other than dwellings and to use them;*
- *visit new dwellings and to use the principal storey.*

(Approved Document M)

6.27.2 Meeting the requirement

Any wall, parapet, balustrade or similar obstruction may serve as a barrier.	K3 (3.8)
If vehicles have access to a floor or roof edge, barriers of at least 375 mm should be provided to any edges that are level with (or above) the adjoining floor or ground.	K3 (3.7)
If vehicles have access to a ramp edge, barriers of at least 610 mm should be provided to any edges that are level with (or above) the adjoining floor or ground.	K3 (3.7)
All barriers should be capable of resisting forces set out in BS 6399.	K3 (3.8)
Loading bays should be provided with at least one exit point from the lower level (preferably near the centre of the rear wall).	K3 (3.9)

Access and facilities for disabled people

Disabled people can reach the principal, or suitable alternative, entrance to the dwelling from the point of access.	M (0.8ii a.)
Access for disabled persons into and within the principal storey of the dwelling is provided.	M (0.8ii b.)
A disabled person should be able to approach and gain access into the dwelling from the point of alighting from a vehicle that may be within or outside the plot. Where possible this should be a level or ramped approach.	M2 (6.1–6.3 and 6.12)

The approach should be as safe and as convenient for disabled people as is reasonable and, ideally, be level or ramped.	M2 (6.6)
If a stepped approach to the dwelling is unavoidable, the aim should be for the steps to be designed to suit the needs of ambulant disabled people.	M2 (6.7 and 6.17)
The surface of an approach available to a wheelchair user should be firm enough to support the weight of the user and his or her wheelchair and smooth enough to permit easy manoeuvre.	M2 (6.9)
The surface should take account of the needs of stick and crutch users.	M2 (6.9)
Loose laid materials, such as gravel or shingle, should not be used as for the approach.	M2 (6.9)
The width of the approach, excluding space for parked vehicles, should take account of the needs of a wheelchair user, or a stick or crutch user.	M2 (6.10 and 6.13)
The point of access should be reasonably level (i.e. with a gradient less than 1 in 20, having a firm surface and width not less than 900 mm) and the approach should not have crossfails greater than 1 in 40.	M2 (6.11)
If the plot gradient is between 1 in 20 and 1 in 15, a ramped approach may be used provided that it:	M2 (6.15)

- has a surface that is firm and even (M2 (6.15));
- has flights whose unobstructed widths are at least 900 mm (M2 (6.15b));
- has individual flights not longer than 10 m for gradients not steeper than 1 in 15, or 5 m for gradients not steeper than 1 in 12 (M2 (6.15c));
- has top and bottom landings and (if necessary), intermediate landings, each of whose lengths is not less than 1.2 m, exclusive of the swing of any door or gate that opens onto it (M2 (6.15d)).

If a stepped approach is used then it should have: M2 (6.17)

- flights whose unobstructed widths are at least 900 mm (M2 (6.17a));
- the rise of a flight between landings is not more than 1.8 m (M2 (6.17b));
- has top and bottom and (if necessary) intermediate landings, each of whose lengths is not less than 900 mm (M2 (6.17c));
- has steps with suitable tread nosing profiles and the rise of each step is between 75 mm and 150 mm and is uniform (M2 (6.17d));
- the going of each step is not less than 280 mm (M2 (6.17e));
- where the flight comprises three or more risers, there is a suitable continuous handrail (i.e. a handrail with a grippable profile between 850 mm and 1000 mm above the pitch line of the flight and extending 300 mm beyond the top and bottom nosings) on one side of the flight (M2 (6.17f)).

If a step into the dwelling is unavoidable, the rise should be no more than 150 mm.   M2 (6.20)

6.28 Yards and passageways

6.28.1 The requirement (Building Act 1984 Section 84)

You are required by the Building Act 1984 to ensure that all courts, yards and passageways giving access to a house, industrial or commercial building (not maintained at the public expense) are capable of allowing satisfactory drainage of its surface or subsoil to a proper outfall.

The local authority can require the owner of any of the buildings to complete such works as may be necessary to remedy the defect.

6.29 Extensions to buildings

6.29.1 The requirement

Conservation of fuel and power

Energy efficiency measures shall be provided that:

(a) • *limit the heat loss through the roof, wall, floor, windows and doors, etc. by suitable means of insulation;*
  • *where appropriate permit the benefits of solar heat gains and more efficient heating systems to be taken into account;*
  • *limit unnecessary ventilation heat loss by providing building fabric that is reasonably airtight;*
  • *limit the heat loss from hot water pipes and hot air ducts used for space heating;*
  • *limit the heat loss from hot water vessels and their primary and secondary hot water connections by applying suitable thicknesses of insulation (where such heat does not make an efficient contribution to the space heating).*

(b) Provide space heating and hot water systems with reasonably efficient equipment such as heating appliances and hot water vessels where relevant, such that the heating and hot water systems can be operated effectively as regards the conservation of fuel and power.

(c) Provide lighting systems that utilize energy-efficient lamps with manual switching controls or, in the case of external lighting fixed to the building, automatic switching, or both manual and automatic switching controls as appropriate, such that the lighting systems can be operated effectively as regards the conservation of fuel and power.

(d) Provide information, in a suitably concise and understandable form (including results of performance tests carried out during the works) that shows building occupiers how the heating and hot water services can be operated and maintained.

(Approved Document L1)

Responsibility for achieving compliance with the requirements of Part L rests with the person carrying out the work. That person may be, for example, a developer, a main (or sub-) contractor, or a specialist firm directly engaged by a private client.

The person responsible for achieving compliance should either themselves provide a certificate, or obtain a certificate from the sub-contractor, that commissioning has been successfully carried out. The certificate should be made available to the client and the building control body.

6.29.2 Meeting the requirement

The fabric U-values given in Table 6.67 can be applied when proposing extensions to dwellings.	L1 (1.11)
The target U-value and carbon index methods can be used only if applied to the whole enlarged dwelling.	L1 (1.11)
The total rate of heat loss is the product of (Area × U-value) for all exposed elements.	L1 (1.12)
For small extensions to dwellings (e.g. porches where the new heated space created has a floor area of not more than about 6 m^2) energy performance should not be less than that in the existing building.	L1 (1.13)
The area-weighted average U-value of windows, doors and rooflights ('openings') in extensions to existing dwellings should not exceed the relevant values in Table 6.69.	L1 (1.14)

Table 6.69 Elemental method – U-values (W/m^{2}K) for construction elements

Exposed element	U-value
Pitched roof with insulation between rafters[1, 2]	0.2
Pitched roof with integral insulation	0.25
Pitched roof with insulation between joists	0.16
Flat roof[3]	0.25
Walls, including basement walls	0.35
Floors, including ground floors and basement floors	0.25
Windows, doors and rooflights (area-weighted average), glazing in metal frames[4]	2.2
Windows, doors and (area-weighted average) rooflights[4], glazing in wood or PVC frames[5]	2.0

Notes:
1. Any part of a roof having a pitch of 70° or more can be considered as a wall.
2. For the sloping parts of a room-in-the-roof constructed as a material alteration, a U-value of 0.3 W/m^{2}K would be reasonable.
3. Roof of pitch not exceeding 10°.
4. Rooflights include roof windows.
5. The higher U-value for metal-framed windows allows for additional solar gain due to the greater glazed proportion.

Figure 6.175 summarizes the fabric insulation standards and allowances for windows, doors and rooflights given in the elemental method.

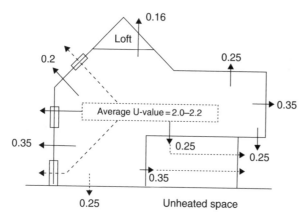

Figure 6.175 Summary of the elemental method

6.30 External balconies

6.30.1 The requirements

Protection from falling

Pedestrian guarding should be provided for any part of a floor (including the edge below an opening window) gallery, balcony, roof (including rooflight and other openings), any other place to which people have access and any light well, basement area or similar sunken area next to a building.

(Approved Document K2)

Requirement K2 (a) applies only to stairs and ramps that form part of the building.

6.30.2 Meeting the requirements

All external balconies and edges of roofs shall have a wall, parapet, balustrade or similar guard at least 1100 mm high.	K3 (3.2)

6.31 Garages

6.31.1 The requirements

As a fire precaution, all materials used for internal linings of a building should have a low rate of surface flame spread and (in some cases) a low rate of heat release.

(Approved Document B2)

6.31.2 Meeting the requirements

If a domestic garage is attached to (or forms an integral part of a house), the garage should be separated from the rest of the house, as shown in Figure 6.176.	B3 (9.14)

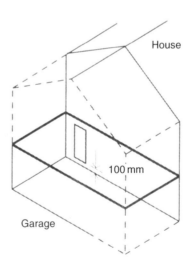

Figure 6.176 Separation between garage and dwellinghouse

The wall and any floor between the garage and the house shall have a 30 minute standard of fire resistance. Any opening in the wall should be at least 100 mm above the garage floor level with an FD30 door.

6.32 Conservatories

6.32.1 Requirements

Conservation of fuel and power

Energy efficiency measures shall be provided that:

(a) • *limit the heat loss through the roof, wall, floor, windows and doors, etc. by suitable means of insulation;*
   • *where appropriate permit the benefits of solar heat gains and more efficient heating systems to be taken into account;*
   • *limit unnecessary ventilation heat loss by providing building fabric that is reasonably airtight;*
   • *limit the heat loss from hot water pipes and hot air ducts used for space heating;*
   • *limit the heat losss from hot water vessels and their primary and secondary hot water connections by applying suitable thicknesses of insulation (where such heat does not make an efficient contribution to the space heating).*

(b) *Provide space heating and hot water systems with reasonably efficient equipment such as heating appliances and hot water vessels where relevant, such that the heating and hot water systems can be operated effectively as regards the conservation of fuel and power.*

(c) *Provide lighting systems that utilize energy-efficient lamps with manual switching controls or, in the case of external lighting fixed to the building, automatic switching, or both manual and automatic switching controls as appropriate, such that the lighting systems can be operated effectively as regards the conservation of fuel and power.*

(d) *Provide information, in a suitably concise and understandable form (including results of performance tests carried out during the works) that shows building occupiers how the heating and hot water services can be operated and maintained.*

(Approved Document L1)

Responsibility for achieving compliance with the requirements of Part L rests with the person carrying out the work. That person may be, for example, a developer, a main (or sub-) contractor, or a specialist firm directly engaged by a private client.

The person responsible for achieving compliance should either themselves provide a certificate, or obtain a certificate from the sub-contractor, that commissioning has been successfully carried out. The certificate should be made available to the client and the building control body.

6.32.2 Meeting the requirement

A conservatory that is not separated from the rest of the dwelling shall be treated as an integral part of the dwelling.	L1
Where the conservatory is separated from the rest of the dwelling and has a **fixed** heating installation, the heating in the conservatory shall have its own separate temperature and on/off controls.	L1
Where an opening is created or enlarged, provision shall be made to limit the heat loss from the dwelling such that it is no worse than before the work was undertaken.	L1
Measurements of thermal conductivity should be made according to BS EN 12664, BS EN 12667, or BS EN 12939.	L1 (0.9)
Measurements of thermal transmittance should be made according to BS EN ISO 8990 or, in the case of windows and doors, BS EN ISO 12567.	L1 (0.9)

Conservatories

For the purposes of the guidance in Part L, a conservatory is considered to not have less than three-quarters of the area of its roof and not less than one-half of the area of its external walls made of translucent material.

If a conservatory is attached to, and built as part of a new dwelling it should be treated as an integral part of the building.	L1 (1.59)
If there is no physical separation between the conservatory and the dwelling, then the conservatory should be treated as an integral part of the dwelling.	L1 (1.59a)
If the conservatory is separated from the dwelling, energy savings can be achieved if the conservatory is not heated.	L1 (1.59b)
If fixed heating installations are used in a conservatory, they should have their own separate temperature and on/off controls.	L1 (1.59b)

When a conservatory is attached to an existing dwelling and an opening is enlarged or newly created as a material alteration, the heat loss from the dwelling should be limited by:

L1 (1.60)

- retaining the existing separation where the opening is not to be enlarged;
- providing separation with an average U-value less than 2.0 W/m²K.

Separating walls and floors between a conservatory and a dwelling should be insulated to at least the same degree as the exposed walls and floors.

L1 (1.61)

Separating windows and doors between a conservatory and a dwelling should have the same U-value and draught-stripping provisions as the exposed windows and doors elsewhere in the dwelling.

L1 (1.61b) AD N

6.33 Regulation 7 – Materials and workmanship

6.33.1 The requirement

Building work shall be carried out –

(a) with adequate and proper materials which –
  (i) are appropriate for the circumstances in which they are used;
  (ii) are adequately mixed or prepared; and
  (iii) are applied, used or fixed so as adequately to perform the functions for which they are designed; and
(b) in a workmanlike manner.

(Regulation 7)

Parts A to K and N of Schedule 1 to these regulations shall not require anything to be done except for the purpose of securing reasonable standards of health and safety for persons in or about buildings (and any others who may be affected by buildings or matters connected with buildings).

6.33.2 Meeting the requirements

Materials (including products, components, fittings, naturally occurring materials, e.g. stone, timber and thatch, items of equipment, and backfilling for excavations in connection with building work) shall be:

- of a suitable nature and quality in relation to the purposes and conditions of their use;
- adequately mixed or prepared (where relevant); and
- applied, used or fixed so as to perform adequately the functions for which they are intended.

Environmental impact of building work

The environmental impact of building work can be minimized by careful choice of materials and the use of recycled and recyclable materials.	Reg. 7 (0.2)
The use of such materials must not have any adverse implications for the health and safety standards of the building work.	Reg. 7 (0.2)

Limitations

Reasonable standards of health or safety for persons in or about the building shall be assured.	Reg. 7 (0.3)
Materials and workmanship shall ensure that fuel and power is conserved.	Reg. 7 (0.3)
Access and facilities for disabled people shall be provided.	Reg. 7 (0.3)

There are no provisions under the Building Regulations for continuing control over the use of materials following the completion of building work.

Fitness of materials

The suitability of material used for a specific purpose shall be assessed using: • British Standards • European Standards (ENs) • other national (and international) technical specifications (from other European Community member states)	Reg. 7 (1.2 to 1.7)

- technical approvals (covered by a national or European certificate issued by a European technical approvals issuing body)

CE Marking (which provides a presumption of conformity with the stated minimum legal requirements)

- independent (UK) product certification schemes (such as those accredited by UKAS)
- tests and calculations (for example, those in conformance with UKAS's Accreditation Scheme for Testing Laboratories)
- past experience (that the material can be shown by experience, to be capable of performing the function for which it is intended)
- sampling (of materials by local authorities).

Short-lived materials and materials susceptible to changes in their properties should only be used in works where these changes do not adversely affect their performance.

Resistance to moisture

Any material that is likely to be adversely affected by condensation, moisture from the ground, rain or snow shall either resist the passage of moisture to the material or be treated or otherwise protected from moisture.

Reg. 7 (1.8)

Resistance to substances in the subsoil

Any material in contact with the ground (or in the foundations) shall be capable of resisting attacks by harmful materials in the subsoil such as sulphates.

Reg. 7 (1.9)

Establishing the adequacy of workmanship

The adequacy (and competence) of workmanship will normally be established by using:

- a British Standard code of practice;
- workmanship specified and covered by a national or European certificate issued by a European technical approvals issuing body;
- the recommendations of an integrated management system (such as ISO 9001: 2000);
- past experience (of workmanship that is capable of performing the function for which it is intended);
- tests (for example, the local authority has the power to test sewers and drains in or in connection with buildings).

Workmanship on building sites

In the main, local authorities will use the codes of practice as contained and detailed in BS 8000 (Workmanship on Building Sites) to monitor and inspect all building work. These codes of practice consist of:

- Part 1: 1989 Code of practice for excavation and filling
- Part 2: Code of practice for concrete work
    Section 2.1: 1990 Mixing and transporting concrete
    Section 2.2: 1990 Sitework with in situ and precast concrete
- Part 3: 1989 Code of practice for masonry
- Part 4: 1989 Code of practice for waterproofing
- Part 5: 1990 Code of practice for carpentry, joinery and general fixings
- Part 6: 1990 Code of practice for slating and tiling of roofs and claddings
- Part 7: 1990 Code of practice for glazing
- Part 8: 1994 Code of practice for plasterboard partitions and dry linings
- Part 9: 1989 Code of practice for cement/sand floor screeds and concrete floor toppings
- Part 10: 1995 Code of practice for plastering and rendering
- Part 11: Code of practice for wall and floor tiling
    Section 11.1: 1989 Ceramic tiles, Terrazzo tiles and mosaics (confirmed 1995)
    Section 11.2: 1990 Natural stone tiles
- Part 12: 1989 Code of practice for decorative wall coverings and painting
- Part 13: 1989 Code of practice for above ground drainage and sanitary appliances
- Part 14: 1989 Code of practice for below ground drainage
- Part 15: 1990 Code of practice for hot and cold water services (domestic scale)
- Part 16: 1997 Code of practice for sealing joints in buildings using sealants.

Appendix A

Access and facilities for disabled people

A.1 The requirement

All precautions shall be taken to ensure that all new dwellings, other buildings and student living accommodation shall be reasonably safe and convenient for disabled people to:

- *gain access to and within buildings other than dwellings and to use them;*
- *visit new dwellings and to use the principal storey.*

(Approved Document M)

A.1.1 New buildings

The requirements apply if a building is newly erected, or has been substantially demolished, leaving only external walls.

A.1.2 Extensions

If an existing building, other than a dwelling (note extensions to dwellings are excluded from Part M), is extended, the requirements of Part M apply to the extension, provided that it contains a ground storey.

When a building is extended, there is no obligation to carry out improvements within the existing building to make it more accessible to and usable by disabled people than it was before.

An extension should be at least as accessible to and usable by disabled people as the building being extended.

A.1.3 Alterations

When a building is altered there is no obligation to improve access and facilities for disabled people.

A.2 Meeting the requirements

Reasonable provision should be made to make sure that dwellings (including any purpose-built student living accommodation, other than a traditional halls of residence providing mainly bedrooms and not equipped as self-contained accommodation) ensure that:

• disabled people can reach the principal, or suitable alternative, entrance to the dwelling from the point of access;	M (0.8iia)
• access for disabled persons into and within the principal storey of the dwelling is provided;	M (0.8iib)
• sanitary accommodation on the principal storey is provided.	M (0.8iic)

A.2.1 Access to a dwelling

A disabled person should be able to approach and gain access into the dwelling from the point of alighting from a vehicle that may be within or outside the plot. Where possible this should be a level or ramped approach.	M2 (6.1–6.3 and 6.12)
The approach should be as safe and as convenient for disabled people as is reasonable and, ideally, be level or ramped.	M2 (6.6)
If a stepped approach to the dwelling is unavoidable, the aim should be for the steps to be designed to suit the needs of ambulant disabled people.	M2 (6.7 and 6.17)
The surface of an approach available to a wheelchair user should be firm enough to support the weight of the user and his or her wheelchair and smooth enough to permit easy manoeuvre.	M2 (6.9)
The surface should take account of the needs of stick and crutch users.	M2 (6.9)
Loose laid materials, such as gravel or shingle, should not be used as for the approach.	M2 (6.9)
The width of the approach, excluding space for parked vehicles, should take account of the needs of a wheelchair user, or a stick or crutch user.	M2 (6.10 and 6.13)

The point of access should be reasonably level (i.e. with a gradient less than 1 in 20, having a firm surface and width not less than 900 mm) and the approach should not have crossfails greater than 1 in 40.

M2 (6.11)

If the plot gradient is between 1 in 20 and 1 in 15 a ramped approach may be used provided that it:

M2 (6.15)

- has a surface that is firm and even (M2 (6.15));
- has flights whose unobstructed widths are at least 900 mm (M2 (6.15b));
- has individual flights not longer than 10 m for gradients not steeper than 1 in 15, or 5 m for gradients not steeper than 1 in 12 (M2 (6.15c));
- has top and bottom landings and (if necessary), intermediate landings, each of whose lengths is not less than 1.2 m, exclusive of the swing of any door or gate that opens onto it (M2 (6.15d)).

If a stepped approach is used then it should have:

M2 (6.17)

- flights whose unobstructed widths are at least 900 mm (M2 (6.17a));
- the rise of a flight between landings is not more than 1.8 m (M2 (6.17b));
- has top and bottom and (if necessary) intermediate landings, each of whose lengths is not less than 900 mm (M2 (6.17c));
- has steps with suitable tread nosing profiles and the rise of each step is between 75 mm and 150 mm and is uniform (M2 (6.17d));
- the going of each step is not less than 280 mm (M2 (6.17e));
- where the flight comprises three or more risers, there is a suitable continuous handrail (i.e. a handrail with a grippable profile between 850 mm and 1000 mm above the pitch line of the flight and extending 300 mm beyond the top and bottom nosings) on one side of the flight (M2 (6.17f)).

If a step into the dwelling is unavoidable, the rise should be no more than 150 mm.

M2 (6.20)

Doors, corridors and passageways

An external door providing access for disabled people has a minimum clear opening width of 775 mm.	M2 (6.23)
Corridors and passageways in the entrance storey should be sufficiently wide to allow convenient circulation by a wheelchair user.	M2 (7.2)
Internal doors need to be of a suitable width to facilitate wheelchair manoeuvre.	M2 (7.4)
A corridor or other access route in the entrance storey or principal storey serving habitable rooms and a room containing a WC (which may be a bathroom) on that level, has an unobstructed width in accordance with Table A1.	M2 (7.5a)
Any local permanent obstruction in a corridor, such as a radiator, should not be longer than 2 m and wider than 750 mm.	M2 (7.5b)
Doors to habitable rooms and/or WCs should have a minimum clear opening width as shown in Figure A1.	M2 (7.5c)

Table A1 Minimum widths of corridors and passageways for a range of doorway widths

Doorway clear opening width (mm)	Corridor/passageway width (mm)
750 or wider	900 (when approach is head on)
750	1200 (when approach is not head on)
775	1050 (when approach is not head on)
800	800 (when approach is not head on)

Stairs

A stair providing vertical circulation within the entrance storey: • should have flights whose clear widths are at least 900 mm (M2 (7.7a)); • should have a suitable continuous handrail on each side of the flight and any intermediate landings where the rise of the flight comprises three or more risers (M2 (7.7b)).	M2 (7.7)

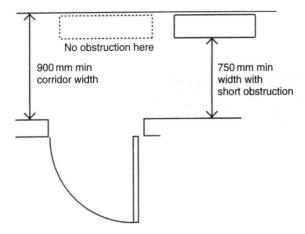

Figure A1 Corridors, passages and internal doors

Electrical fittings

• switches and socket outlets for lighting and other equipment should be located so that they are easily reachable.	M2 (8.2)
• switches and socket outlets (for lighting) should be installed between 450 mm and 1200 mm from the finished floor level (see Figure A2).	M2 (8.3)

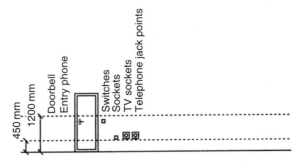

Figure A2 Heights for switches and sockets

Lifts

Where a lift is provided, it should be suitable for an unaccompanied wheelchair user.	M2 (9.1–9.4)

Passenger lifts

Passenger lifts should:

• have a minimum load capacity of 400 kg;	M2 (9.6)
• have a clear landing at least 1500 mm wide and at least 1500 mm long in front of its entrance;	M2 (9.7a)
• have door(s) wide enough to provide a clear opening width of at least 800 mm;	M2 (9.7b)
• have a car whose width is at least 900 mm and whose length is at least 1250 mm;	M2 (9.7c)
• have landing and car controls that are not less than 900 mm and not more than 1200 mm above the landing and the car floor at a distance of at least 400 mm from the front wall;	M2 (9.7d)
• be provided with visual and voice indication of the floor reached;	M2 (9.7e)
• be provided with tactile indication of the floor level;	M2 (9.7f)
• incorporate a signalling system that the lift is answering a landing call.	M2 (9.7g)

Requirements for common stairs

If a passenger lift is not installed, an internal stair may be used provided that:

• all step nosings are distinguishable through contrasting brightness;	M2 (9.5a)
• there is a continuous handrail on each side of the flights (and the landings) if the overall rise of the stair is two or more risers;	M2 (9.5f)
• the rise of each step is uniform and not more than 170 mm;	M2 (9.5c)
• the going of each step is uniform and not less than 250 mm;	M2 (9.5d)
• the risers are not open.	M2 (9.5e)

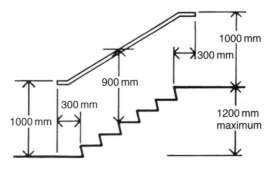

Figure A3 Common stairs in blocks of flats

Provision of WCs

Whenever possible a WC should be provided in the entrance storey or the principal storey of a dwelling.	M2 (10.1–10.3)
The door to the WC compartment should open outwards, and be positioned so as to enable wheelchair users to access the WC and for the WC to have a washbasin positioned so that it does not impede access.	M2 (10.3)
To enable transfer, the wheelchair should be able to approach within 400 mm of the front of the WC.	M2 (10.3)

It is appreciated that in most circumstances, it will not always be practical for the wheelchair to be fully accommodated within the WC compartment.

A.2.2 Basic requirements for buildings other than dwellings

Requirements for common stairs

If a passenger lift is not installed, an internal stair may be used provided that:

• all step nosings are distinguishable through contrasting brightness;	M2 (9.5a)
• there is a continuous handrail on each side of the flights (and the landings) if the overall rise of the stair is two or more risers;	M2 (9.5f)

• the rise of each step is uniform and not more than 170 mm;	M2 (9.5c)
• the going of each step is uniform and not less than 250 mm;	M2 (9.5d)
• the risers are not open.	M2 (9.5e)

Disabled people can reach the principal entrance to the building and other entrances, from the edge of the site curtilage and from car parking within the curtilage. M1 (0.8ia and c)

Parts (e.g. corners) of the building do not constitute a hazard for a person with an impairment of sight. M1 (0.8ib)

Disabled people should be able to use the building's facilities. M1 (0.8id)

Disabled people should be able to use the principal entrance provided for visitors or customers and an entrance which is intended, exclusively, for members of staff. M2 (1.2)

Sanitary accommodation should be available for disabled people. M2 (0.8ie)

Suitable accommodation should be available for disabled people in audience or spectator seating. M2 (0.8if)

Aids to communication for people with an impairment of hearing or sight in auditoria, meeting rooms, reception areas and ticket offices should be provided. M2 (0.8ig)

Gradients should be as gentle as the circumstances allow. M2 (1.7)

💡 Whilst a 'level approach' will satisfy Requirement M2 if it has a surface width of at least 1.2 m and its gradient is not steeper than 1 in 20, for steeper slopes a ramped approach should be provided.

Where possible, a 'level' approach with a surface width of at least 1.2 m and a gradient not steeper than 1 in 20 should be provided. M2 (1.9)

Slopes steeper than 1 in 20 require a ramped approach. M2 (1.10)

There should be space for people passing in the opposite direction.

M2 (1.8)

When a pedestrian route to the building is being used by disabled people:

M2 (1.11)

- a tactile warning should be provided for people with impaired vision where the route crosses a carriageway and at the top of steps;
- dropped kerbs should be provided for wheelchair users (see Figure A4).

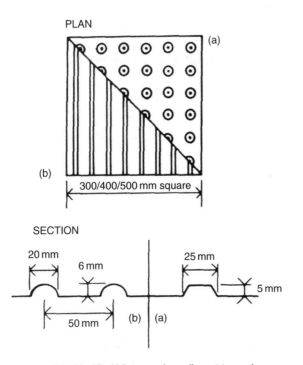

(a) Modified blister paving adjacent to carriageway crossings.
(b) Corduroy paving adjacent to steps

Figure A4 Tactile paving slabs

Figure A4 illustrates paving slabs with tactile warning services. The blister type paving is considered suitable for use at pedestrian crossing points. The corduroy type paving is considered suitable for use at the top of external stairs.

Ramped (internal and external) approaches

A ramped approach (see Figure A5) should:

• have a gradient as gentle as possible;	M2 (1.12)
• provide adequate space to enable wheelchair users to stop on landings and to open and pass through doors without needing to reverse into circulation routes or face the risk of rolling back down slopes;	M2 (1.14)
• have kerbs or solid balustrades on open sides;	M2 (1.17)
• where practicable, have easy going steps;	M2 (1.18)
• have a surface which reduces the risk of slipping;	M2 (1.19a)
• have flights whose surface widths are at least 1.2 m (if unobstructed this can be reduced to 1 m);	M2 (1.19b)
• not be steeper than 1 in 15, if individual flights are not longer than 10 m (or 1 in 12 if individual flights are not longer than 5 m);	M2 (1.19c)
• have top and bottom landings at least 1.2 m in length and (if necessary) intermediate landings not less than 1.5 m clear of any door swing;	M2 (1.19d)

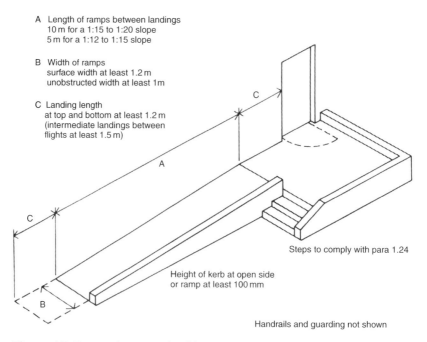

A Length of ramps between landings
   10 m for a 1:15 to 1:20 slope
   5 m for a 1:12 to 1:15 slope

B Width of ramps
   surface width at least 1.2 m
   unobstructed width at least 1m

C Landing length
   at top and bottom at least 1.2 m
   (intermediate landings between
   flights at least 1.5 m)

Steps to comply with para 1.24

Height of kerb at open side
or ramp at least 100 mm

Handrails and guarding not shown

Figure A5 Ramped approach with complementary steps

- have a raised kerb at least 100 mm high on any open side of a flight or a landing;   M2 (1.19e)
- have a continuous suitable handrail on each side of flights and landings, if the length of the ramp exceeds 2 m.   M2 (1.19f)

Stepped approach

Although similar design considerations apply to stepped as do to ramped approaches (see Figure A6), there are some additions.

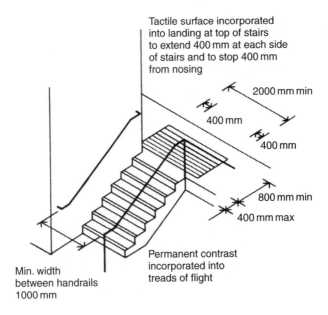

Tactile surface incorporated into landing at top of stairs to extend 400 mm at each side of stairs and to stop 400 mm from nosing

2000 mm min

400 mm

400 mm

800 mm min

400 mm max

Permanent contrast incorporated into treads of flight

Min. width between handrails 1000 mm

Figure A6 Tactile and visual warnings

The existence of individual steps, on their own or within a flight, should be made apparent.   M2 (1.21)

Tread dimensions should allow both feet to be placed square onto it.   M2 (1.21)

The top landing shall have a tactile surface, to give advance warning of the change in level.   M2 (1.24a)

All step nosings should be distinguishable through contrasting brightness.

The minimum width between handrails shall be 1000 mm.	M2 (1.24b)
The minimum width between flights shall be at least 1 m.	M2 (1.24c)
The rise of a flight between landings shall be less than 1.2 m.	M2 (1.24d)
Intermediate landings shall have at least 1.2 m clearance for door swings.	M2 (1.24e)
The rise of each step shall be uniform and not more than 150 mm.	M2 (1.24f)
The going of each step shall not be less than 280 mm.	M2 (1.24g)
The risers shall not be open.	M2 (1.24)
If the rise of the stepped approach comprises two or more risers, then there shall be a continuous handrail on each side of the flight and landings.	M2 (1.24j)

Handrails

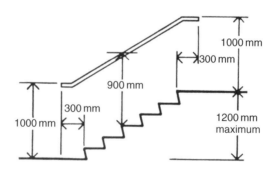

Figure A7 External steps and handrails

Grippable and well supported handrails shall be available for those who have physical difficulty in negotiating changes of level.	M2 (1.25a)
The top of a handrail shall be 900 mm above the surface of a ramp or the pitch line of a flight of steps and 1000 mm above the surface of a landing.	M2 (1.25b)

The handrail shall extend at least 300 mm beyond the top and bottom of a ramp (or the top and bottom nosings of a stepped approach) and terminate in a closed end that does not project into a route of travel.	M2 (1.25c)

Internal stairs

Internal stairs should have:

• flights with unobstructed widths of at least 1000 mm;	M2 (2.21a)
• all step nosings that are distinguishable through contrasting brightness;	M2 (1.21b)
• top and bottom and, if necessary, intermediate landings, each of whose lengths is not less than 1200 mm clear of any door swing into it;	M2 (1.21d)
• a minimum width between handrails of 1000 mm;	M2 (1.24b)
• the rise of a flight between landings of less than 1.2 m;	M2 (1.21c)
• a continuous handrail on each side of the flights (and the landings) if the overall rise of the stair is two or more risers;	M2 (1.21h)
• the rise of each step that is uniform and not more than 170 mm;	M2 (1.21e)
• the going of each step that is uniform and not less than 250 mm;	M2 (1.21f)
• risers that are not open.	M2 (1.21g)
If the rise of the stepped approach comprises two or more risers, then there shall be a continuous handrail on each side of the flight and landings.	M2 (1.24j)

Access routes

Windows or doors in general use that open outwards should not cause an obstruction on a path that runs along the face of a building.	M2 (1.28)
Convenient access into the building should be provided for disabled people whether they are visitors or staff who work there and whether they arrive in a wheelchair or are on foot.	M2 (1.30, 1.31a and 1.30b)

Principal entrance doors

The principal entrance door should:

• include a space alongside the leading edge of a door to reduce the risk of a wheelchair user being prevented from reaching the door handle;	M2 (1.33)
• allow people approaching the other side of entrances to see wheelchair users;	M2 (1.34)
• be of sufficient width to enable a wheelchair to manoeuvre;	M2 (1.32)
• contain a leaf which provides a minimum clear opening width of not less than 800 mm;	M2 (1.35a)
• have an unobstructed space on the side next to the leading edge for at least 300 mm;	M2 (1.35b)
• have a glazed panel giving a zone of visibility from a height of 900–500 mm from the finished floor level wherever the opening action of the door could constitute a hazard.	M2 (1.35c)

Internal doors

Internal doors should:

• include a space alongside the leading edge of a door to reduce the risk of a wheelchair user being prevented from reaching the door handle;	M2 (1.33)
• allow people approaching the other side of entrances to see wheelchair users;	M2 (1.34)
• be of sufficient width to enable a wheelchair to manoeuvre;	M2 (1.32)
• contain a leaf that provides a minimum clear opening width of not less than 750 mm;	M2 (2.4a)
• have an unobstructed space on the side next to the leading edge for at least 300 mm;	M2 (2.4b)
• have a glazed panel giving a zone of visibility from a height of 900–500 mm from the finished floor level for each doorway across an accessible corridor or passageway.	M2 (2.4c)

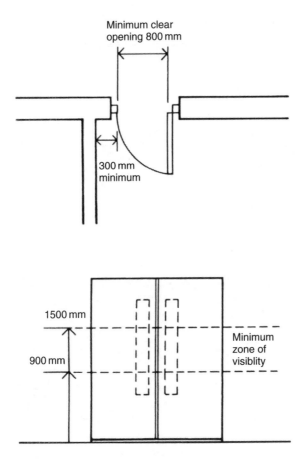

Figure A8 Entrance doorways. (a) Minimum clear opening: single doors, at least one of a pair, should provide the minimum clear opening. (b) Glazed doors: doors across circulation routes should have visibility glazing

Revolving doors

Whilst some larger types of revolving doors are considered suitable for use by disabled people on their own, they should, nevertheless:

• be capable of accommodating several people at the same time;	M2 (1.37)
• revolve very slowly;	M2 (1.37)
• be equipped with mechanisms to stop them as soon as they feel resistance.	M2 (1.37)

Corridors and passageways

In locations required to be accessible to wheelchair users, corridors and passageways need to be wide enough to allow for wheelchair manoeuvre and for other people to pass.

Wheelchair users should have access to an unobstructed width of 1200 mm (or 1000 mm if access is via a stairway or through an existing building).	M2 (2.6a and b)

Internal lobbies

A wheelchair user should be able to move clear of one door before using the next one.	M2 (2.7)
If there is no passenger lift to provide vertical access, a stair should be designed to satisfy the needs of ambulant disabled people.	M2 (2.10)
Stairs should be designed to be suitable for people with impaired sight.	M2 (2.10)

Passenger lifts

Wheelchair users need sufficient space and time to manoeuvre into a lift and, once in, should not be restricted for space. Passenger lifts should:

• have a clear landing at least 1500 mm wide and at least 1500 mm long in front of the entrance;	M2 (2.14a)
• have door(s) wide enough to provide a clear opening width of at least 800 mm;	M2 (2.14b)
• have a car whose width is at least 1100 mm and whose length is at least 1400 mm;	M2 (2.14c)
• have landing and car controls that are not less than 900 mm and not more than 1200 mm above the landing and the car floor at a distance of at least 400 mm from the front wall;	M2 (2.14d)

- be provided with visual and voice indication of   M2 (2.14g)
  the floor reached;

- be provided with tactile indication of the floor   M2 (2.14f)
  level;

- incorporate a signalling system to indicate that the   M2 (2.14h)
  lift is answering a landing call.

Wheelchair users should be able to reach the controls   M2 (2.12)
that summon and direct the lift.

Suitable means of access should be provided from   M2 (2.13c)
the lift to the remainder of the storey.

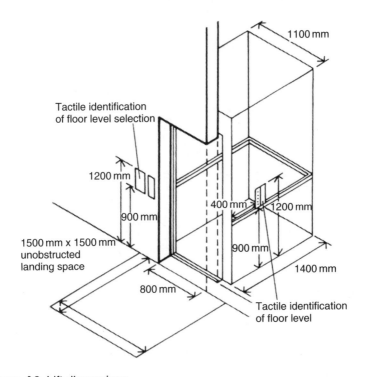

Figure A9 Lift dimensions

In some buildings it may be impractical to provide a passenger lift. In such circumstances, a wheelchair stairlift (in accordance with BS 5776: 1996 Specification for powered stairlifts) or a platform lift (in accordance with BS 6440: 1983 Powered lifting platforms for use by disabled people) is considered a reasonable alternative.

A.2.3 Other buildings

The design of 'other buildings' should reflect the fact that disabled people should be able to visit them, independently or with companions.

Common facilities (such as canteens and cloakrooms, doctors' and dentists' consulting rooms or other health facilities) should be located in a storey to which wheelchair users have access.	M2 (3.3a)
Restaurants and bars should reflect the fact that disabled people should be able to visit them, independently or with companions (e.g. access should be available to the full range of services offered and at least half the area where seating is provided should be accessible to wheelchair users).	M2 (3.4–3.6)
Hotel and motel bedrooms – at least one guest room out of every twenty should be suitable in terms of size, layout and facilities for wheelchair users.	M2 (3.7–3.9)
Recreational facilities – changing rooms (in particular) need special consideration.	M2 (3.10–3.11)
Auditoriums and churches – where appropriate, loop induction or infrared systems should be fitted to enable people with impaired hearing to take part.	M2 (3.12–3.17)
Educational establishments – in schools or other educational establishments, the requirements for disabled persons are covered by Design Note 18, 1984 'Access for Disabled Persons to Educational Buildings'.	

Sanitary facilities

A wheelchair user should not have to travel more than one storey to reach a suitable WC.	M3 (4.3)
The design of the WC compartments should reflect ease of access and use at any time.	M3 (4.4)
Sanitary accommodation for wheelchair users can be provided on a 'unisex' or 'integral' basis.	M3 (4.5)
A 'unisex' facility should be approached separately from other sanitary accommodation.	M3 (4.5a)

WCs should be capable of allowing frontal, lateral,   M3 (4.6)
diagonal and backward transfer onto the WC and should
have facilities for hand washing and hand drying within
reach from the WC, prior to transfer back onto the
wheelchair.

Audience or spectator seating

In theatres, cinemas, concert halls, sports stadia etc., 1% or 6 (whichever is greater) of fixed audience or spectator seats available to the public, should be dedicated as 'wheelchair spaces' (M4 (5.4)).

A 'wheelchair space' is defined as a clear space (one that is kept clear or one that can readily be provided for the occasion by removing a seat) at least 900 mm wide by 1400 mm deep that is easily accessible to a wheelchair user and provides a clear view of the event (M4 (5.6)).

Appendix B

Conservation of fuel and power

B.1 Main changes in the 2002 edition

The main changes made by the Building Regulations (Amendment) Regulations 2001 primarily concerned the legal requirements for:

- the conservation of fuel and power in buildings (Domestic L1 and non-domestic L2);
- heating and hot water systems (including boiler efficiency);
- efficient internal and external lighting systems fixed to the dwelling;
- replacement work on windows (and other glazed elements), boilers and hot water vessels.

These changes provoked an improved method for calculating the U-values by bringing them in line with European standards and by setting lower (i.e. better) values. For example:

- The nominal average U-value for windows and other glazed elements has been improved and is now based on sealed double units with low-emissivity inner panes and 16 mm air gaps.
- Metal windows are given some credit for their thinner framing members and hence the increased useful solar gain.
- The energy rating method that used SAP ratings has been replaced by the new carbon index method.
- The guidance on reducing thermal bridging has been improved and there is new guidance on reducing unwanted air leakage.
- New guidance is given on complying with the requirements for commissioning heating and hot water systems, for lighting and for the provision of information to householders.

- The guidance on conservatories has been extended to cover conservatories attached to existing dwellings.
- Guidance on material alterations and changes of use remains unchanged but has been placed in a new section along with new guidance work on controlled services in existing dwellings and the conservation and restoration work in historic buildings.

B.2 The requirement

B.2.1 Dwellings

Reasonable provision shall be made for the conservation of fuel and power in dwellings by:

(a) *limiting the heat loss:*
    (i)   *through the fabric of the building;*
    (ii) *from hot water pipes and hot air ducts used for space heating;*
    (iii) *from hot water vessels;*
(b) *providing space heating and hot water systems which are energy efficient;*
(c) *providing lighting systems with appropriate lamps and sufficient controls so that energy can be used efficiently;*
(d) *providing sufficient information with the heating and hot water services so that building occupiers can operate and maintain the services in such a manner as to use no more energy than is reasonable in the circumstances.*

💡 The requirement for sufficient controls in requirement L1 (c) applies only to external lighting systems fixed to the building.

💡 In the context of Approved Document L1, 'building work' consists of the provision of a window, rooflight, roof window, door (being a door which together with its frame has more than 50% of its internal face area glazed), a space heating or hot water service boiler, or a hot water vessel.

B.3 Meeting the requirement

Energy efficiency measures shall be provided that:

(a)   • limit the heat loss through the roof, wall, floor, windows and doors, etc. by suitable means of insulation;
     • where appropriate permit the benefits of solar heat gains and more efficient heating systems to be taken into account;

- limit unnecessary ventilation heat loss by providing building fabric that is reasonably airtight;
- limit the heat loss from hot water pipes and hot air ducts used for space heating;
- limit the heat losss from hot water vessels and their primary and secondary hot water connections by applying suitable thicknesses of insulation (where such heat does not make an efficient contribution to the space heating).

(b) Provide space heating and hot water systems with reasonably efficient equipment such as heating appliances and hot water vessels where relevant, such that the heating and hot water systems can be operated effectively as regards the conservation of fuel and power.

(c) Provide lighting systems that utilize energy-efficient lamps with manual switching controls or, in the case of external lighting fixed to the building, automatic switching, or both manual and automatic switching controls as appropriate, such that the lighting systems can be operated effectively as regards the conservation of fuel and power.

(d) Provide information, in a suitably concise and understandable form (including results of performance tests carried out during the works) that shows building occupiers how the heating and hot water services can be operated and maintained.

Responsibility for achieving compliance with the requirements of Part L rests with the person carrying out the work. That person may be, for example, a developer, a main (or sub-) contractor, or a specialist firm directly engaged by a private client.

The person responsible for achieving compliance should either themselves provide a certificate, or obtain a certificate from the sub-contractor, that commissioning has been successfully carried out. The certificate should be made available to the client and the building control body.

B.3.1 General

A conservatory that is not separated from the rest of the dwelling shall be treated as an integral part of the dwelling.	L1
Where the conservatory is separated from the rest of the dwelling and has a **fixed** heating installation, the heating in the conservatory shall have its own separate temperature and on/off controls.	L1
Where an opening is created or enlarged, provision shall be made to limit the heat loss from the dwelling such that it is no worse than before the work was undertaken.	L1

Measurements of thermal conductivity should be made   L1 (0.9)
according to BS EN 12664, BS EN 12667, or BS EN 12939.

Measurements of thermal transmittance should be made   L1 (0.9)
according to BS EN ISO 8990 or, in the case of windows
and doors, BS EN ISO 12567.

📖 **Thermal conductivity** (i.e. the λ-value) of a material is a measure of the rate at which that material will pass heat and is expressed in units of watts per metre per degree of temperature difference (W/mK).

Thermal transmittance (i.e. the U-value) is a measure of how much heat will pass through 1 m^2 of a structure when the air temperatures on either side differ by 1 °C.

B.3.2 Calculation of U-values

U-values should be calculated using the methods given in: L1 (0.11)

- for walls and roofs: BS EN ISO 6946;
- for ground floors: BS EN ISO 13370;
- for windows and doors: BS EN ISO 10077–1 or prEN (SO 10077–2);
- for basements: BS EN ISO 13370 or the BCA/NHBC Approved Document.

💡 Appendix A to Approved Document L1 contains tables of U-values and examples of their use, which provide a simple way to establish the amount of insulation needed to achieve a given U-value for some typical forms of construction.

When calculating U-values the thermal bridging effects of   L1 (0.13)
timber joists, structural and other framing, normal mortar
bedding and window frames should generally be taken
into account.

Thermal bridging can be disregarded where the difference   L1 (0.13)
in thermal resistance between the bridging material and
the bridged material is less than 0.1 m^{2}K/W.

Normal mortar joints need not be taken into account in   L1 (0.13)
calculations for brickwork.

Walls containing in-built meter cupboards, and ceilings   L1 (0.13)
containing loft hatches, recessed light fittings, etc. should
be calculated using area-weighted average U-values.

B.3.3 Design and construction

Under separate provisions in the Building Regulations a new dwelling created by building work, or by a material change of use in connection with which building work is carried out, must be given an energy rating, using the SAP edition having the Secretary of State's approval at the relevant time in the particular case; and the rating must be displayed in the form of a notice. Administrative guidance on producing and displaying SAP ratings is given in DETR Circular No 07/2000 dated 13 October 2000.

Three methods are currently used for demonstrating reasonable provision for limiting heat loss through the building fabric:

1. An elemental method;
2. A target U-value method;
3. A carbon index method.

✎ **SAP** means the Government's Standard Assessment Procedure for Energy Rating of Dwellings. The SAP provides the methodology for the calculation of the carbon index, which can be used to demonstrate that dwellings comply with Part L.

B.3.4 Elemental method

The Elemental Method can be used only when the heating system is either a gas boiler, oil boiler, heat pump, community heating with CHP, biogas or biomass fuel, but **not** for direct electric heating or other systems.

The elemental method is suitable for alterations and extension work, and for newbuild work when it is desired to minimize calculations.	L1 (1.3)
When using the elemental method, the requirement will be met for new dwellings by selecting construction elements that provide the U-value thermal performances given in Table B1.	L1 (1.3)

✎ **Exposed element** means an element exposed to the outside air (including a suspended floor over a ventilated or unventilated void, and elements so exposed indirectly via an unheated space), or an element in the floor or basement in contact with the ground. In the case of an element exposed to the outside air via an unheated space (previously known as a 'semi-exposed element') the U-value should be determined using the method given in the SAP 1998 (replaced by SAP 2001 in 2001). Party walls, separating two dwellings or other premises that can reasonably be assumed to be heated to the same temperature, are assumed not to need thermal insulation.

Table B1 Elemental method – U-values (W/m²K) for construction elements

Exposed element	U-value
Pitched roof with insulation between rafters[1, 2]	0.2
Pitched roof with integral insulation	0.25
Pitched roof with insulation between joists	0.16
Flat roof[3]	0.25
Walls, including basement walls	0.35
Floors, including ground floors and basement floors	0.25
Windows, doors and rooflights (area-weighted average), glazing in metal frames[4]	2.2
Windows, doors and (area-weighted average) rooflights[4], glazing in wood or PVC frames[5]	2.0

Notes:
1. Any part of a roof having a pitch of 70° or more can be considered as a wall.
2. For the sloping parts of a room-in-the-roof constructed as a material alteration, a U-value of 0.3 W/m²K would be reasonable.
3. Roof of pitch not exceeding 10°.
4. Rooflights include roof windows. (A roof window is a window in the plane of a pitched roof and may be considered as a rooflight for the purposes of this Approved Document.)
5. The higher U-value for metal-framed windows allows for additional solar gain due to the greater glazed proportion.

> Single-glazed panels can be acceptable in external doors provided that the heat loss through all the windows, doors and rooflights matches the relevant figure in Table B1 and the area of the windows, doors and rooflights together does not exceed 25% of the total floor area.   **L1 (1.5)**
>
> Boiler efficiency would be demonstrated by using a boiler with SEDBUK not less than the appropriate entry in Table B2.   **L1 (1.7)**

SEDBUK is the Seasonal Efficiency of a Domestic Boiler in the UK, defined in the Government's Standard Assessment Procedure for the Energy Rating of Dwellings.

Table B2 Minimum boiler SEDBUK corresponding to Table B1 for use with the target U-value

Central heating system fuel	SEDBUK percentage
Mains natural gas	78%
LPG	80%
Oil	85%

The average U-value of windows, doors and rooflights   L1 (1.7)
should match the relevant figure in Table B1 and the area
of the windows, doors and rooflights together does not
exceed 25% of the total floor area.

💡 Examples of how the average U-value is calculated
are given in Appendix D to approved Document L1.

B.3.5 Extensions to dwellings

The fabric U-values given in Table B1 can be applied   L1 (1.11)
when proposing extensions to dwellings.

The target U-value and carbon index methods can be   L1 (1.11)
used only if applied to the whole enlarged dwelling.

The total rate of heat loss is the product of   L1 (1.12)
(area × U-value) for all exposed elements.

For small extensions to dwellings (e.g. porches where   L1 (1.13)
the new heated space created has a floor area of not
more than about 6 m²) energy performance should
not be less than those in the existing building.

The area-weighted average U-value of windows,   L1 (1.14)
doors and rooflights ('openings') in extensions to
existing dwellings should not exceed the relevant
values in Table B1.

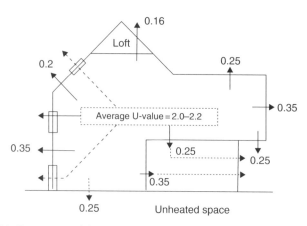

Figure B1 Summary of the elemental method

Figure B1 summarizes the fabric insulation standards and allowances for windows, doors and rooflights given in the elemental method.

Examples of the procedures used in this method are given in Appendices A to C of Approved Document L1.

B.3.6 Target U-value

This method allows greater flexibility than the elemental method in selecting the areas of windows, doors and rooflights, and the insulation levels of individual elements in the building envelope, taking into account the efficiency of the heating system and enabling solar gain to be addressed. It can be used for any heating system.

The requirement would be met if the calculated average U-value of the dwelling does not exceed the target U-value, corrected for the proposed method of heating, as determined from the following equation:

$$U_T = [0.35 - 0.19(A_R/A_T) - 0.10(A_{GF}/A_T) + 0.413(A_F/A_T)]$$

where:

U_T is the target U-value prior to any adjustment for heating system performance or solar gain

A_R is the exposed roof area

A_T is the total area of exposed elements of the dwelling (including the ground floor)

A_{GF} is the ground floor area

A_F is the total floor area (all storeys).

The target U-value equation assumes equal distribution of glazed openings on north and south elevations.

For dwellings whose windows have metal frames (including thermally broken frames) the target U-value can be increased by multiplying by a factor of 1.03, to take account of the additional solar gain due to the greater glazed proportion.

B.3.7 Carbon index method

The aim in this method is to provide more flexibility in the design of new dwellings whilst achieving similar overall performance to that obtained by following the elemental method. The carbon index method can be used with any heating system.

The carbon index for a dwelling (or each dwelling in a block   L1 (1.27)
of flats or converted building) should not be less than 8.0.
(Examples of dwellings with a carbon index of 8.0 or more
are given in Appendix F to Approved Document L1.)

B.3.8 Limiting thermal bridging at junctions and around openings

The building fabric should be constructed so that there   L1 (1.30)
are no significant thermal bridges or gaps in the insulation
layer(s) within the various elements of the fabric, at the
joints between elements, and at the edges of elements
such as those around window and door openings.

B.3.9 Limiting air leakage

Reasonable provision should be made to reduce unwanted   L1 (1.33)
air leakage.

A continuous barrier to air movement around the habitable   L1 (1.34)
space (including separating walls and the edges of
intermediate floors) that is in contact with the inside
of the thermal insulation layer should be maintained
wherever possible.

B.3.10 Space heating system controls

Heating systems other than space heating provided by individual solid fuel, gas and electric fires or room heaters should be provided with:

- zone controls,
- timing controls, and
- boiler control interlocks.

For electric storage heaters the system should have an automatic charge control that detects the internal temperature and adjusts the charging of the heater accordingly.

B.3.11 Zone controls

Hot water central heating systems, fan controlled electric storage heaters and electric panel heaters should control the temperatures independently in areas (such as separate sleeping and living areas that have different heating needs).	L1 (1.38)
Room thermostats and/or thermostatic radiator valves (or any other suitable temperature sensing devices) should be used where appropriate.	L1 (1.38)
In most dwellings one timing zone divided into two temperature control sub-zones would be sufficient.	L1 (1.39)
Large dwellings should be divided into zones with floor area no greater than $150\,m^2$.	L1 (1.39)

B.3.12 Timing controls

For gas-fired and oil-fired systems and for systems with solid-fuel-fired boilers (where forced-draught fans operate when heat is required), timing devices should be provided to control the periods when the heating systems operate.	L1 (1.40)
Separate timing control should be provided for space heating and water heating, except for combination boilers or solid fuel appliances.	L1 (1.40)

B.3.13 Boiler control interlocks

Gas- and oil-fired hot water central heating system controls should switch the boiler off when no heat is required whether control is by room thermostats or by thermostatic radiator valves.	L1 (1.41)
The boiler in systems controlled by thermostats should operate only when a space heating or vessel thermostat is calling for heat.	L1 (1.41a)
A room thermostat or flow switch should also be provided to switch off the boiler when there is no demand for heating or hot water.	L1 (1.41b)

B.3.14 Hot water systems

Systems incorporating integral or separate hot water storage vessels should:

• have an insulating vessel with a 35 mm thick, factory-applied coating of PU-foam having a minimum density of 30 kg/m^3.	L1 (1.43)
Unvented hot water systems need additional insulation to control the heat losses through the safety fittings and pipework.	L1 (1.43)

B.3.15 Operating and maintenance instructions for heating and hot water systems

The building owner and/or occupier should be given information on the operation and maintenance of the heating and hot water systems.	L1 (1.51)
The instructions should be directly related to the system(s) in the dwelling and should explain to householders how to operate the systems so that they can perform efficiently, and what routine maintenance is advisable for the purposes of the conservation of fuel and power.	L1 (1.51)

B.3.16 Insulation of pipes and ducts

Pipes and ducts should be insulated to conserve heat and hence maintain the temperature of the water or air heating service.	L1 (1.52)
Space heating pipework located outside the building fabric insulation layer(s) should be wrapped with insulation material having a thermal conductivity at 40 °C not exceeding 0.035 W/m^{2}K and a thickness equal to the outside diameter of the pipe up to a maximum of 40 mm.	L1 (1.52a)
Warm air ducts should be insulated in accordance with BS 5422: 2001.	L1 (1.52b)

| Hot pipes connected to hot water storage vessels (including the vent pipe, and the primary flow and return to the heat exchanger, where fitted) should be insulated for at least 1 m from their points of connection (or up to the point where they become concealed). | L1 (1.52c) |
| To protect against freezing, central heating and hot water pipework in unheated areas may also need increased insulation thicknesses. | L1 (1.53) |

B.3.17 Internal lighting

Table B3 gives an indication of recommended number of locations (excluding garages, lofts and outhouses) that need to be equipped with efficient lighting (L1 (1.55–1.56)).

| In locations where lighting can be expected to have most use, fixed lighting (e.g. fluorescent tubes and compact fluorescent lamps – but not GLS tungsten lamps with bayonet cap or Edison screw bases) with a luminous efficacy greater than 40 lumens per circuit-watt should be available. | L1 (1.54) |

Circuit-watts means the power consumed in lighting circuits by lamps and their associated control gear and power factor correction equipment.

Table B3

Number of rooms created (see Note)	Recommended minimum number of locations
1–3	1
4–6	2
7–9	3
10–12	4

Note: Hall, stairs and landing(s) count as one room, as does a conservatory.

B.3.18 External lighting fixed to the building

External lighting (including lighting in porches, but not lighting in garages and carports) should:

- automatically extinguish when there is enough   L1 (1.57a)
  daylight, and when not required at night;

- have sockets that can only be used with lamps having   L1 (1.57b)
  an efficacy greater than 40 lumens per circuit watt
  (such as fluorescent or compact fluorescent lamp
  types, and **not** GLS tungsten lamps with bayonet cap
  or Edison screw bases).

B.3.19 Conservatories

For the purposes of the guidance in Part L, a conservatory is considered to not have less than three-quarters of the area of its roof and not less than one-half of the area of its external walls made of translucent material.

If a conservatory is attached to and built as part of a new dwelling it should be treated as an integral part of the building.	L1 (1.59)
If there is no physical separation between the conservatory and the dwelling, then the conservatory should be treated as an integral part of the dwelling.	L1 (1.59a)
If the conservatory is separated from the dwelling, energy savings can be achieved if the conservatory is not heated.	L1 (1.59b)
If fixed heating installations are used in a conservatory, they should have their own separate temperature and on/off controls.	L1 (1.59b)
When a conservatory is attached to an existing dwelling and an opening is enlarged or newly created as a material alteration, the heat loss from the dwelling should be limited by: • retaining the existing separation where the opening is not to be enlarged; • providing separation with an average U-value less than $2.0\,W/m^2K$.	L1 (1.60)
Separating walls and floors between a conservatory and a dwelling should be insulated to at least the same degree as the exposed walls and floors.	L1 (1.61)
Separating windows and doors between a conservatory and a dwelling should have the same U-value and draught-stripping provisions as the exposed windows and doors elsewhere in the dwelling.	L1 (1.61b) AD N

B.3.20 Material alterations

Depending on the circumstances, material alterations with respect to the requirements for conserving fuel and power as shown in Approved Document L could be consided to be:

- **Roof insulation**: when substantially replacing any of the major elements of a roof structure in a material alteration, providing insulation to achieve the U-value for new dwellings.
- **Floor insulation**: where the structure of a ground floor or exposed floor is to be substantially replaced, or re-boarded, providing insulation in heated rooms to the standard for new dwellings.
- **Wall insulation**: when substantially replacing complete exposed walls or their external renderings or cladding or internal surface finishes, or the internal surfaces of separating walls to unheated spaces, providing a reasonable thickness of insulation.
- **Sealing measures**: when carrying out any of the above work, including reasonable sealing measures to improve airtightness.
- **Controlled services and fittings**: when replacing controlled services and fittings.

B.3.21 Replacement of controlled services or fittings

For the purpose of L1, 'building work' consists of the provision of a window, rooflight, roof window, door (being a door which together with its frame has more than 50% of its internal face area glazed), a space heating or hot water service boiler, or a hot water vessel.

Replacement work on controlled services or fittings (whether replacing with new but identical equipment or with different equipment) depends on the circumstances in the particular case and would also need to take account of historic value (L1 (2.3)).

Windows, doors and rooflights: when windows, etc.   L1 (2.3a)
are being replaced with new draught-proofed ones either
with an average U-value not exceeding the appropriate
entry in Table B1, or with a centre-pane U-value not
exceeding $1.2\,\mathrm{W/m^2K}$.

This requirement does not apply to repair work on
parts of these elements, such as replacing broken glass
or sealed double-glazing units or replacing rotten
framing members.

Heating boilers: replacement heating boilers in dwellings having a floor area greater than 50 m² shall be treated as if it were a new dwelling and in the case of:	L1 (2.3a)
• ordinary oil or gas boilers, shall be by a boiler with a SEDBUK not less than the approproate entry in Table B2	L1 (2.3a(1))
• back boilers, shall be by a boiler having a SEDBUK of not less than three percentage points lower than the appropriate entry in Table B2	L1 (2.3a(2))
• solid fuel boilers, shall be by a boiler having an efficiency not less than that recommended for its type in the HETAS certification scheme.	L1 (2.3a(3))
Hot water vessels: replacements shall all be new equipment as if for a new dwelling.	L1 (2.3c)
Boiler and hot water storage controls: the work may also need to include replacement of the time switch or programmer, room thermostat, and hot water vessel thermostat, and provision of a boiler interlock and fully pumped circulation.	L1 (2.3d)
Commissioning and providing operating and maintenance instructions: where heating and hot water systems are to be altered as in paragraphs (a) to (e), reasonable provision would also include appropriate commissioning and the provision of operating and maintenance instructions.	L1 (2.3f)

B.3.22 Material changes of use

Depending on the circumstances, material changes of use for the purpose of conserving fuel and power as required by Approved Document L, could be satisfied by:

- **Accessible lofts**: when upgrading insulation in accessible lofts, providing additional insulation to achieve a U-value not exceeding 0.25 W/m²K where the existing insulation provides a U-value worse than 0.35 W/m²K.
- **Roof insulation**: when substantially replacing any of the major elements of a roof structure, providing insulation to achieve the U-value considered reasonable for new dwellings.
- **Floor insulation**: when substantially replacing the structure of a ground floor, providing insulation in heated rooms to the standard for new dwellings.

- **Wall insulation**: when substantially replacing complete exposed walls or their internal or external renderings or plaster finishes or the internal renderings and plaster of separating walls to an unheated space, providing a reasonable thickness of insulation.
- **Sealing measures**: when carrying out any of the above work, including reasonable sealing measures to improve airtightness.

B.3.23 Historic buildings

Historic buildings include:

- listed buildings,
- buildings situated in Conservation Areas,
- buildings that are of architectural and historical interest and that are referred to as a material consideration in a local authority's development plan,
- buildings of architectural and historical interest within national parks, areas of outstanding natural beauty, and World Heritage sites.

The need to conserve the special characteristics of such historic buildings needs to be recognized. In such work, the aim should be to improve energy efficiency where practically possible, always provided that the work does not prejudice the character of the historic building, or increase the risk of long-term deterioration to the building fabric or fittings. In arriving at an appropriate balance between historic building conservation and energy conservation, it would be appropriate to take into account the advice of the local planning authority's conservation officer.

B.3.24 Additional guidance

Attached to Approved Document L1 are a number of appendices that provide the reader with additional guidance and worked examples concerning fuel and power conservation and methods for obtaining U-values. These consist of:

- Appendix A Tables of U-values
- Appendix B Calculating U-values
- Appendix C U-values of ground floors
- Appendix D Determining U-values for glazing
- Appendix E Target U-value examples
- Appendix F SAP ratings and carbon indexes
- Appendix G Carbon Index.

Bibliography

Standards referred to

Title	Standard
Building components and building elements – Thermal resistance and thermal transmittance – Calculation method	BS EN ISO 6946: 1997
Building materials and products – Hydrothermal properties – Tabulated design values	BS EN 12524: 2000
Capillary and compression tube fittings of copper and copper alloy: Part 2: 1983 Specification for capillary and compression fittings for copper tubes	BS 864
Cast iron pipes and fittings, their joints and accessories for the evacuation of water from buildings. Requirements, test methods and quality assurance	BS EN 877: 1999
Chimneys. Clay/Ceramic Flue Blocks for Single Wall Chimneys. Requirements and Test Methods	BS EN 1806: 2000
Chimneys. Clay/Ceramic Flue Liners. Requirements and Test Methods	BS EN 1457: 1999
Chimneys. General Requirements	BS EN 1443: 1999
Chimneys. Metal Chimneys. Test Methods	BS EN 1859: 2000
Code of practice for accommodation of building services in ducts	BS 8313: 1989
Code of practice for building drainage	BS 8301: 1985
Code of practice for design and installation of damp-proof courses in masonry construction	BS 8215: 1991
Code of practice for design and installation of natural stone cladding and lining	BS 8298
Code of practice for design and installation of small sewage treatment works and cesspools	BS 6297: 1983

Code of practice for design of non-load bearing external vertical enclosures of buildings	BS 8200: 1985
Code of practice for drainage of roofs and paved areas	BS 6367: 1983
Code of practice for fire door	BS 8214: 1990
Code of Practice for Flues and Flue Structures in Buildings	BS 5854: 1980 (1996)
Code of practice for foundations	BS 8004: 1986
Code of practice for mechanical ventilation and air-conditioning in buildings	BS 5720: 1979
Code of Practice for Oil Firing: Part 1: 1977 Installations up to 44 kW Output Capacity for Space Heating and Hot Water Supply Purposes Part 2: 1978 Installations of 44 kW or Above Output Capacity for Space Heating, Hot Water and Steam Supply Purposes	BS 5410
Code of practice for powered lifting platforms for use by disabled persons	BS 6440: 1983
Code of practice for protection of structures against water from the ground	BS 8102: 1990
Code of practice for protective barriers in and about buildings	BS 6180: 1995
Code of practice for sanitary pipework	BS 5572: 1978
Code of practice for sheet roof and wall coverings: Part 14: 1975 Corrugated asbestos-cement	BS 5247
Code of practice for site investigation	BS 5930: 1981
Code of practice for stone masonry	BS 5390: 1976 (1984)
Code of practice for the control of condensation in buildings	BS 5250: 1989
Code of practice for the storage and on-site treatment of solid waste from buildings	BS 5906: 1980 (1987)
Code of practice for thermal insulation of cavity walls (with masonry or concrete inner and outer leaves) by filling with urea-formaldehyde (UF) foam systems	BS 5618: 1985
Code of practice for use of masonry: Part 1: 1978 Structural use of unreinforced masonry Part 3: 1985 Materials and components, design and workmanship	BS 5628
Code of practice for ventilation principles and designing for natural ventilation	BS 5925: 1991

Title	Standard
Code of practice. External rendered finishes	BS 5262: 1976
Components for smoke and heat control systems: Part 2: 1990 Specification for powered smoke and heat exhaust ventilators	BS 7346
Components of automatic fire alarm systems for residential premises: Part 1: 1990 Specification for self-contained smoke alarms and point-type smoke detectors	BS 5446

Concrete:
Part 1: 1990 Guide to specifying concrete
Part 2: 1990 Method for specifying concrete mixes
Part 3: 1990 Specification for the procedures to be
   used in producing and transporting concrete
Part 4: 1990 Specification for the procedures
   to be used in sampling, testing and assessing
   compliance of concrete
	BS 5328

Construction and testing of drains and sewers	BS EN 1610: 1998

Copper and copper alloys. Plumbing fittings:
Part 1: Fittings with ends for capillary soldering
   or capillary brazing to copper tubes
Part 2: Fittings with compression ends for use
   with copper tubes
Part 3: Fittings with compression ends for use
   with plastic pipes
Part 4: Fittings combining other end connections
   with capillary or compression ends
Part 5: Fittings with short ends for capillary brazing
   to copper tubes
	BS EN 1254: 1998

Copper and copper alloys. Seamless, round copper tubes for water and gas in sanitary and heating applications	BS EN 1057: 1996
Copper indirect cylinders for domestic purposes. Specification for double feed indirect cylinders	BS 1566-1: 1984
Discharge and ventilating pipes and fittings, sand-cast or spun in cast iron: Part 1: 1990 Specification for spigot and socket systems Part 2: 1990 Specification for socketless systems	BS 416
Drain and sewer systems outside buildings: Part 1: 1996 Generalities and definitions Part 2: 1997 Performance requirements	BS EN 752

Part 3: 1997 Planning
Part 4: 1997 Hydraulic design and environmental
    aspects
Part 5: 1997 Rehabilitation
Part 6: 1998 Pumping installations
Part 7: 1998 Maintenance and operations

Ductile iron pipes, fittings, accessories and their   BS EN 598: 1995
joints for sewerage applications. Requirements and
test methods

Emergency lighting: BS 5266
Part 1: 1988 Code of practice for the emergency
    lighting of premises other than cinemas and
    certain other specified premises used
    for entertainment

Factory-Made Insulated Chimneys: BS 4543
Part 1: 1990 (1996) Methods of Test, AMD 8379
Part 2: 1990 (1996) Specification for Chimneys
    with Stainless Steel Flue Linings for Use with
    Solid Fuel Fired Appliances
Part 3: 1990 (1996) Specification for Chimneys
    with Stainless Steel Flue lining for Use with
    Oil Fired Appliances

Fibre Cement Flue Pipes, Fittings and Terminals: BS 7435
Part 1: 1991 (1998) Specification for Light
    Quality Fibre Cement Flue pipes,
    Fittings and Terminals.
Part 2: 1991 Specification for heavy quality cement
    flue pipes, fittings and terminals

Fire detection and alarm systems for buildings: BS 5839
Part 1: 1988 Code of practice for system design,
    installation and servicing
Part 2: 1983 Specification for manual call points
Part 6: 1995 Code of practice for the design and
    installation of fire detection and alarm systems
    in dwellings
Part 8: 1998 Code of practice for the design,
    installation and servicing of voice alarm
    systems

Fire extinguishing installations and equipment   BS 5306
    on premises:
Part 1: 1976 (1988) Hydrant systems,
    hose reels and foam inlets
Part 2: 1990 Specification for sprinkler
    systems.

Title	Standard
Fire precautions in the design, construction and use of buildings:	BS 5588

Part 0: 1996 Guide to fire safety codes of practice for particular premises

Part 1: 1990 Code of practice for residential buildings

Part 4: 1998 Code of practice for smoke control using pressure differentials

Part 5: 1991 Code of practice for firefighting stairs and lifts

Part 6: 1991 Code of practice for places of assembly

Part 7: 1997 Code of practice for the incorporation of atria in buildings

Part 8: 1999 Code of practice for means of escape for disabled people

Part 9: 1989 Code of practice for ventilation and air conditioning ductwork

Part 10: 1991 Code of practice for shopping complexes

Part 11: 1997 Code of practice for shops, offices, industrial, storage and other similar buildings

| Fire safety signs, notices and graphic symbols: | BS 5449 |

Part 1: 1990 Specification for fire safety signs

| Fire tests on building materials and structures: | BS 476 |

Part 3: 1958 External fire exposure roof tests

Part 4: 1970 (1984) Non combustibility Test for Materials

Part 6: 1981 Method of test for fire propagation for products

Part 6: 1989 Method of test for fire propagation for products.

Part 7: 1971 Surface spread of flame tests for materials.

Part 7: 1987 Method for classification of the surface spread of flame of products

Part 7: 1997 Method of test to determine the classification of the surface spread of flame of products. BS 476: Fire tests on building materials and structures

Part 8: 1972 Test methods and criteria for the fire resistance of elements of building construction

Part 11: 1982 (1988) Method for Assessing the Heat Emission from Building Materials

Part 20: 1987 Method for determination of the fire resistance of elements of construction (general principles)

Part 21: 1987 Methods for determination of the fire
   resistance of loadbearing elements of construction
Part 22: 1987 Methods for determination
   of the fire resistance of non-loadbearing
   elements of construction
Part 23: 1987 Methods for determination of the
   contribution of components to the fire resistance
   of a structure
Part 24: 1987 Method for determination of the fire
   resistance of ventilation ducts
Part 31: Methods for measuring smoke penetration
   through doorsets and shutter assemblies:
   Section 31.1: 1983 Measurement under ambient
   temperature conditions

Flue Blocks and Masonry Terminals for Gas   BS 1289-1: 1986
Appliances:
Part 1: 1986 Specification for Precast Concrete
   Flue Blocks and Terminals
Part 2: 1989 Specification for Clay Flue Blocks
   and Terminals

Fuel Oils for Non-Marine Use: BS 2869: 1998
Part 2: 1988 Specification for Fuel Oil for
   Agricultural and Industrial Engines and Burners
   (Classes A2, C1, C2, D, E, F, G and H).

Glossary of Terms Relating to Solid Fuel Burning   BS 1846
Equipment:
Part 1: 1994 Domestic Appliances

Gravity drainage systems inside buildings: BS EN 12056: 2000
Part 1 Scope, definitions, general and performance
   requirements
Part 2 Wastewater systems, layout and calculation
Part 3 Roof drainage layout and calculation
Part 4 Effluent lifting plants, layout and calculation
Part 5 Installation, maintenance and user instructions

Guide for design, construction and maintenance of BS 6661: 1986
single-skin air supported structures

Guide to assessment of suitability of external   BS 8208
cavity walls for filling with thermal insulants
Part 1: 1985 Existing traditional cavity construction

Guide to development and presentation of fire tests BS 6336: 1998
and their use in hazard assessment

Guide to the choice, use and application of wood   BS 1282: 1975
   preservatives

Title	Standard
Heating Boilers. Heating Boilers with Forced Draught Burners. Terminology, General Requirements, Testing and Marketing	BS EN 303-1: 1999
Installation and Maintenance of Flues and Ventilation for Gas Appliances of Rated Input not exceeding 70 kW net: Part 1: 2000 Specification for Installation and Maintenance of Flues Part 2: 2000 Specification for Installation and Maintenance of Ventilation for Gas Appliances	BS 5440
Installation of Chimneys and Flues for Domestic Appliances Burning Solid Fuel (Including Wood and Peat): Part 1: 1984 (1998) Code of Practice for Masonry Chimneys and Flue Pipes	BS 6461
Installation of Domestic Heating and Cooking Appliances Burning Solid Mineral Fuels: Part 1: 1994 Specification for the Design of Installations Part 2: 1994 Specification for Installing and Commissioning on Site Part 3: 1994 Recommendations for Design and on Site Installation	BS 8303
Installation of Factory-Made Chimneys to BS 4543 for Domestic Appliances: Part 1: 1992 (1998) Method of Specifying Installation Design Information Part 2: 1992 (1998) Specification for Installation Design Part 3: 1992 Specification for Site Installation Part 4: 1992 (1998) Recommendations for Installation Design and Installation	BS 7566
Installations for separation of light liquids (e.g. petrol or oil): Part 1 Principles of design, performance and testing, marking and quality control	BS EN 858: 2001
Internal and external wood doorsets, door leaves and frames: Part 1: 1980 (1985) Specification for dimensional requirements	BS 4787

Lifts and service lifts:	BS 5655
Part 1: 1986 Safety rules for the construction and installation of electric lifts (Part 1 to be replaced by BS EN 81-1, when published)	
Part 2: 1988 Safety rules for the construction and installation of hydraulic lifts (Part 2 to be replaced by BS EN 81-2, when published)	
Part 5: 1989 Specifications for dimensions for standard lift arrangements	
Part 7: 1983 Specification for manual control devices, indicators and additional fittings	
Loading for buildings:	BS 6399
Part 1: 1996 Code of practice for dead and imposed loads	
Measurement of sound insulation in buildings and of building elements:	BS 2750
Part 1: 1980 Recommendations for laboratories	
Part 3: 1980 Laboratory measurement of airborne sound insulation of building elements	
Part 4: 1980 Field measurement of airborne sound insulation between rooms	
Part 6: 1980 Laboratory measurement of impact sound insulation of floors	
Part 7: 1980 Field measurements of impact sound insulation of floors	
Method for specifying thermal insulating materials for pipes, tanks, vessels, ductwork and equipment operating within the temperature range $-40\,°C$ to $+70\,°C$	BS 5422: 2001
Method of test for ignitability of fabrics used in the construction of large tented structures	BS 7157: 1989
Methods for rating the sound insulation in building elements:	BS 5821
Part 1: 1984 Method for rating the airborne sound insulation in buildings and interior building elements	
Part 2: 1984 Method for rating the impact sound insulation	
Methods of test for flammability of textile fabrics when subjected to a small igniting flame applied to the face or bottom edge of vertically oriented specimens, Test 2	BS 5438: 1989

Title	Standard
Methods of testing plastics: Part 1: Thermal properties: Methods 120A to 120E: 1990 Determination of the Vicat softening temperature of thermoplastics	BS 2782
Oil Burning Equipment, Part 5: 1987 Specification for Oil Storage Tanks	BS 799
Plastic piping systems for non-pressure underground drainage and sewerage. Unplasticized poly(vinylchloride) (PVC-U). Specifications for pipes, fittings and the system	BS EN 1401-1: 1998
Plastic piping systems for soil and waste (low and high temperature) within the building structure. Acryionitrilebutadiene-styrene (ABS). Specifications for pipes, fittings and the system	BS EN 1455-1: 2000
Plastic piping systems for soil and waste discharge (low and high temperature) within the building structure. Chlorinated polyvinyl chloride (PVC-C). Specification for pipes, fittings and the system	BS EN 1566-1: 2000
Plastic piping systems for soil and waste discharge (low and high temperature) within the building structure. Unplasticized polyvinyl chloride (PVC-U). Specifications for pipes, fittings and the system	BS EN 1329-1: 2000
Plastic piping systems for soil and waste discharge (low and high temperature) within the building structure. Polypropylene (PP). Specifications for pipes, fittings and the system	BS EN 1451-1: 2000
Plastic piping systems for soil and waste discharge (low and high temperature) within the building structure. Polyethylene (PE). Specifications for pipes, fittings and the system	BS EN 1519-1: 2000
Plastic piping systems for soil and waste discharge (low and high temperature) within the building structure	BS EN 1565-1: 2000
Precast concrete masonry units: Part 1: 1981 Specification for precast concrete masonry units	BS 6073
Precast concrete pipes fittings and ancillary products: Part 2: 1982 Specification for inspection chambers and street gullies Part 100: 1988 Specification for un-reinforced and reinforced pipes and fittings with flexible joints	BS 5911

Part 101: 1988 Specification for glass composite
    concrete (GCC) pipes and fittings with flexible
    joints
Part 120: 1989 Specification for reinforced jacking
    pipes with flexible joints
Part 200: 1989 Specification for un-reinforced and
    reinforced manholes and soakaways of circular
    cross section

Pressure sewerage systems outside buildings   BS EN 1671: 1997

Safety and control devices for use in hot water   BS 6283
systems:
Part 2: 1991 Specification for temperature relief
    valves for pressures from 1 bar to 10 bar
Part 3: 1991 Specification for combined
    temperature and pressure relief valves
    for pressures from 1 bar to 10 bar

Sanitary installations: BS 6465
Part 1: 1984 Code of practice for scale of provision,
    selection and installation of sanitary appliances

Sanitary tapware. Waste fittings for basins, bidets   BS EN 274: 1993
and baths. General Technical Specifications

Small wastewater treatment plants less than 50 PE   BS EN 12566-1: 2000

Specification for aggregates from natural sources   BS 882: 1983
for concrete

Specification for asbestos-cement pipes,   BS 3656: 1981 (1990)
joints and fittings for sewerage and drainage

Specification for calcium silicate (sandlime and   BS 187: 1978
flintlime) bricks

Specification for cast iron spigot and socket drain   BS 437: 1978
pipes and fittings

Specification for cast iron spigot and socket flue   BS 41: 1973 (1981)
or smoke pipes and fittings

Specification for clay and calcium silicate modular   BS 6649: 1985
bricks

Specification for clay bricks   BS 3921: 1985

Specification for Clay Flue Linings and Flue   BS 1181: 1999
Terminals

Specification for copper and copper alloys. Tubes: BS 2871
Part 1: 1971 Copper tubes for water, gas and sanitation

Specification for copper direct cylinders for   BS 699: 1984
    domestic purposes

Title	Standard
Specification for copper hot water storage combination units for domestic purposes	BS 3198: 1981
Specification for Dedicated Liquified Petroleum Gas Appliances. Domestic Flueless Space Heaters (including Diffusive Catalytic Combustion Heaters)	BS EN 449: 1997
Specification for design and construction of fully supported lead sheet roof and wall coverings	BS 6915: 1988
Specification for design, installation, testing and maintenance of services supplying water for domestic use within buildings and their curtilages	BS 6700: 1987
Specification for electrical controls for household and similar general purposes	BS 3955: 1986
Specification for fabrics for curtains and drapes: Part 2: 1980 Flammability requirements	BS 5867
Specification for fibre boards	BS 1142: 1989
Specification for flexible joints for grey or ductile cast iron drain pipes and fittings (BS 437) and for discharge and ventilating pipes and fittings (BS 416)	BS 6087: 1990
Specification for impact performance requirements for flat safety glass and safety plastics for use in buildings	BS 6206: 1981
Specification for Installation in Domestic Premises of Gas-Fired Ducted-Air Heaters of Rated Input Not Exceeding 60 kW	BS 5864: 1989
Specification for Installation of Domestic Gas Cooking Appliances (lst, 2nd and 3rd Family Gases)	BS 6172: 1990
Specification for Installation of Gas Fired Catering Appliances for Use in All Types of Catering Establishments (lst, 2nd and 3rd Family Gases)	BS 6173: 2001
Specification for Installation of Gas Fires, Convector Heaters, Fire/Back Boilers and Decorative Fuel Effect Gas Appliances: Part 1: 2001 Gas Fires, Convector Heaters and Fire/Back Boilers and heating stoves (lst, 2nd and 3rd Family Gases)	BS 5871

Part 2: 2001 Inset Live Fuel Effect Gas Fires of
    Heat Input Not Exceeding 15 kW (2nd and 3rd
    Family Gases)
Part 3: 2001 Decorative Fuel Effect Gas
    Appliances of Heat Input Not Exceeding 20 kW
    (2nd and 3rd Family Gases), AMD 7033

Specification for installation of gas-fired hot water boilers of rated input not exceeding 60 kW	BS 6798: 2000
Specification for Installation of Hot Water Supplies for Domestic Purposes, Using Gas Fired Appliances of Rated Input not Exceeding 70 kW	BS 5546: 2000
Specification for ladders for permanent access to chimneys, other high structures, silos and bins	BS 4211: 1987
Specification for Metal Flue Pipes, Fittings, Terminals and Accessories for Gas-Fired Appliances with a Rated Input Not Exceeding 60 kW, AMD 8413	BS 715: 1993
Specification for metal ties for cavity wall construction	BS 1243: 1978
Specification for modular co-ordination in building	BS 6750: 1986
Specification for Open Fireplace Components	BS 1251: 1987
Specification for performance requirements for cables required to maintain circuit integrity under fire conditions	BS 6387: 1994
Specification for Performance Requirements for Domestic Flued Oil Burning Appliances (including Test Procedures)	BS 4876: 1984
Specification for plastics inspection chambers for drains	BS 7158: 2001
Specification for plastics waste traps	BS 3943: 1979 (1988)
Specification for Portland cements	BS 12: 1989
Specification for powered stairlifts	BS 5776: 1996
Specification for prefabricated drainage stack units in galvanized steel	BS 3868: 1995
Specification for safety aspects in the design, construction and installation of refrigerating appliances and systems	BS 4434: 1989
Specification for sizes of sawn and processed softwood	BS 4471: 1987
Specification for softwood grades for structural use	BS 4978: 1988

Title	Standard
Specification for the use of structural steel in building: Part 2: 1969 Metric units	BS 449
Specification for thermoplastics waste pipe and fittings	BS 5255: 1989
Specification for thermostats for gas-burning appliances	BS 4201: 1979 (1984)
Specification for tongued and grooved softwood flooring	BS 1297: 1987
Specification for un-plasticized PVC soil and ventilating pipes, fittings and accessories	BS 4514: 1983
Specification for un-plasticized polyvinyl chloride (PVC-U) pipes and plastics fittings of nominal sizes 110 and 160 for below ground drainage and sewerage	BS 4660: 1989
Specification for unplasticized PVC pipe and fittings for gravity sewers	BS 5481: 1977 (1989)
Specification for unvented hot water storage units and packages	BS 7206: 1990
Specification for urea-formaldehyde (UF) foam systems suitable for thermal insulation of cavity walls with masonry or concrete inner and outer leaves	BS 5617: 1985
Specification for Vitreous-Enamelled Low-Carbon-Steel Fluepipes, Other Components and Accessories for Solid-Fuel-Burning Appliances with a Maximum Rated Output of 45 kW	BS 6999: 1989 (1996)
Specification for vitrified clay pipes, fittings and ducts, also flexible mechanical joints for use solely with surface water pipes and fittings	BS 65: 1991
Stainless Steels. List of Stainless Steels	BS EN 10 088-1: 1995
Stairs, ladders and walkways: Part 1: 1977 Code of practice for stairs Part 2: 1984 Code of practice for the design of helical and spiral stairs Part 3: 1985 Code of practice for the design of industrial type stairs, permanent ladders and walkways	BS 5395

Steel Plate, Sheet and Strip: Part 2: 1983 Specification for Stainless and Heat-Resisting Steel Plate, Sheet and Strip	BS 1449
Steel plate, sheet and strip. Carbon and carbon manganese plate, sheet and strip. General specifications	BS 1449-1: 1991
Structural design of buried pipelines under various conditions of loading. General requirements	BS EN 1295-1: 1998

Structural use of concrete:
Part 1: 1985 Code of practice for design
  and construction
Part 2: 1985 Code of practice for special
  circumstances
Part 3: 1985 Design charts for single reinforced
  beams, doubly reinforced beams and rectangular
  columns
	BS 8110

Structural use of concrete. Code of practice for design and construction	BS 8110-1: 1997

Structural use of steelwork in buildings:
Part 1: 1990 Code of practice for design in simple
  and continuous construction: hot rolled sections
Part 2: 1992 Specification for materials, fabrication
  and erection: hot rolled sections
Part 3: Design in composite construction:
  Section 3.1: 1990 Code of practice for design
  of simple and continuous composite beams
Part 4: 1982: Code of practice for design of floors
  with profiled steel sheeting
	BS 5950

Structural use of timber: Part 2: 1991 Code of practice for permissible stress design, materials and workmanship Part 3: 1985 Code of practice for trussed rafter roofs Part 6: Code of practice for timber framed walls Part 6.1: 1988 Dwellings not exceeding three storeys	BS 5268
The principles of the conservation of historic buildings	BS 7913: 1998
Thermal bridges in building construction – Calculation of heat flows and surface temperatures: Part 1: General methods	BS EN ISO 10211-1: 1996
Thermal bridges in building construction – Calculation of heat flows and surface temperatures: Part 2: Linear thermal bridges	BS EN ISO 10211-2: 2001

Title	Standard
Thermal insulation – Determination of steady-state thermal transmission properties – Calibrated and guarded hot box	BS EN ISO 8990: 1996
Thermal insulation of cavity walls by filling with blown man-made mineral fibre: Part 1: 1982 Specification for the performance of installation systems Part 2: 1982 Code of practice for installation of blown man-made mineral fibre in cavity walls with masonry and/or concrete leaves	BS 6232
Thermal performance of building materials and products – Determination of thermal resistance by means of guarded hot plate and heat flow meter methods – Dry and moist products of low and medium thermal resistance	BS EN 12664: 2001
Thermal performance of building materials and products – Determination of thermal resistance by means of guarded hot plate and heat flow meter methods – Products of high and medium thermal resistance	BS EN 12667: 2000
Thermal performance of building materials and products – Determination of thermal resistance by means of guarded hot plate and heat flow meter methods – Thick products of high and medium thermal resistance	BS EN 12939: 2001
Thermal performance of buildings – Heat transfer via the ground – Calculation methods	BS EN ISO 13370: 1998
Thermal performance of windows and doors – Determination of thermal transmittance by hot box method: Part 1: Complete windows and doors	BS EN ISO 12567-1: 2000
Thermal performance of windows, doors and shutters – Calculation of thermal transmittance: Part 1: Simplified methods	BS EN ISO 10077-1: 2000
Thermoplastics piping systems for non-pressure underground drainage and sewerage – Structure walled piping systems of unplasticized poly(vynyl chloride) (PVC-U), polypropylene (PP) and polyethylene (PE) – Part 1: Specification for pipes, fittings and the system:	BS EN 13476-1: 2001

Part 1: 1997 Guide to specifying concrete
Part 2: 1997 Methods for specifying concrete
   mixes
Part 3: 1990 Specification for the procedures
   to be used in producing and transporting
   concrete
Part 4: 1990 Specification for the procedures
   to be used in sampling, testing and assessing
   compliance of concrete

Vacuum drainage systems inside buildings	BS EN 12109: 1999
Vacuum sewerage systems outside buildings	BS EN 1091: 1997

Vitrified clay pipes and fittings and pipe joints   BS 295
for drains and sewers:
Part 1: 1991 Test requirements
Part 2: 1991 Quality control and sampling
Part 3: 1991 Test methods

Vitrified clay pipes and fittings and pipe joints   BS EN 295
for drains and sewers:
Part 1: 1991 Test requirements
Part 2: 1991 Quality control and sampling
Part 3: 1991 Test methods
Part 6: 1996 Requirements for vitrified clay
   manholes

Wastewater lifting plants for buildings and sites –   BS EN 12050: 2001
principles of construction and testing:
Part 1 Lifting plants for wastewater containing
   faecal matter
Part 2 Lifting plants for faecal-free wastewater
Part 3 Lifting plants for wastewater containing
   faecal matter for limited application

Windows, doors and rooflights: BS 8213: Part 1: 1991
Part 1 Code of practice for safety in use and during
   cleaning of windows and doors (including
   guidance on cleaning materials and methods)

Wood stairs Part 1: 1989 Specification for stairs   BS 585
with closed risers for domestic use, including
straight and winder flights and quarter and half
landings

Workmanship on building sites. Code of practice   BS 8000-13: 1989
for above ground drainage and sanitary
appliances

Workmanship on building sites. Code of practice   BS 8000-14: 1989
for below ground drainage

Other publications

Building Research Establishment Good Building Guide No 42 Reed Beds.
Part 1 Application and Specification. BRE GBG 42 Part 1 (2000).
ISBN 1 86081 436 0
Part 2 Design Construction and Maintenance. BRE GBG 42 Part 2 (2000).
ISBN 1 86081 4379

Building Research Establishment. Soakaway design. BRE Digest 365

CIRIA Infiltration drainage – Manual of good practice. CIRIA Report 156

CIRIA Sustainable urban drainage systems – design manual for England
and Wales. CIRIA Report C522 2000. ISBN 0 8601 7 522 7 Drainage and
waste disposal

Cladding systems: a test method, BRE Fire Note 9 (BRE, 1999)

Code of practice. Hardware essential to the optimum performance
of fire-resisting timber doorsets. Association of Builders Hardware
Manufacturers, 1993 (available from the ABHM, 42 Heath Street,
Tamworth, Staffs B79 7JH)

Construction Products Directive (CPD). The Council Directive reference
89/106/EEC dated 21 December 1988 and published in the Official Journal
of the European Communities No L40/12 dated 11.2.89. The CE Marking
Directive (93/68/EEC) amends the CPD

Department of the Environment. Development on Unstable Land: Landslides
and Planning. Planning Policy Guidance Note 14 annex 1. March 1996
ISBN 0 11 753259 2 (www.planning.dtir.gov.uk/ppg/index.htm)

Design methodologies for smoke and heat exhaust ventilation. BR 368,
BRE 1999. (Revision of design principles for smoke ventilation in
enclosed shopping centres. BR 186, BRE, 1990.)

Design, construction, specification and fire management of insulated
envelopes for temperature controlled environments. The International
Association of Cold Storage Contractors (European Division), 1999
(available from the IACSC, Downmill Road, Bracknell, Berks RG12 1GH)

DOE Circular 12/92. Houses in multiple occupation. Guidance to local housing
authorities on standards of fitness under section 352 of the Housing Act
1985. HMSO, 1992

DOE Circular 12/93. Houses in multiple occupation. Guidance to local
housing authorities on managing the stock in their area. HMSO, 1993

Draft guide to fire precautions in existing residential care premises.
Home Office/Scottish Home and Health Department, 1983

Environment Agency. Pollution Prevention Guidelines No 3, Use and Design
of Oil Separators in Surface Water Drainage Systems. Environment
Agency 1999 (www.environmentagency.gov.uk)

Environment Agency. Pollution Prevention Guidelines No 4, Disposal of Sewage Where no Mains Drainage is Available. Environment Agency 1999 (www.environment-agency.gov.uk)

Fire and steel construction: The behaviour of steel portal frames in boundary conditions, 1990. 2nd edition (available from the Steel Construction Institute, Silwood Park, Ascot, Berks, SL5 7QN)

Fire performance of external thermal insulation for walls of multi-storey buildings. BR 135, BRE 1988

Fire Precautions (Workplace) Regulations 1997 (SI 1997 No 1840), as amended by the Fire Precautions (Workplace) (Amendment) Regulations 1999

Fire Precautions Act 1971. Guide to fire precautions in existing places of work that require a fire certificate. Factories, offices, shops and railway premises. (Home Office/Scottish Office) HMSO, 1993

Fire protection for structural steel in buildings, 2nd edition. ASFP/SCI/FTSG, 1992 (available from the ASFP, Association House, 235 Ash Road, Aldershot, Hants GU12 4DD and the Steel Construction Institute, Silwood Park, Ascot, Berks SL5 7QN)

Fire safety in construction work, HSG 168. ISBN 0-7176-1332-1

Fire safety of PTFE-based materials used in buildings. BR 274, BRE 1994

Firecode. HTM 81. Fire precautions in new hospitals. (NHS Estates) HMSO, 1996

Firecode. HTM 88. Guide to fire precautions in NHS housing in the community for mentally handicapped (or mentally ill) people. (DHSS) HMSO, 1986

Gas Safety (Installation and Use) Regulations 1998, SI 1998 No 2451

Good Building Guide 25, Radon and Buildings (1996) ISBN 1 86081 0705

Guide to Safety at Sports Grounds. (Department of National Heritage/Scottish Office) HMSO, 1997

Guidelines for the construction of fire-resisting structural elements. BR 128, BRE, 1988

Heat radiation from fires and building separation. Fire Research Technical Paper 5. HMSO, 1963

Hydraulics Research Ltd, Report SR 463 Performance of Syphonic Drainage Systems for Roof Gutters. Hydraulic Research Ltd, Wallingford

Increasing the fire resistance of existing timber floors. BRE Digest 208, 1988

Information Guidance Note 09-02-05 Marking and Identification of Pipework for Reclaimed and Grey Water Systems (available from: WRAS, Fern Close, Pen-y-Fan Industrial Estate, Oakdale, Gwent, NP11 3EH, Tel: 01 495 248454, Fax: 01495 249234) (www.wras.co.uk)

Pollution Prevention Guidelines PPG2 – Above Ground Oil Storage Tanks. The Control of Polution (Oil Storage) (England) Regulations (2001).

Masonry Bunds for Oils Storage Tanks Concrete Bunds for Oils Storage Tanks
Above documents available without charge from DMS Mailing House:
Fax 0151 604 1222; email environment-agency@dmsltd.co.uk

prEN 12380 Ventilating pipework, air admittance valves

prEN 13564 Anti flooding devices for buildings

prEN 1825 Installations for separation of grease

prEN 1857: 1995 Chimneys – Components Concrete Flue Liners

prEN 858 Installations for separation of light liquids (e.g. petrol or oil)
Part 2 Selection of nominal size, installation operation and maintenance

prEN ISO 10077–2 Thermal performance of windows, doors and shutters –
Calculation of thermal transmittance
Part 2: Numerical method for frames

Radon: guidance on protective measures for new dwellings, BR 211, 1999,
ISBN 1 86081 3283

Safety in the installation and use of gas systems and appliances, Approved
Code of Practice and Guidance L56. HSE Books, ISBN 0 7176 1635 5

Safety signs and signals. The Health and Safety (Safety signs and signals)
Regulations 1996. Guidance on Regulations. (HSE, L64). HSE Books,
1996

Sewers for Adoption Water UK/WRc, 5th edition, 2001, ISBN 1 898920 43 5

Spillage of flue gases from solid-fuel combustion appliances. BRE IP 7/94

Thatched buildings. New properties and extensions [the 'Dorset Model']
(can be found on www.dorset-technical-committee.org.uk;
email: info@dorset-technical-committee.org.uk)

Water Regulations Advisory Scheme Information Guidance Note 09-02-05
Marking and Identification of Pipework for Reclaimed and Grey Water
Systems. (Available from: WRAS, Fern Close, Pen-y-Fan Industrial
Estate, Oakdale, Gwent, NP11 3EH, Tel: 01 495 248454, Fax: 01495
249234) (www.wras.co.uk)

Water Regulations Advisory Scheme leaflet No 09-02-04. Reclaimed Water
Systems. Information about installing, modifying or maintaining reclaimed
water systems. (Available from: WRAS, Fern Close, Pen-y-Fan Industrial
Estate, Oakdale, Gwent, NP11 3EH, Tel: 01 495 248454, Fax: 01495
249234) (www.wras.co.uk)

Welsh Office Circular 25/92. Local Government and Housing Act 1989.
Houses in multiple occupation: standards of fitness. Welsh Office, 1992

Welsh Office Circular 55/93. Houses in multiple occupation. Guidance on
management strategies. Welsh Office, 1993

Workplace health, safety and welfare. The Workplace (Health, Safety and
Welfare) Regulations 1992, Approved Code of Practice and Guidance; The
Health and Safety Commission, L24; HMSO 1992; ISBN 0-1 1-886333-9

Books by the same author

ISO 9001: 2000 for Small Businesses (2nd edition). Butterworth-Heinemann, ISBN: 0 7506 4882 1.

Fully revised and updated, ISO 9001: 2000 for Small Businesses explains the new requirements of ISO 9001: 2000 and helps businesses draw up a quality plan to allow them to meet the challenges of the marketplace.

ISO 9001: 2000 Audit Procedures. Butterworth-Heinemann, ISBN: 0 7506 5436 8.

A complete set of audit check sheets and explanations to assist auditors in completing internal, external and third party audits of **newly implemented** ISO 9001: 2000 Quality Management Systems, **existing** ISO 9001: 1994, ISO 9002: 1994 and ISO 9003: 1994 compliant QMSs and **transitional** QMSs.

ISO 9001: 2000 in Brief. Butterworth-Heinemann, ISBN: 0 7506 4814 7.

A 'hands on' book providing practical information on how to cost-effectively set up an ISO 9001: 2000 Quality Management System.

Optoelectronic and Fiber Optic Technology. Butterworth-Heinemann, ISBN: 0 7506 5370 1.
A 'user-friendly' introduction to this increasingly important topic.

CE Conformity Marking. Butterworth-Heinemann, ISBN: 0 7506 4813 9.

Essential information for any manufacturer or distributor wishing to trade in the European Union. Practical and easy to understand.

Environmental Requirements for Electromechanical and Electronic Equipment.
Butterworth-Heinemann, ISBN: 0 7506 3902 4.

Definitive reference containing all the background guidance, ranges, test specifications, case studies and regulations worldwide.

MDD Compliance using Quality Management Techniques.
Butterworth-Heinemann, ISBN: 0 7506 4441 9.

Easy to follow guide to the Medical Devices Directive, enabling the purchaser to customize the Quality Management System to suit his own business.

Quality and Standards in Electronics. Butterworth-Heinemann, ISBN: 0 7506 2531 7.

Ensures that manufacturers are aware of all the UK, European and international necessities, know the current status of these regulations and standards, and where to obtain them.

Useful contact names and addresses

Asbestos specialists

**Asbestos Information
Centre Ltd**
PO Box 69
Widnes
Cheshire
WA8 9GW
051 420 5866

Concrete specialists

**British Ready Mixed
Concrete Association**
The Bury
Church Street
Chesham
Buckinghamshire
HP5 1JE
0494 791050

Damp, rot, infestation

**British Wood Preserving
& Damp Proofing
Association**
Building No. 6
The Office Village
4 Romford Road
Stratford
London
E15 4EA
081-519 2588

English Nature
Northminster House
Northminster Road
Peterborough
PE1 1VA
0733 340345

Countryside Council for Wales
Plaspenrhos
Penrhos Road
Bangor
Gwynedd
LL57 2LG
0248 370444

Scottish Natural Heritage
12 Hope Terrace
Edinburgh
EH9 2AS
031 447 4784

**Local Department of
Environmental Health**
Refer to your local *Yellow Pages*

Decorators

British Decorators Association
32 Coton Road
Nuneaton
Warwickshire
CV11 STW
0203 353776

Scottish Decorators Federation
41A York Place
Edinburgh
EH1 3HT
031 557 9345

Electricians

**National Inspection Council
for Electrical Installation
Contracting**
37 Albert Embankment
London
SE1 7UJ
071 582 7746

Fencing erectors

Fencing Contractors Association
St John's House
304–310 St Albans Road
Watford
WD2 SPE
0923 248895

Glazing specialists

Glass and Glazing Federation
44–48 Borough High Street
London
SE1 1XB
071 403 7177

Heating installers

British Gas Regional Office
Refer to your local *Yellow Pages*

Electricity Supply Company
Refer to your local *Yellow Pages*.

British Coal Corporation
Hobart House
Grosvenor Place
London

SWIX 7AE
071 235 2020

**Heating and Ventilating
Contractors Association**
Esca House
34 Palace Court
London
W2 4JG
071 229 2488

**National Association of Plumbing,
Heating and Mechanical Services
Contractors,**
Ensign House
Ensign Business Centre
Westwood Way
Coventry
CV4 8JA
0203 470626

Home security

Local crime prevention officer
Refer to your local *Yellow Pages*

Local fire prevention officer
Refer to your local *Yellow Pages*

**National Approval Council for
Security Systems**
Queensgate House
14 Cookham Road
Maidenhead
S16 8AJ
0628 37512

Master Locksmiths Association
Units 4–5
Woodford Halse Business Park
Great Central Way
Woodford Halse
Daventry
NN1 6PZ
0327 62255

British Security Industry Association
Security House
Barbourne Road
Worcester
WR1 1RS
0905 21464

Insulation installers

Draught Proofing Advisory Association Ltd
External Wall Insulation Association
National Cavity Insulation Association
National Association of Loft Insulation Contractors
PO Box 12
Haslemere
Surrey
GU27 3AH
0428 654011

Plasterers

Federation of Master Builders
14 Great James Street
London
WC1N 3DP
071 242 7583

Plumbers

Institute of Plumbing
64 Station Lane
Hornchurch
Essex
M12 6NB
0708 472791

National Association of Plumbing, Heating and Mechanical Services Contractors
Ensign House
Ensign Business Centre

Westwood Way
Coventry
CV4 8JA
0203 470626

Roofers

Builders' Merchants Federation
15 Soho Square
London
W1V SFB
071 439 1753

National Federation of Roofing Contractors
24 Weymouth Street
London
1N 3FA
071 436 0387

Ventilation

Heating and Ventilating Contractors Association
Esca House
34 Palace Court
London
W2 4JG
071 229 2488

Other Useful contacts

British Board of Agrément (BBA)
PO Box 195
Bucknalls Lane, Garston, Watford
WD2 7NG
Tel: 01923 665300
Fax: 01923 665301
E-mail: bba@btinternet.com
Internet: http://www.bbacerts.co.uk

BSI
British Standards Institution
389 Chiswick High Road, London
W4 4AL
Tel: 0181 996 9001
Fax: 0181 996 7001
E-mail: info@bsi.org.uk
Internet: www.bsi.org.uk

The Stationery Office
The Publications Centre
PO Box 29, Norwich NR3 1GN
Telephone orders/general enquiries
0870 600 5522
Fax orders 0870 600 5533
www.thestationeryoffice.com

UKAS
United Kingdom Accreditation
Service

21–47 High Street
Feltham
Middlesex
TW3 4UN
Tel: 0181 917 8400
Fax: 0181 917 8500

WIMLAS
WIMLAS Limited
St Peter's House
6–8 High Street
Aver
Buckinghamshire
SL0 9NG
Tel: 01753 737744
Fax: 01753 792321
E-mail: wimlas@compuserve.com

Index